THE SEVEN SISTERS OF THE PLEIADES

Stories from around the world

MUNYA ANDREWS

SPINIFEX

First published in 2004 by Spinifex Press, Australia.
Reprinted 2013, 2014, 2018

Spinifex Press Pty Ltd
PO Box 5270
North Geelong, Victoria 3215
and
PO Box 105
Mission Beach Qld 4852
Australia
women@spinifexpress.com.au
www.spinifexpress.com.au

Edited by Averil Lewis, Melbourne
Typeset in *Adobe Garamond* by Claire Warren, Melbourne
Cover design by Deb Snibson, MAPG

Cataloguing-in-Publication

Andrews, Munya, 1960–.
The seven sisters of the Pleiades: stories from around the world.
Bibliography
ISBN: 9781876756451 (paperback)
1. Pleiades - Mythology. 2. Mythology. Aboriginal Australian.
I. Title.
202.12

This project has been assisted by the Commonwealth Government through the Australia Council, its Arts funding and advisory body.

For my beloved Pleiadian grandmother, Canice Cox Ishiguchi,
her great granddaughter, Cordelia Andrews, and
her great, great, granddaughters, Paige and Georgia Newton,
for them, to keep the 'Dreaming of the Seven Sisters' alive
and to pass on the traditions.

ACKNOWLEDGEMENTS

Heartfelt thanks to the following people:

Christine Franks, for her wonderful friendship and support

Crystal, Levi and Amanda Bok, for sharing their home and resources

Carrie Maddison, for her assistance with picture research

Kylie Toomey, for retyping the manuscript following a computer virus!

John Ley, for his usual brilliance and generosity of spirit, for his editorial assistance in reworking the Greek chapter, and as someone with whom to share and discuss intellectual ideas and theories

Chris Sitka, for introducing me to the work of Marija Gimbutas and shared discussions of the 'Bird Goddess' in European cultures

Ashirirea San in Nibutani, Hokkaido, for her wonderful Ainu hospitality

Georgie Stevens, for her translation of Japanese and Ainu starlore

Riteria Nikora, for information on Maori starlore

Averil Lewis, for her superb editing skills

Nigel Andrews, Jean Gardiner, Maureen and Phil Newton for sharing their computer resources

Will Bon, Blanche Bowles, Nytunga Phillips, Diana Scifleet and Jan Testro, for their encouraging words of love and support

To all the cultural traditions featured in the book, for sharing their stories of the Pleiades so that the world may come to realise our common humanity and origins

To my Aboriginal elders and teachers in this and Spirit world — Aunty Lorraine Mafi-Williams, Robert Mate-Mate, David Mowaljarlai, Violet Newman and Daisy Utemorrah

To Susan Hawthorne — for believing in me and taking the risk

And to my kantrimin — the Pleiades

CONTENTS

A Grandmother's Tale

Seven Stars for Seven Sisters

As a little girl growing up in the bush in the Kimberley region of Western Australia, I would spend many nights with my family looking up at the sheer majesty and glory of the heavens. It is here, far away from the light pollution of the towns and cities that the night skies are the clearest and brightest you will ever see. It was here that my adoptive grandmother, Canice Cox Ishiguchi, a Nyigina woman from Noonkanbah,[1] would often relate Dreamtime stories of the stars and our relationship with them. Our favourite story was that of the Seven Sisters, the Kungakungaranga, otherwise known in Western astronomy as the Pleiades star cluster in the constellation of Taurus.[2]

'Look up there Munya,' she would say, 'up there in the night skies — the Seven Sisters. Can you see them?'

I would look up to the 'saucepan', as Orion is sometimes irreverently described, to that familiar celestial region where I knew the Sisters were located nearby. 'Where are they Granny?'

'Over there,' she would reply, 'to the left of the hunter's belt, low in the north-west.' Then she would add, quite excitedly, 'there they are — the Girls. Your kantrimin, your relatives.'[3]

'But how are they our kantrimin Granny?'

'Because we are the same *mob* as them Munya. We are the *same* people. *One* people. We come from the same *country*.'[4]

Ah yes, country.

Now I understood.

Whenever Aboriginal people speak about *country*, they are not necessarily just referring to a sovereign nation state like Australia, Canada or the United States of America, which is the general meaning of the term in Standard English. In an Aboriginal English context, the word 'country' takes on a more significant, cultural meaning. It describes and encompasses the overall spiritual, physical and emotional *connection* that an Aboriginal person has with the land.[5] It is this relationship that gives Aboriginal people their identity.

Whenever my grandmother spoke of 'the Girls', as she affectionately called the Pleiades, she would do so with such warmth and love, I half expected an aunt, cousin or niece to literally drop out of the sky to visit us. It was not uncommon to have family and friends drop in unannounced from nearby Broome, Noonkanbah or Fitzroy Crossing. These townships are, at the very least, 200 kilometres or more from the small town of Derby in the West Kimberley, where I grew up. Which is much closer than the estimated 410 light-years that the Pleiades are from Earth![6]

My grandmother would ask, 'How many stars can you see Munya?'

I would begin to count. 'One . . . two . . . three . . . four . . . five . . . six. I can see six stars Granny.'

'Well if you look closer you will see that there is another fainter star.'

'*Another* star Granny?'

'Yes Munya, but most people can only see six. That's because the seventh star is the youngest sister. She doesn't shine so bright because she is lost. She's got a little behind and is trying to catch up with her older sisters.'

'But why is she trying to catch up to them Granny?'

'Ah, Munya, that's because of what happened in the Dreamtime.'

'What happened in the Dreamtime, Granny?'

'Well that is the *Dreaming of the Seven Sisters*.'

She would pause and then begin with the familiar phrase: 'Long, long ago in the Dreamtime . . .' Her voice would fill the Kimberley night air, her words lighting up the dark skies like Lejmorro, the Milky Way, as my grandmother retold the ancient, timeless tale. A tale of the earthly and celestial exploits of the Seven Sisters who came down to Earth from the Pleiades.

In grandmother's story, the Seven Sisters would often come down from the sky and always landed on a high hill.[7] This was no ordinary hill for it was hollow inside as it contained a cave. A secret passageway leading into the cavern from the outside enabled the Sisters to come and go between the worlds. This cave served as their temporary home while they were on Earth.

On one of these visits, the Sisters went hunting for food in the bush. They were excellent hunters and soon gathered enough meat and other bush foods to eat. On their way back to the cave, an old man saw them but the Sisters were too busy collecting food and other things and did not notice him at all. The old man decided to follow the young women, as he wanted a wife. When they were camped by a creek he jumped out from behind a bush and grabbed the youngest sister. The other Sisters started running toward the cave in the hill to escape. They ran into the secret passageway and climbed to the top of the hill. The remaining Sisters flew off into the sky with their digging sticks.

In the meantime the youngest sister was still struggling with the old man, trying to escape. She called out to her older sisters to come and help her but did not realise they had already left.

'Tjitja (Sisters), please help me,' she cried as she fought with the old man. The youngest sister started to hit the old man; she kicked and punched him as hard as she could until she managed to break free. She ran into the cave and took off after her sisters. The old man gave chase and followed her.

She called out once more, 'Tjitja (Sisters), an old man is chasing me.' He followed her up into the sky, back to their country in the stars.

If you look hard you can see her in the distance trying to catch up with her older sisters. Sometimes you cannot see her at all and when that happens it is because she has lost her way back to her country. You can still see that old man in the sky, still chasing the girls, still trying to grab the seventh sister and make her his wife. According to our elders, you can still see that old man in the night skies as they point to the evening and morning star (Venus). There he goes, they say, still chasing the Seven Sisters.

You see him as the Evening and Morning star because he comes and goes in the night skies as he continues to pursue the Seven Sisters.

Notes

1. One of several hundred Aboriginal tribes or nations of Bandaiyan (Australia) that each in turn have their own languages as opposed to a dialect; hence there are literally hundreds of different Aboriginal language names for the Pleiades.

2. This story is similar to many western desert stories of the Seven Sisters, in particular that of the Kukatja myth recorded in *Tjarany Roughtail Lizard.* The simple explanation for this is that my adoptive grandmother's tribe, the Nyigina people of the Fitzroy River basin area of Western Australia, have strong cultural and linguistic ties with several western desert peoples such as the Waldmadjarri and the Kukatja near Balgo. On one of my visits to Balgo in 1987 I met with one of my grandmother's relatives who fondly remembered her and was equally thrilled to meet two of Canice's granddaughters — myself and my cousin, Colleen Sariago. Quite clearly, the similarity between the stories indicates the extent of these linguistic and cultural affiliations in this region of Australia. My biological mother's people are the Bardi and Nyul Nyul peoples of the Dampier Peninsula north of Broome in the Kimberley region of Western Australia. See also the similarities with *The Legend of the Seven Sisters: A Traditional Aboriginal Story from Western Australia* by May O'Brien.

3. Kantrimin is an Aboriginal English, or Kriol term, that derives from the English term 'country men' to signify one's kin. In Aboriginal terms kinship can be based on a number of factors besides biology — including geographical ties — whereby people who come from the same locale or elsewhere are connected to particular stretches of land via various Dreamings or sacred sites.

4. 'Country' or Kantri (in its kriolised form) is another of those Aboriginal English terms, which — although derived from the English language — is used in another context, as explained further on in the story.

5. I use the term 'connection' in its broadest sense, to encompass the spiritual and material realms. In no way is it restricted to its legal sense, such as in the leading Australian Native Title case of *Mabo*, where it was decided that Indigenous people had to show a continuous *connection* to their land in order to claim Native Title.

6. David Levy, *Skywatching*, p. 215. A light-year in astronomical terms is the distance that light travels in one year in a vacuum; that is, a mere nine-and-a-half trillion kilometres, or six trillion miles.

7. This particular story was told to me as a child and is essentially a children's story, as opposed to the more intricate and complex secret sacred stories of women's business. As an Aboriginal child reaches adolescence she is introduced to more complicated elements of the myth. In some regions of Bandaiyan there is an emphasis on eroticism in stories of the Seven Sisters that is not suitable for young children. I have chosen to share this particular children's version with readers that provided my introduction as a child to the mystery of the Dreaming of the Seven Sisters.

CHAPTER ONE

The Sweet Influence of the Pleiades

Unravelling the Mystery

Many a night from yonder ivied casement, ere I went to rest,
Did I look on great Orion, sloping slowly to the west.
Many a night I saw the Pleiads. Rising through the mellow shade,
Glitter like a swarm of fireflies tangled in a silver braid.
Here about the beach I wander'd, nourishing a youth sublime
With the fairy tales of science, and the long result of Time;
When the centuries behind me like a fruitful land reposed;
When I clung to all the present for the promise that it closed:
When I dipt into the future far as human eye could see;
Saw the Vision of the World, and all the wonder that would be.

— from *Locksley Hall* by Lord Alfred Tennyson[1]

No other stars in the passage of time seem to have captivated and enthralled our imaginations quite like those of the Pleiades. Revered and worshipped by many diverse peoples, cultures and civilisations, this small cluster of stars has had an enormous influence on the human psyche and on our collective unconscious, where they continue to charm and fascinate. Throughout millennia their gentle glow in the night skies has inspired and guided sailors over the seven seas and other explorers on land in search of their hopes and dreams during the endless migrations of humanity across the globe. People looked to the Pleiades to tell them when to sow and harvest their produce, when the important rains would come and when to keep their sacred ceremonies. Poets, priests, prophets, shamans, storytellers, singers and historians have all sung their praises down through the ages from Homer to Hesiod, Mohammed to Milton, Plato to Edgar Allen Poe.[2] Other acclaimed writers moved and mused by their presence include the Romantic poets Byron, Keats, and Tennyson. Artists have depicted these famous stars on bark paintings, in caves, on petroglyphs, in sculptures, on canvas and, in modern times, in cyberspace. Many important buildings, temples and other ancient monuments were aligned to the Pleiades including the Temple of the Sun in Mexico, the Great Pyramid in Egypt, the Parthenon in Greece, the 'Golden Enclosure' of the ancient Mayan capital in Peru and the pyramid of Chichén Itzá in the Yucatan, to name just a few.[3] These faint, gentle stars have touched all our lives on a multitude of levels. Their celestial influence in all spheres of life is prolific while their esoteric, spiritual nature in world mythology is profound. Beyond their symbolic meaning, the practical application of the Pleiades in the sciences — especially in measurements, geodesics, geometry, architecture and navigation — is considerable.

Their association with timekeeping in particular is multitudinous and legendary. In the fifth century the Greek dramatist Euripides referred to them as 'nocturnal timekeepers', and a century before the poet Sappho noted the passage of time during the night while observing the Pleiades, which she recollects in a melancholic poem.[4] Such was their reputation that the 26,000-year cycle of precession was named in their honour, where it was once known as the 'Great Year of the Pleiades' in the ancient world. Elsewhere their rise and setting marked the seasons of the calendar year, including the end of the old and commencement of the New Year. Many well known festivals owe their origins to the observation and worship of the Pleiades, including Halloween and other feasts of the dead.[5] Even Japanese and Indian lantern festivals can be traced back to earlier celebrations involving these stars.[6] Their influence on the development of world calendars, especially the acclaimed Mayan Calendar, is only just beginning to be realised largely through the writings of Mayan scholar John Major Jenkins who has identified the key role that the stars of the Pleiades played in Mesoamerica.[7] Exactly thirty-four years earlier, Gertrude and James Jobes went so far as to suggest that a Pleiades Calendar 'may have preceded the lunar and solar calendars.'[8] If proven to be true then it would establish the Pleiades Calendar as one of the world's oldest calendars. This connection with time meant that in some cultures the Pleiades took on a prophetic aspect, as in ancient Egypt where they were regarded as the Seven Fates who foretold the destiny of every newborn child,[9] or in India where they govern the world ages or *yugas* embodied in the game of dice.[10] These connections with fate, time and destiny are explored in more detail in the Egyptian and Hindu chapters. The last chapter examines the role the Pleiades played in many world calendars and prophecies.

More popularly known as the 'Seven Sisters' in world mythology, their official name in astronomy comes from Greek legends where they were known as the Seven Daughters of Atlas and Pleione.[11] Atlas, one of Seven Titans who plotted and fought against Zeus the king of the Greek gods and his Olympian associates, was severely punished and made to bear the burden of the world upon his shoulders for eternity.[12] The underlying symbolism of this heroic act is looked at in more detail in that chapter. In the meantime the Sisters, who were in train to the goddess Artemis as young nymphs, each went on to influence the course of human history by marrying kings and giving birth to gods and heroes who laid the foundation of many civilisations, including the ancient city-state of Troy.

The world's leading theosophist of the nineteenth century, Helena Blavatsky or Madame Blavatsky as she is more popularly known, has much to say about the role of the Pleiades in history and cosmology in her celebrated treatise *The Secret Doctrine*. Often referring to the Pleiades as the Atlantides (after their father Atlas and because of their connections with Atlantis), she maintains that the Seven Sisters play a vital role in the unfolding of human destiny and in the karma of nations and individuals.[13] Just what this outcome entails is considered throughout the book, although given the sometimes encryptic disposition of her writings much of it remains conjecture and therefore subject to different interpretations. Whatever the true nature of her claims there can be no denying the prophetic roles ascribed to the stars of the Pleiades throughout history by a diverse range of peoples and cultures.

In honour of their special role in navigation, the ancient Greeks referred to the Pleiades as the 'sailing stars' and designated their Oceanid mother Pleione the 'sailing queen.'[14] This naval tradition continues to be observed in Germany, where they are still called by their maritime nickname Schiffahrts Gestirn (sailor's stars) even though their official name is Plejaden.[15] At other times they are simply referred to as Das Siebengestirn or 'seven stars'.

Of all the sailing nations, however, including the Phoenicians who were famed for their seafaring prowess, none could match the extraordinary maritime achievements of the Polynesians who turned sailing by the stars into an exact science. Although they relied on several individual bright stars and constellations besides the Pleiades to guide them across the vast Pacific, their love of these illustrious stars is reflected in their cultures, languages and especially in their chants and songs. The valuable role which the Pleiades played in Polynesian navigation is looked at in more detail in the chapter on Matariki, as they are known in Aotearoa (New Zealand). In Bandaiyan (Australia) they are known by many different names to reflect the prolific number of Aboriginal languages (fully fledged and complete languages in themselves, not 'dialects' as often mistakenly thought).

Navigation aside, the Greek legends of the Pleiades have given us words like electron, electrum, electricity and atlas. Not only is their father commemorated in the collection of maps that bears his name[16] but his memory is evoked whenever we speak of the Atlas Mountains in northern Africa, the Atlantic Ocean or even the lost continent of Atlantis.[17] The cultural affiliations of these stars in fashion and the media are present in ancient and contemporary times. In England the only street in London that turns itself into a 'lane' solely on weekends, known for its wares and collectibles —

Petticoat Lane — actually derives its name from the Pleiades because the Romans named the garment hanging outside brokers' shops for this star cluster.[18] Kate Winslet and Leonardo DiCaprio, in the popular cinematic love story *Titanic*, sailed as fictitious young lovers Rose and Jack on the ill-fated ship. The great vessel, which received its unfortunate epithet 'unsinkable,' derived its name from the very race of giants from whom the Pleiades are descended. Even our name for the month of May comes from these stars after Maia, the eldest and most beautiful of the Seven Sisters in the Greek legend.[19] In one way or another, either directly or indirectly, and irrespective of our ethnic origins, the stars of the Pleiades have had an enormous influence on human cultures and languages. An examination and analysis of world mythology surrounding them reveals many universal themes, which suggest a very likely common human origin. At the very least, we are more alike than some of us care to admit and perhaps our so-called cultural differences may largely be of our own making.

Many great works of literature including the various mystical traditions, philosophies, codices and other religious writings such as the Kabbalah, Koran, Hermetica, Rig Veda and the Zohar all contain references to these stars. They are mentioned several times throughout the Bible, especially in the Book of Revelations, where they are implicated through their special relationship with the magical number seven. The Book of Job, in particular, asks the rhetorical question 'Canst thou bind the sweet influence of the Pleiades or loosen the bands of Orion?[20] Intrigued by the riddle of this biblical phrase, many academics, writers and theologians have sought to understand its hidden meaning. Academy-award winning actress and New Age author Shirley MacLaine seizes upon this passage to pose the question, 'Why is the influence of the Pleiades denoted as *sweet* when Orion's depiction is *constricting*?'[21] Although the biblical phrase does not refer to the Pleiades as female and Orion as male, their representation in world mythology suggests there is a sexual division based on gender to which these characteristics may be perceived in traditional terms. Thus Orion is often portrayed as a man or more importantly as a warrior or hunter and the Pleiades as a group of young maidens.

According to Lloyd Motz and Carol Nathanson there may be an astronomical explanation for the phrase 'loosening' the bands of Orion. This is because one of the Belt Stars, Alnitak (*Zeta Orionis*) 'is moving away from both Alnilam and Mintaka,' (*Epsilon* and *Delta Orionis*) along with other stars in the Orion system.[22] What this means effectively is that 'the entire constellation will alter its shape, owing to the stars' changing positions; and an equal factor in Orion's altered appearance will be the evolutionary development of those stars.'[23] Therefore, say the authors, 'the Lord will indeed, one day hundreds of thousands of years hence,' loosen the bands of Orion.[24] The stars of the Pleiades, on the other hand, are all relatively speaking the same astronomic age and were born from the same starry womb.[25] And while their individual stars will one day grow apart from one another and disperse themselves across the night skies, they are at the very least all travelling through space in the same direction.[26] So far as the alleged sweet nature of the Pleiades is concerned, an essential clue is what I have identified as the 'honey theme' in all these stories.

Taurean star clusters

In astronomy, as in mythology, the Seven Sisters of the Pleiades continue to impress and mystify. One of the most celebrated star clusters in the sky and designated on star maps as M 45, they can be seen in the constellation of Taurus the Bull.[27] Their alphabetic and numeric designation refers to their astronomic classification in the *Messier Catalogue*, named for the eighteenth-century French astronomer whose inventory of 110 celestial objects largely includes star clusters, nebulae and distant galaxies.[28] As their name suggests, star clusters are a group of stars bunched together in a relatively small area of the sky. Basically there are two kinds of star clusters — open and closed (or globular).[29] Taurus contains two sets of *open* clusters, the Pleiades and their lesser known celestial neighbours and siblings, the Hyades.[30] The identification of this particular region of the night sky with a bull or cow is widespread throughout Europe, including India and the earlier civilisations of Mesopotamia.

Just how far back in time this association goes is not entirely certain, but some writers like cosmologist Frank Edge and Michael Rappenglueck suggest a far more remote period stretching back thousands of years before the appearance of the early Mesopotamian civilisations of ancient Sumer, Akkadia and Babylon. In his research paper *Aurochs in the Sky*, Edge examines the prehistoric cave paintings of aurochs (a prehistoric animal related to our modern bull) in Lascaux in southern France, and argues that the cluster of six dots above an auroch's back may in fact represent the Pleiades.[31] He points out that not only are they the same number of visible stars (as seen from that location) but their configuration closely resembles the same 'spatial relationships' of individual stars within the cluster.[32] As well, 'they have approximately the same relationship as the Pleiades to the head and face of the related bull.'[33] The Lascaux caves house one of the oldest prehistoric cave paintings in Europe that is estimated to have been painted more than 17,000 years ago.[34]

What this effectively means, says Graham Hancock in *Heaven's Mirror*, is that the cave dwellers would have painted the aurochs 'more than 14,000 years before the supposed first invention and naming of the twelve constellations of the zodiac by the ancient Babylonians and Greeks.'[35] If correct, this would make the cave paintings one of the oldest representations of Taurus and the Pleiades in Europe and possibly one of the earliest star maps of that region from which other European traditions, including ancient Greece, followed. In early Northern Hemisphere European drawings and in modern star charts, the Pleiades represent the Bull's shoulder.[36] The Hyades, on the other hand, are the set of stars that form the distinctive V-shaped pattern of Taurus that depicts the bull's head with the beautiful orange star, Aldebaran, marking the eye of the bull.[37] The symbolism behind this asterism will become much clearer when we consider some of the universal themes found in an assortment of different cultural stories of the Pleiades.

Universal Pleiadian themes

Despite the existence of a number of common Pleiadian themes among world mythologies, only a select handful of writers have attempted to draw parallels between the various Pleiadian legends and fewer still have made any kind of inter-tribal comparison of the Aboriginal legends. The late nineteenth century Australian author, Katherine Langloh Parker, who published a general collection of Aboriginal Dreamtime stories in *Australian Legendary Tales* (1896) and *More Australian Legendary Tales* (1898), paved the way for such a comparison.[38] Her views on Aboriginal people and their cultures were unusually enlightened for the times. For one, she regarded Aboriginal myths to be on an equal par with ancient Greece and believed they were just as complex and sophisticated in their storytelling as other ancient civilisations. As a consequence, she was one of the first white Australian authors to recognise and comment on the similarities between Aboriginal Dreamtime stories and ancient Greek mythology.

In *Wise Women of the Dreamtime*, Johanna Lambert draws upon the well of Aboriginal stories collected by Langloh Parker and analyses them from an intercultural, anthropological and spiritual perspective. Her intention, she states, was to follow Langloh Parker's 'insight and interpret her translations comparatively with other world mythologies.'[39] This she does most eloquently throughout her book, and especially in the chapter 'Where the Frost Comes From', where she examines one particular Aboriginal legend of the Pleiades — that of the Bundjalung peoples of northern New South Wales on the east coast of Australia.[40] This is the story of the Maimai women of the Pleiades and the Berai-Berai men of Orion who fell in love with them. In this chapter, Lambert makes several connections between ancient Greek, Indian and Egyptian legends and those of the Bundjalung stories of the Seven Sisters of the Pleiades.

Twelve years prior to Lambert, Jennifer Isaacs, the editor of *Australian Dreaming*, included a general, brief discussion on the Pleiades with reference to three different Aboriginal legends.[41] Although her comparative analysis is limited, she makes some interesting observations from a broader, bigger-picture perspective. She notes, for instance, that the majority of the stories are essentially about young women, seven in number, with one sister who is either missing or lost, and that they are pursued either by an older man or else a group of men. Lambert, on the other hand, identifies specific mythic aspects including the Sisters' relationship with honey in the Australian Aboriginal and Greek legends, and their description as female judges in ancient Hindu, Egyptian and Greek mythologies. Noting similar attributes among the Maimai, she focuses on their portrayal as strong warrior women in Aboriginal mythology who act as role models for young Aboriginal girls approaching womanhood.

Lambert also comments on the connection between the Pleiades and music, in particular the use of drums in ancient Mesopotamia. This is particularly significant because of the involvement of sound with creation in world mythology and science, and the fact that the playing of this instrument was once exclusively the domain of women in several cultures, including Aboriginal Australia. Other writers such as Richard Allen, William Tyler Olcott and Robert Burnham Jr have commented on individual related themes such as the link between the Pleiades and rain or their depiction as birds, but have done no more with their observations. More recently, Diane Bell has written on the significance of the Seven Sisters Dreaming in South Australia, specifically in relation to the Ngarrindjeri people's beliefs of the waters surrounding Kumarangk (Hindmarsh Island). In *Ngarrindjeri Wurruwarrin* she touches on some of these Pleiadian themes, namely their association with water, their role in denoting the seasons and their directives or 'sacred orders' to young Aboriginal women, especially during initiation ceremonies.[42] I have taken these observations, along with my own, expanded on each in the following sections and incorporated these familiar themes throughout this book and within specific chapters.

Women's Dreaming

By far the most prevalent theme is the correspondence of the Pleiades with what Aboriginal people refer to as 'women's business' or 'women's dreaming', for almost everywhere they are universally portrayed as young women. There are some exceptions to this general rule, like some Native American legends that refer to seven young boys instead of girls.[43] Despite this slight mythic variation, the significance of the Pleiades to women's dreaming remains essentially intact. Given that women are the primary caregivers and nurturers of children, it makes sense that young boys are allowed within the sphere of women's influence, so long as they have not become men, either in the biological or initiatory sense. Bell suggests the stars of the Pleiades are largely portrayed as young women because of their astronomical nature. Their mythological status as maidens, she says, is due to the fact that, scientifically speaking, the Pleiades are relatively young adolescent stars in comparison to our middle-aged Sun and other more elderly stars in the cosmos.[44] The interesting corollary to this scientific fact is that in the Ngarrindjeri tale of the Seven Sisters, the mother, as befitting a middle-aged woman, elected to stay behind on Earth after sending her daughters up into the sky.[45] Is this yet another example of science verifying mythology? It makes you wonder just how much scientific knowledge and wisdom the Ancients may have possessed, and what may have been lost along the way.

More intriguing is the Quiché Mayan reference to the Pleiades in their sacred text *Popul Vuh* as representing 400 heavenly youths that once fought down here on Earth before returning to their homeland in the skies.[46] If we take the line of argument in *Hamlet's Mill* that myths are encoded with astronomical and other scientific data, one interpretation of this particular choice of number may be a reference to the number of light-years distance that the Pleiades are from Earth. Previous estimates by astronomers like David Levy in *Skywatching*[47] have placed the star cluster at about 410 light-years distance from us, but more recently Robert Burnham and others in *Astronomy: The Definitive Guide*, suggest the distance is exactly 400 light-years.[48] This latest estimate would bring mythology and science more in line with one another. But even if the latest estimate is wrong and the Pleiades are more than 400 light-years away, then the Ancients might still be right, given the possible existence of wormholes in space-time that lessen enormous distances involved in space travel. Either way, it's not a bad estimate. Another interpretation might suggest that the 400 youths represent the actual number of suns in the Pleiades star system. But once again, because these estimates vary between scientists who say there are 400 to 500 or *more* stars in the Pleiades (possibly as many as 3,000 stars!), this theory is less likely.[49] In any event, the myth serves to emphasise the notion that only young boys are associated with the Pleiades, and that supports the basic mythological premise that the region is largely a women's domain.

Sisters seven

Although many of the stories refer to *seven* sisters in the Pleiades more than any other number, some writers such as American astronomer Edwin Krupp suggest the number seven has no significance at all. Because some cultures have counted *more* than seven stars in the cluster — some as many as ten, thirteen or even sixteen — he believes these so-called 'contradictory accounts' divest the numeral seven of any credibility.[50] As proof of this observation, Krupp refers to the Australian Aboriginal bark painting 'Orion and the Pleiades' by Minimini Mamarika that clearly shows thirteen stars and not seven. However, for reasons that are outlined in this book, and with due respect to Krupp, I believe the argument is somewhat flawed. While it is certainly true that some cultures have seen more than the obligatory seven stars, it does not diminish or take away from the numerological, spiritual or scientific significance ascribed to the number seven and its association with the Pleiades by so many others.

Where more than seven stars in the cluster have been recorded (such as in the less populated northern and desert regions of Australia and in the mountainous terrain of South America) this is largely because conditions for optimal stargazing are more enhanced. As Anthony Aveni points out in *Stairways to the Stars*, thirteen Pleiades stars is 'not an unrealistic number to be seen at this high altitude in the rarefied Andean air.'[51] Certainly many other Australian Aboriginal tribes have reported seeing more than seven stars in the cluster, yet despite these observations, the Pleiades are still referred to as the *Seven Sisters*, not the eleven, twelve or thirteen sisters. The reasons for this are not, as Margaret Simons suggests in *The Meeting of the Waters*, a direct result of European colonisation and import of the universal term 'Seven Sisters' into Aboriginal mythology.[52] Nor is it because Aboriginal languages supposedly 'don't have a word for "seven", or for any numerals above three,'[53] which is the standard anthropological line. She is correct in stating that the cluster 'is often referred to as *The Girls*,'[54] as related in my grandmother's story, but it is not true that they were never traditionally referred to as the Seven Sisters.

This unchallenged anthropological authoritative voice on Aboriginal people and their cultures often ignores Aboriginal realities and experiential ways of being. These sorts of perceptions and claims about Aboriginal numeracy fail to comprehend how number operates within Indigenous cultures and sciences, especially mathematics. As David Peat points out in *Blackfoot Physics*, 'number and mathematics have always played a special role' in Indigenous cultures.[55] Also, because number is tied to spiritual and ritual significance in some instances 'its importance is so high that number may not be used for secular or commercial purposes.'[56] This is especially true of esoteric lore surrounding the Dreaming of the Seven Sisters in Australia. Even the ancient Inca civilisation, which depicted thirteen Pleiades stars in their artwork and written astronomical codes, still referred to the Pleiades as the 'Seven Eyes of Viracocha' — their god of thunder and creation.[57] This suggests the number seven was of enormous significance, spiritually and otherwise. It tells us that the number seven is not coincidental to world mythology surrounding the Pleiades but is, in fact, intrinsic to the legend of the Seven Sisters.

This preoccupation with seven is not just a matter of numeric convenience. Neither is it because people felt they had to assign this particular number to the Pleiades to match or imitate the seven stars of the Big Dipper, or Ursa Major, as Krupp suggests.[58] While it is true that the Pleiades star pattern does appear as a miniature Big Dipper in the Northern Hemisphere, and in this sense may be said to 'mimic' its shape, the argument is not convincing for viewers in the Southern

Hemisphere on two accounts. Firstly the Pleiades, like so many of the other stars visible to both hemispheres, appears 'upside down' in the Southern Hemisphere. Therefore any possible resemblance to the Big Dipper in shape or form is not immediately apparent to observers in the Southern Hemisphere.[59] Secondly, as amateur astronomer Patrick Moore points out in *Stars of the Southern Skies*, while the seven prominent stars of the Big Dipper (for there are more that comprise the constellation of Ursa Major) may be a familiar, distinctive constellation to viewers in the Northern Hemisphere, in large areas of Australia and South Africa they can only be partially seen in places below 30° S latitude, and cannot be seen from New Zealand at all.[60] The seven stars of the Pleiades, therefore, hold their own symbolism in at least half of the world that either knows nothing or else very little of the seven stars of Ursa Major. Thus Krupp's argument does not satisfactorily explain why the number seven is given more emphasis to the Pleiades within the cultures of that region.

An ancient Egyptian calendar found buried with a mummy clearly illustrated seven stars as part of a twelve-star column that did not represent the seven stars of the Big Dipper but those of the Pleiades.[61] At this latitude in Egypt, the stars of Ursa Major are just as visible as the Pleiades and, to distinguish between the two, ancient Egyptians (whose astronomical knowledge and expertise is well documented) drew the constellation of Ursa Major as a single bull's thigh.[62] By assigning the primary motif of seven stars as the sole emblem of the Pleiades and not that of Ursa Major, ancient Egyptians clearly afforded a central role to the Pleiades. Obviously they regarded them as important in their own right, not merely as a carbon copy of Ursa Major.

The relationship of the Pleiades with number seven has a far deeper cosmological and spiritual significance than is immediately apparent. Occasionally the myths refer to *six* young people, not seven, but this merely reflects aspects of yet another familiar theme — that of the lost or missing sister. A number of mythical, historical and scientific theories have been offered to explain the persistence of this idea, some satisfactorily, others less so. Predictably, Blavatsky offers a more esoteric explanation. Writing on the significance of seven in astronomy, magic and science in *The Secret Doctrine* she says this number is closely connected with the occult significance of the Pleiades, particularly in regard to the lost sister, 'the six present, the seventh hidden.'[63] This theme, she argues, is not coincidental to astronomy but has a much deeper, hidden meaning relating to the various cycles of time. These cycles govern our universe and include the Kali Yuga, the Age of Aquarius and the Apocalypse of St John's Revelation.[64] The spiritual and scientific aspects of the Lost Pleiad in relation to time are explored throughout this book, especially in the Hindu chapter on the Krittika.

Seven, the sacred number of creation*

Number plays an important role in many cultural traditions. On a practical level, it enables us to measure and quantify our world to assist in carrying out a wide variety of tasks. These range from the ordinary, mundane chores of daily life through to more complex computations of science, which facilitate technological advances leading to exploration of spatial domains from the macrocosm to nano realities. From a spiritual perspective, number takes on a more profound, consecrated dimension. This was particularly true of the early Hermetic and Pythagorean teachings, which gave rise to the sacred sciences of numerology and geometry. As David Peat eloquently says, within these traditions, as in Indigenous sciences, 'number is seen in a profoundly different way; not as dry, abstract, and dehumanising, but as alive, real, and immediate.'[65] Within this context numbers symbolise the 'fundamental principle from which the whole objective world proceeds,'[66] and because they possess 'a quality and a spirit of (their) . . . own,' numbers therefore serve as epiphanies of spiritual and scientific revelation, perhaps none more so than the mystical number seven.[67]

Throughout many cultures and in many of the world's leading religious texts from ancient Egypt's sacred canon, the Hermetica to the Jewish Zohar, the Muslim Koran and the Christian Bible, seven is the number *par excellence*. Its application to music, mythology, science and religion is prolific, as evidenced by the myriad references to its numerical value in popular sayings and phrases. These include the seven days of the week, seven colours of the rainbow, seven root chakras of the body, seven rays of the sun, seven notes of the musical scale, seven gates of Ishtar, seven heavens, seven seas, seven major planets, seven wonders of the world and more. Seven represents 'the number of the universe, the macrocosm,' says Jeanne Cooper in her *Illustrated Encyclopaedia of Traditional Symbols.*[68] It exemplifies completeness, totality and perfection, safety and security, rest and plenty, reintegration and synthesis. Its significance is drawn from the fact that it combines 'the three of the heavens (the soul) and the four of the earth (the body).'[69] Therefore 'it is the first number which contains both the spiritual and temporal.'[70]

Seven also stood for virginity, says Cooper, as in the seven vestal virgins, and it represents 'the number of the Great Mother.'[71] Exploring the symbolism behind its parthenogenic nature Blavatsky notes Pythagoreans believed that number seven was neither born of a mother or a father but that it proceeded 'from the Monad directly' and was therefore considered 'to be a religious and perfect number.'[72] In mathematical terms this means 'it is not generated as the product of 3 x 2, nor does it give birth, as do 2 and 4 in producing 8, 2 and 5 in producing 10, and so on,' says John Michell and Christine Rhone in *Twelve-Tribe Nations.*[73] In this sense, seven represents eternity for 'Only that which has not been born can be called eternal.'[74] Writing on the secret numbers of the gods in *The Cosmic Code*, Zecharia Sitchin points out that the number seven played a central role in the Christian creation story, which explains why it is a key digit in the Book of Revelation.[75] In ancient Sumer seven not only represented the code number for Enlil (the chief commander of the Annunaki) but it also stood for the 'planetary number' of Earth.[76] In other words, visitors entering our solar system would count Earth as the seventh planet as opposed to the third planet from the Sun, if they were coming from the other direction. Enlil's other official duties included overseeing treaties and administering oaths says Sitchin, who points out that the English word 'swear' (as in the undertaking of an oath) is derived from the Hebrew root word for seven.[77]

Many people once believed 'that the existence of the universe depends on the harmony of the seven,' say the Jobes, largely because of its relation to the biblical 'seven days of creation' that 'control the cycle of the week.'[78] Others relate this dependence to the Pythagorean theory of the 'Harmony of the Spheres,' whose seven tones or 'voice of nature' lies at the heart of all creation.[79] These musical symphonies are seen to govern the endless cycle of world ages, which bring with them cataclysmic changes to our physical environment and spiritual consciousness. Ancient beliefs in the creative powers of number seven have been recently affirmed by science. In *Just Six Numbers: The Deep Forces that Shape the Universe*, astronomer Martin Rees has identified one number as the source of our creation, and that is 0.007. What is so remarkable about this number, says Rees, 'is that no carbon-based biosphere could exist if this number had been 0.006 or 0.008 rather than 0.007.'[80]

In other words, as Robert Temple points out in *The Crystal Sun*, we simply 'wouldn't even exist if the tiny number had varied slightly.'[81] In an involved and detailed discussion of the laws of physics, Temple equates 0.007 close to one-half of a process and equation known as the 'Particle of Pythagoras' which makes up the 'Comma of Pythagoras' — the 'tiny gap' that exists between measuring or distinguishing 'between the ideal and real.'[82] This mathematical equation, says Temple, lies at the heart of the Pythagorean harmony of the spheres. In the sacred science of numerology, zero is not counted for it is whole and complete in itself.[83] This means that the tiny sum of 0.007 therefore translates to the magical number seven that lies at the heart of music, science and creation.

Curiously, as with my grandmother's people, many other Indigenous peoples around the world claim to be of Pleiadian origin and because of the emphasis placed on the Seven Sisters in their mythologies, the number seven is central to this creation. Hindu mythology, for instance, refers to the Pleiades as the Seven Mothers of the World.[84] What could this possibly mean? Were there just seven women who gave birth to humanity and if so, were they the legendary Seven Sisters of the Pleiades? If the recent genetic research conducted on people of

European ancestry by Professor Bryan Sykes of Oxford University is anything to go by, it is possible that science may one day confirm the beliefs of many Indigenous peoples about the stars of the Pleiades. Sykes' research led to the discovery that the gene pool of European peoples is centred on 'seven major genetic clusters.'[85] An even more exciting breakthrough was the fact that just one woman in each case carried 'the single founder sequence at the root of each of the seven clusters.'[86] What this means effectively is that almost everyone of European descent in the world 'can trace an unbroken genetic link . . . way back into the remote past, to one of only seven women.'[87] Referring to these women as 'Clan Mothers', Sykes was able to trace their existence through the mitochondrial gene sequence in our DNA, which is only inherited maternally.[88]

Although these women lived in different regions of Europe and at different time periods from one another (and not simultaneously in the same generation, as one would expect of siblings), nonetheless these amazing genetic discoveries affirm the septenary nature of creation. If Indigenous claims about our beginnings are true, then it is possible that this star seeding may have occurred over an equally longer period of time, which parallels the genetic history of these female progenitors or the *Seven Daughters of Eve*. Regardless of the outcome, Sykes' work is enormously exciting for what it can potentially tell us about our human origins, and to what extent it may confirm or negate human mythologies, including Blavatsky's more controversial 'Seven Root Races' theory espoused in the late nineteenth century. This aspect of the sevening of creation is looked at in more detail in the Greek chapter, for Blavatsky sees within the story of the Seven Sisters of the Pleiades allegories of the Seven Rounds of creation.

Seven Fates

An additional association between the Pleiades and number seven concerns matters of fate and destiny. At the beginning of this chapter we saw that the Pleiades were known as the Seven Fates in ancient Egypt. Their prophetic influences are further revealed in the game of dice, which was often used as a divinatory tool in India and other

parts of the ancient world because of its numerical arrangement and the perceived magical qualities of number seven.[89] And as we have seen, Madame Blavatsky has much to say about the occult significance of number seven and its connection with these stars, including their celestial relationship with the cyclical creation, destruction and reconstruction of the universe. The Pleiades are implicated through their relationship with the black goddess Kali and the numeric symbolism of the die where the throws represent the four ages of time.[90] Although six-sided, its numerals are arranged so that its opposite faces all add to the mystical *seven* — six plus one, five plus two, four plus three, three plus four, two plus five and one plus six.

What can we make of this observation? Quite apart from their symbolic representation of the world ages or *yugas*, how might we read the different combinations of these equations? A possible clue provided by Peat suggests that gambling games involving chance are not designed purely for entertainment purposes at all but are a form of 'sacred ceremony that acknowledges the basic metaphysics of the cosmos.'[91] Their existence, he says, 'implies that such people have a knowledge of the laws of probability and are able to compute the odds of various outcomes.'[92]

It may be that the range of combinations speaks to eternal mysteries relating to time and the creation of the universe, including that of the Lost Pleiad. The numeric layout of the die with its six sides and opposing numbers, which add up to the magic seven, is a perfect visual illustration of the interplay between the numbers six and seven that feature in the aspect of the missing sister. This interaction is repeated in the popular saying to be 'at sixes and sevens' with someone or something. According to *The Oxford Dictionary of English Proverbs*, the saying means 'to be careless of consequences, or let things go to disorder', and derives from the game of dice.[93] The original expression, says the author, was to be '*on* six and seven'.[94] Given the connection of the Pleiades with dice and time, the six sides of the die may be seen to represent the group less the lost sister, who is incorporated into the dice through the above additions. She is simultaneously both visible and invisible, there and not there.[95]

The Lost Pleiad is now immortalised in the company logo of an international car manufacturer, who chose the emblem of six stars to represent their organisation which is named for the popular and better known Japanese name for the Pleiades — *Subaru*.[96] Strangely, even though their legends refer to Seven Sisters, Subaru decided to depict six stars as their insignia, on the basis that most people can only ever see this number of stars, not because the legends refer to a lost sister. No matter the reasons for their decision, the logo naturally incorporates this mythic aspect into the design — which happens to be astronomically accurate — for the individual stars are placed in their correct celestial alignment and spatial distance from one another.[97]

The Sevenfold Deity, One in the Many and Many in the One

The stars of the Pleiades are mentioned at least twice in the Book of Job, where they are referred to as *Kimah*, a Hebrew term for 'cluster' or 'heap'.[98] Many cultures have emphasised the collective, bunched nature of the Pleiades star cluster. The Pawnee Indians of the North American plains look on them as symbols of unity and pray to these stars to teach their people how to be united as them.[99] Likewise, the Aboriginal people of Bandaiyan (Australia) explain the cluster's closeness 'by their being close kin.'[100] The emphasis of these stories, says Bell, is that sisters 'should stay together for safety and through affection, shared rights in land, and shared responsibilities.'[101] Those who wander or stray from the group's safety are considered lost or endangered in a spiritual, emotional and physical sense, which explains Aboriginal people's concern for the lost or missing sister.

A Polynesian myth tells how the Great Star of Matariki once formed a single, bright, big star before breaking up into the individual stars of the Pleiades.[102] Although their names for the star cluster vary among the different Polynesian languages there remains obvious linguistic similarities between them. Maori people of Aotearoa (New Zealand) still refer to this cluster as Matariki collectively, and individually as Matariki and her six daughters.[103] The Arapaho native people

of Turtle Island (America) tell a remarkably similar tale of an exploding star, Alcyone, which they refer to as the 'Broken Chest Star,' which created the star cluster.[104] As the brightest star in the Pleiades this may explain her selection as the original intact star from which the others may have emerged.[105] Oddly, Robert Graves refers to Alcyone as the 'mystic' leader of the cluster.[106] What he was alluding to is not known, but the possible meaning of this epithet is explored in the Greek chapter.

This tale of a single star giving birth to other suns in a starburst is a wonderful visual image from nature that perfectly illustrates the popular notion of the Godhead as the 'One in the Many and Many in the One'. This phrase refers to the emanation of divinity from zero (the source of all) to two, three, four and even more multiple deities. The idea of a sevenfold deity in particular appears to have been a common phenomenon in the ancient world and in some contemporary civilisations. In ancient Mesopotamia the Pleiades are referred to as Mul in their religious texts *Mul Apin*, which 'literally means star,' says Krupp.[107] Another name for the star cluster refers to them as the Sevenfold One.[108] The seven stars of the Pleiades were viewed in a similar manner in ancient Egypt where they were seen as the stars of Hathor the Cow Goddess, hence their epithet as the Seven Cows.[109] The Blackfoot native peoples of the American and Canadian plains refer to their god as '*Ekitsikuno* the Seven One'.[110] Although they place their Supreme Being in Ursa Major, the seven stars of the Pleiades still play an equally important role in their initiation ceremonies and sacred mysteries.[111] As the Seven Cows, Hathor represents the embodiment of the multiple goddess as expressed and understood by this spiritual axiom. Their Arabic name Al Thuraya or 'the many little ones', bears a remarkable similarity to their Egyptian name Athurai or Atauria, says Allen.[112] Also known as Chu or Chow they were said to represent either the goddess Nit or Neith.[113]

This is not as contradictory as it first appears, for she is part of the sevenfold deity I have already identified as the Seven Sisters in ancient Egyptian cosmology. The idea that the Godhead can represent more than one personality is common to many religions, hence the popular

saying, 'All the Gods are One God.'[114] One of the better known divisions is the Triple Deity of Brahma, Shiva and Krishna of Hindu tradition, or the Father, Son and Holy Spirit in Christianity, or the Triple Goddess of Crone, Mother and Maiden in Goddess worship and pagan religions. In *The Language of the Goddess*, Marija Gimbutas says that notions of the sevenfold deity or the multiple deity are 'aspects of the one great Goddess with her core functions: life-giving, death-wielding, regeneration and renewal.'[115] Nature itself is multiple, she stresses, which explains the Goddess' many manifestations, especially in her aspect as Mother Nature.[116]

Water girls, ice maidens, winged bulls and bird goddesses of the Pleiades

The painting that Krupp refers to comes from Groote Eylandt in Arnhem Land in the Northern Territory of Australia. It portrays an artist's interpretation of a traditional story of the Wutarinja fisher-women of the Pleiades and their hunter husbands in the Orion constellation. In this part of Australia, Orion is seen as a canoe in which the Burumburumrunja fishermen travel to fish and hunt for food for their families, including their Pleiadian wives.[117] In this regard, it is remarkably similar to the Maori star legends of the Tainui waka (canoe) of which the three belt stars of Orion make up its stern and the stars of the Pleiades its bow.[118] Although the 'theme of pursuit' is not present in this particular story, as Isaacs points out,[119] both myths highlight another familiar theme — the association of the Pleiades with water in all its various forms: rain, frost, ice, snow, rivers, creeks, lakes and oceans. In Greek and Aboriginal mythology the Pleiades are often referred to as ocean or sea nymphs or as water girls and ice maidens. Their relationship with water is multi-layered and multi-faceted and we see numerous connections of the Pleiades with the weather, agriculture, navigation and sailing.

A passage in the *Talmud* connects the Pleiades with floods and rain: when the Almighty Creator wished to send a flood to Earth he simply removed two stars from the Pleiades to cause the Deluge, only to replace them with two stars from Ursa Major as celestial bath plugs

when he wished the waters to abate.[120] In a similar vein, an Aboriginal Australian legend from Ooldea on the Nullarbor Plains of South Australia tells how the Minmara (the Bird Women of the Pleiades) stemmed the floodwaters of the Southern Ocean from 'eroding the mainland.'[121] The flood started after a kangaroo skin bag full of fresh water burst during a disagreement between two brothers. The Sisters promptly set about building a barrier made from the roots of the kurrajong tree to stop the ocean waters flooding the plains. 'This explains why *ngalda* roots contain fresh water and the kurrajong is the *water tree*,' say Ronald and Catherine Berndt in *The Speaking Land*.[122] A vital clue to the Pleiades' connection with water lies in their representation as birds, particularly aquatic birds.

In many Aboriginal Australian legends, the Pleiades are often interchangeably described as birds and women, like the seven emu women who are chased by dingo men in the story of the Magara. The application of avian imagery to the Pleiades in many world cultures is a prevalent motif, says Richard Allen, who points out that in many eastern European countries the Pleiades are often depicted as 'a hen with her chickens.'[123] In her account of Polynesian astronomy, Maud Makemson relates that the Pukapukan Islanders in the northern Cook Islands, who see themselves as the descendants of the Pleiades, also refer to their people as Te Manu Mataliki, 'the Birds of the Pleiades.'[124] In the Greek legends the Sisters are turned into doves and one of them, the second eldest, is linked with the European kingfisher that gave rise to the legendary halcyon days in Greek mythology.[125]

Very few writers have attempted to provide an adequate explanation for this avian association with the Pleiades. One of the first was the nineteenth-century British ornithologist D'Arcy Wentworth Thompson, who speculated on the existence of some kind of astronomical relationship in the Greek legend but does little more with this observation.[126] Robert Graves goes one step further in exploring their astronomic and spiritual symbolism, but only in relation to the halcyon legend and not to their wider application.[127] Similarly, Marija Gimbutas writes of the significance of bird imagery in the art

and spiritual traditions of what she calls 'Old Europe' in regard to Goddess worship, but does not relate this discovery to the stars generally or to the Pleiades specifically.[128] Nonetheless, her diligent research and probing insight into the underlying symbolism of birds, in particular their aquatic aspects, has enormous implications for understanding their connection to the stars of the Pleiades.

The 'Bird Woman' is another aspect of the Mother Goddess as supreme creator and is one of her most ancient insignia, says Gimbutas, who spent years analysing old European artefacts with these images.[129] She believes many of the decorative icons on these relics, which included sculptures and clay pottery, may be read as a kind of 'language' of the Goddess and she identifies various types of abstract or hieroglyphic symbols that comprise its iconography. For instance some markings are clearly alphabetic, such as the letters V, X and M. Others are more glyph-like, such as diamonds and triangles. Some images are more representational in that they show the Goddess in zoological form as a bird, snake, pig, frog or bee, having real features such as breasts, eyes, hands or bird's feet.[130] V markings, or the chevron in particular, says Gimbutas, are typical of waterbirds and other aquatic items such as mussels whose significance in Australian Aboriginal culture is part of their sacred laws.[131] Over time the chevron has become 'the designating mark of the bird goddess.'[132] Its significance is that it marks the 'pubic triangle' and therefore represents the vulva — the 'life-giving moisture of the Goddess' body.'[133] Other signs besides the chevron that embrace the aquatic sphere include 'zigzags, wavy or serpentine bands, net and checkerboard' of streams, rain, creeks, rivers, lakes and oceans.[134] All these glyphs are associated with the life waters of the Goddess and 'with her functions as Life Giver.'[135] They embrace the aquatic realm because of the prevailing belief that 'all life comes from water.'[136]

This so-called language of the Bird Goddess not only gives added insight to the aquatic themes of the Pleiades in various world mythologies but it provides an analytical framework for deeper analysis. It is especially relevant to the Dreaming of the Seven Sisters among the Ngarrindjeri people of South Australia, whose spiritual beliefs were viciously denigrated, then vindicated, in the highly publicised Hindmarsh Island case, but only after an exhausting and damaging legal battle with Australian courts who failed to recognise and value the richness and wisdom of their Indigenous cultural heritage.[137] Perhaps some day the legal fraternity and the wider Australian society will understand and appreciate the significance of these Dreamings and what they mean to our world heritage.

The idea that all life comes from water is central to Ngarrindjeri beliefs relating to the creation of their world of all living beings, human and otherwise. Although the publicised legend (as opposed to that which remains secret) does not specifically refer to the Seven Sisters as birds, nonetheless flight of some sort is implicated. Flight enables the Sisters to come down from the heavens to visit their mother who lives in the waters surrounding Kumarangk (Hindmarsh Island).[138] Having stayed behind to invest the waters with life, the mother's presence is central to the spiritual wellbeing of the people and the environment, without which there can be no existence, no connections and no communications between people and their ngatji — totemic animals, fish and plants.[139]

The confluence of the River Murray and its tributaries, Lake Alexandrina and the Southern Ocean, where saltwater mixes with freshwater, is the centre of Ngarrindjeri creation.[140] Because this area is connected with conception, reproduction and fertility it involves women's business generally, and also specifically, with the Dreaming of the Seven Sisters. For it is the Mantjingga who issue divine orders to Ngarrindjeri women by giving them 'directives about life, about preparing for womanhood, marriage and childbearing.'[141] Every winter when the stars of the Pleiades are no longer visible in the evening skies, the Ngarrindjeri believe the Sisters are visiting their mother in the sacred waters. But, says Ngarrindjeri elder Veronica Brodie, 'there has to be a clearway, so they can return.'[142] The necessity to ensure this clearway resulted in the legal battle at Hindmarsh Island, but to no avail and the bridge was built against the wishes of the Ngarrindjeri women. To what extent it has obstructed the flow of celestial energies from the Pleiades or their impact on the Dreaming of the Seven Sisters in this area is not known, but it remains an important Dreaming place for those who choose to believe.

Besides their immediate association with the Pleiades, aquatic symbolism is used to represent women or femaleness in Aboriginal Australia. Among the threefold Aboriginal nation of the Ngarinyin, Worora and Wunambal in the Kimberley, the sign for a woman is the water mussel or darrul darrul.[143] In this region of Bandaiyan, Aboriginal Law forbids young boys to eat the water mussel because 'it symbolises a woman's body' and is therefore sacred or mahmah.[144] In *Yorro Yorro*, one of my Aboriginal elders and teacher, the late David Mowaljarlai, explains why this is so: when the Wandjina creator 'made the little girl, he put a water mussel between her legs so that man would find it when she grew up, to revere and look after his companion.'[145] The splay of two fingers to form a V sign is used to indicate how 'Wandjina drew that on earth' as part of his plan to create women.[146] It had never occurred to me until I read Gimbutas' work, why so many ancient cultures placed the Mother Goddess in Taurus. The answer is now obvious, for Taurus is the only constellation in the night skies that is marked by a distinctive V-shaped asterism or star pattern. As Gimbutas states, this V pattern is the very mark of the Goddess in her bird and bovine aspects that represents her horns and feet. The pattern is also played out in the skies above us in the flight path formation of birds.

The connection of the V with Taurus, in particular the shape of its horns, 'provides the key to understanding why the bull is linked with regeneration,' says Gimbutas.[147] Drawing on the earlier work of Dorothy Cameron in *Symbols of Birth and Death in the Neolithic Era*, she says the head and horns of the bull bear an 'extraordinary likeness' to 'the female uterus and fallopian tubes'; the perfect symbol of the female reproductive organs.[148] This interplay between birds and bulls as interchangeable signs for Taurus and the Pleiades comes from direct observations of the symbiotic relationship between both animals, which is a direct result of this regeneration process. Cows and bulls, like birds, eat grass seeds, which in turn are further dispersed and pollinated through the flight of birds. Their relationship is further strengthened by the mutual benefits they bring to one another; the bull offers protection from predators while birds remove ticks and other parasites, which threaten it with infestation and the possibility of disease.

Some writers suggest that the universal terrestrial image of birds sitting on the backs of bulls may have served as an astronomical symbol for the stars of Taurus and the Pleiades in several ancient civilisations, including Greece, Egypt and Mesopotamia. These images adorned the walls and ceilings of palaces and pyramids and in some instances were imprinted on coins. Richard Allen suggests that an ancient coin from the Greek Isle of Samos showing a single dove on a bull's back, estimated to be about 2,600 years old, might represent the Pleiadian doves.[149] Today in Egypt the same image can be seen on the ceiling of the tomb of Set I in the Valley of the Kings.[150] Although some would argue this bird represents the falcon-god Horus, it is worth remembering that his mother was Hathor the Cow Goddess and Pleiadian Queen in ancient Egyptian mythology.[151] The winged bull was a popular icon in ancient Mesopotamia where it represented Taurus or the 'Bull of Heaven' as it is referred to in the classic Gilgamesh epic. In some 'rare illustrations', say the Jobes, it was depicted with a lone bird on its back, which may have indicated the Pleiades.[152] On the other hand, the composite image of bird and bull may have emphasised the joint relationship between Taurus and the Pleiades.

Apart from the life-giving functions of the Bird Goddess, another aspect to the spiritual dimension of birds relating to their shamanistic nature is particularly revealing of their connection with the stars of the Pleiades. In many cultures birds are seen as sacred messengers because of their ability to fly into the heavens where they communicate with gods and spirits. As a consequence the flight of birds takes on a deeper meaning, which represents spiritual transcendence and release from earthly bondage. 'Magical flights' and other kinds of 'ascent into the sky' are the special trademarks of shamanism, says world-renowned historian of religion Mircea Eliade, especially when they are tied to dreams and ecstatic ritual practices.[153] The use of feathers in ceremonial rituals is therefore designed to emulate the perceived qualities and talents of birds and to call on their totemic powers to assist shamans and other ritual participants. Besides the use of hallucinogenic substances, music — particularly drumming — is an integral component of inducing altered states of consciousness that enhance the shamanic experience. The authors of *Hamlet's Mill* take the analysis to another level. They argue that quite apart from sky travels and heavenly ladders, a closer inspection of shamanic practices reveals 'very ancient patterns' that are associated with universal rhythm and the maintenance of time through sound and vibration.[154]

The idea that sound is fundamental to the process of creation is familiar to modern science, as it was to the Ancients and expressed in Plato's 'Harmony of the Spheres'. Several writers, including Madame Blavatsky and Barbara Hand Clow, have linked the stars of the Pleiades to the Harmony of the Spheres and the creation of our universe through the manipulation of sound and vibration. Hand Clow's reference to the birdsong of the Pleiades is alluring in this regard. In *The Pleiadian Agenda* she claims that birdsong is a seven-dimensional sound that 'causes human language to be so sound coded.'[155] It is intriguing that some Aboriginal Australian Dreamtime legends tell of a time in the distant past when birds once taught human beings how to speak! Although Blavatsky's theories and Hand Clow's writings are largely ignored by mainstream academia because they are perceived as being too obscure or New Age, on the other hand the more erudite and scholarly (and perhaps more academically acceptable) *Hamlet's Mill* is effectively saying the same thing. In examining the motifs and designs of drums in ancient Mesopotamia, Giorgio De Santillana and Hertha Von Dechend identify the region surrounding Taurus and the Pleiades as *the* central celestial location governing time and creation. They note in particular that the 'tuning' and 'untuning' of the skies is specifically connected to the changing of the World Ages.[156]

On a grander cosmic scale, Barbara Hand Clow offers an alternative explanation for the special relationship that exists between birds, humans and the Pleiades. As in Hindu beliefs and other older spiritual traditions like the Aboriginal Dreamtime, which teaches that every-thing is consciousness, including stars, Hand Clow maintains that animals are the primary source of 'star consciousness' or 'star wisdom' for human beings.[157] Specifically this translates to cats living 'the star consciousness of Sirius, the birds of the Pleiades, and the bears of the Andromeda Galaxy.'[158] So in *The Pleiadian Agenda* she refers to the Seven Sisters as our 'Pleiadian bird teachers', whom she claims taught us many things during human evolution.[159] Many Indigenous peoples, including Aboriginal Australians, share the idea that animals and stars are our greatest teachers. Of these, the birds of the Pleiades play a leading role in the initiation of young women. These shamanic practices lie at the heart of all Aboriginal women's rituals and cere-monies, especially those relating to the Dreaming of the Seven Sisters. Who could forget the Seven Sisters corroboree performed at the opening ceremony of the Sydney Olympic Games in September 2000, when Australia's Indigenous women spectacularly broadcast the *Dance of the Pleiades* to the rest of the world?

Birds of flight, divination and navigation

Beyond these spiritual aspects, the flight of birds, their individual traits and biological characteristics have contributed enormously to our understanding of navigation, which led to world exploration and colonisation of lands that enabled the survival of our species. The in-built navigation systems of birds, in particular, made them reliable companions to ancient mariners where certain species indicated the presence of landfall.[160] The recognition of the avian navigator and their role in piloting sailors to safety led to the ancient Greek custom of releasing pigeons at the commencement of the sailing season.[161] Given that the stars of the Pleiades heralded this auspicious event, and the fact that Zeus turned them into doves (a species of pigeon), then it is likely these birds symbolised the Sisters in zoological form. Here we can make all sorts of connections between the Sisters with the celestial, atmospheric and nautical realms. We can trace these associations in sailing, the flight of birds, and the fullness of seasonal cycles, numerical systems and even sexual relations.

In his well known poem *Works and Days*, the Greek historian Hesiod left instructions to Mediterranean mariners to avoid sailing when the stars of the Pleiades fell below the horizon, indicating stormy weather. This is reflected in the turning of a full seasonal cycle from the Pleiades heliacal rising in May and their setting in late autumn, which marked the start and end of the maritime season.[162] Even the word *plias* means 'to sail'; hence the star cluster was known as the 'sailing stars'.[163] Another word *pleos*, meaning 'full', connects with the idea of a full sail, while the plural of *pleos* or 'many', suggests the combination of forces at work — wind, sky, sea and ship.[164] To these connections we can add the Greek word *peleiades*, or a flock of doves, because for aeons the appearance of birds has shown sailors that land is near.[165] This idea of fullness brings us back to Zeus and the amorous gods above ancient Greece, and the word 'satiated' which derives from the Hebrew root word for seven.[166] Finally, as Allen argues, 'as Pleione was the mother of the Seven Sisters,' it is more than likely that these celestial maidens were named for her.[167]

Whatever the true etymological explanation, doves were particularly sacred to the Goddess, says Barbara Walker.[168] This may explain in part why the planet Venus rules the zodiacal sign of Taurus in astrological lore, for that constellation houses her favourite birds. Some writers have likened the Pleiades to a bunch of grapes but the association may derive from their original perception as doves. For instance, the same ornithologist, Thompson, traced the origin of the medieval Latin term for bunch of grapes, or *butrum*, to a species of dove known as *columba oenas*. He noted in particular that these birds migrated at the time of vintage and that their purplish-red breast colouring corresponded to certain dark wines of similar hue.[169] This connection of doves with grapes, says Richard Allen, is confirmed by the designs of ancient coins from the Grecian city of Mallos in Cilicia that show 'doves with bodies formed by bunches of grapes,' later superseded by drawings of grapes alone.[170] These ancient Pleiadian images of doves with grapes still survive in Christian symbolism today.

Another link between doves and grapes, other than the fruit's colouring or time of harvest, may be the relationship between alcohol and other forms of magic potions. The Greek philosopher Homer mentions the Pleiades in *The Odyssey* where he describes them as 'seven doves that start out from the west with ambrosia for the infant Zeus.'[171] As the birds flew to the Wandering Islands they entered the gateway to the Black Sea known as the Symplegades where the notorious twin rocks closed in and crushed one of them. Allen suggests the story may have originated from tales of the single white dove, which guided Jason and the Argonauts through the precarious passageway on their ship *Argo* in search of the Golden Fleece. This argument does not hold sway in view of other world mythologies that relate the Sisters to a flock of birds, usually seven.[172] It also fails to comprehend the underlying symbolism of the divine connection between the Pleiades, and their portage (in zoological form as birds) of powerful elixirs such as ambrosia and soma to Indo-European storm gods. These mythic references to sweet drinks of immortality reveal hidden aspects of our creation beyond the meteorological meaning ascribed to them as 'seasonal symbols of thunder, lightning, and rain.'[173] Could Homer's tale allude to the Lost Pleiad instead?

However this may be interpreted, the lone dove endures as a symbol of loss, where deviation from the course leads to emotional or physical demise, while returning to the fold brings fulfilment. This aspect of the potential spiritual loss of a tribal member provides the focal point for sense of community among Aboriginal Australians that evoke deep feelings of concern and compassion for the Lost Sister, whoever she is and wherever she may be. She need not be Aboriginal; rather she represents any lost soul searching for community and a sense of belonging. Compassion, as the Dalai Lama states, is one of the greatest of human virtues, and this philosophy is equally shared by Aboriginal peoples whose use of the Aboriginal English term 'sorry' expresses this feeling of deep sympathy or empathy for someone or something such as land or 'country'.[174]

Oracle doves, navel stones and geodesics

Doves played vital roles as sacred messengers at oracle centres such as Delphi and Dodona, where women priests known as 'doves' interpreted divine messages.[175] Originally built to honour the prophetic wisdom of the Mother Goddess, these oracular shrines were later taken over by patriarchal forces.[176] Threatened by feminine powers, the appointment of male priests paved the way for eradicating women from playing an active role in spiritual matters. A deeper historical examination of the ancient civilisations of Greece and Egypt reveals some fascinating connections between doves, oracular shrines and navigation. Herodotus tells of two black doves that flew away from Thebes in Egypt with the message that oracular shrines should be built at two particular locations, Dodona in Greece and Ammon in Egypt (now Siwa).[177] Graham Hancock tells another story in *Heaven's Mirror* of Zeus releasing two eagles 'from opposite ends of the earth' and directing them to 'fly towards the centre — where they naturally met at Delphi.'[178] The flight path of these two birds running east and west around the globe established a line of latitude to pinpoint the exact orientation of the Delphic oracle. Whether they were eagles or doves is not really in dispute, says Hancock, for both were interchangeable in ancient architectural iconography and measurements. What is more revealing is that the Pleiades are astronomically aligned along an east–west direction of the Earth in relation to our Sun, and therefore many buildings in the ancient world were oriented to this star cluster.[179] At the centre of these oracles stood the foundation stone, the *omphalos*, or 'navel' in Greek. Ovoid in shape, carved lattice type, basket-weaving mesh patterns decorated the stones and two doves were etched into their top surfaces or sides.[180] Interestingly, a pair of doves set the surveying standard for parallels, meridians and measurements for the Great Pyramid, according to Roberto Stecchini,[181] who argues that the mesh covering represents 'a net of meridians and parallels.'[182]

How did the Ancients come to symbolise architectural and surveying measurements with doves? What information did they possess in those times that led to iconography of this sort? The association of doves with geometric measurements is intriguing in the light of scientific research into the eyesight of pigeons and their cognition, which suggests the birds' flight is determined by a combination of visual grids and sensitivity to the earth's magnetic field.[183] Though little was known about these ancient forms of computation until recently, the science of geodesics shows that the celestial alignment of ancient structures involved a highly sophisticated surveyor system. The Delphi temple, for instance, featured a roulette wheel device that operated as an abacus to calculate angles for geometric positioning, and the navel stone's mesh patterns indicated latitude and longitude somewhat like the grid pattern used in modern day maps.[184] Thus the omphalos stone at Delphi served as a geodesic pointer to energy patterns in the environment and astronomic realm. Shared by many different cultures, this complex system involves the so-called 'ley lines' traversing the British Isles, the 'dragon paths' in China, and 'songlines' in Australia.

Although Delphi is often referred to as 'the navel of the world', several writers point out there were many such navels, often positioned at the exact spot where meteorites fell to earth. The stars of the Pleiades, as we have seen, are connected to many of these sacred sites through mythology and cultural artefacts. The remarkable feature of meteorites, as Robert Bauval and Adrian Gilbert point out in *The Orion Mystery*, is that they all enter the Earth's atmosphere from the east heading in a westerly direction as they follow the course of the Sun.[185] Given the close proximity of the Pleiades to the ecliptic, this may explain in part why ancient buildings were aligned to these stars along an east–west axis.[186] To mark these centres of creation and increased fertility, rods or pillars were sunk into the ground or other objects erected. The energy lines emanating from these navel sites are like umbilical cords linking our embryonic creation to the universe.[187] In the central Australian desert the monolithic red rock Uluru is viewed by Aboriginal people in the Kimberley as the navel of the bisexual being Bandaiyan or Terra Australis.[188] This androgynous ancestral creator lies on its back straddling the continental shelf between the Indian and Pacific oceans. Not only does Bandaiyan cross those oceans, it also appears in zoological form as the world-supporting turtle of Eastern philosophy and in its anthropomorphic form as Purusha, the supreme bisexual creator being in Hindu mythology.[189]

Cosmic mountains, pyramids, knolls, hillocks and mounds

Apart from sacred waters and a presence in the sky, the Pleiades have other places on Earth that allow communication between the worlds. In many of the stories we see the Seven Sisters associated with some kind of high hill, rock or tower, or a series of seven hills. These geographical features are used in the legends as virtual launch pads and landing platforms that enable the Sisters to visit and leave the Earth and to serve as a refuge when being chased. In Aboriginal Australia, as in other parts of the world, the locations of these hills are regarded as sacred sites by the Indigenous peoples because of their relationship with Dreaming Beings. In chapter four we see this similarity in the Native American Kiowa legend of the seven star girls who take refuge on a rock tower from a bear that chases them. In some areas of Bandaiyan (Australia) the 'Seven Sisters Dreaming' is centred on seven hills, such as on the Atherton Tablelands west of Cairns in northern Queensland.[190]

This mythic connection of the Pleiades with seven hills in Australia raises questions about their possible existence in other regions of the world. One example is the legendary Seven Hills of Rome on which the city is founded.[191] Could there be a story of the Seven Sisters involving these particular hills? Or has the passage of time eroded their presence from early Roman history? What of other countries, do they have certain geological features and landmarks connected with the Seven Sisters of the Pleiades? We see, for instance in southern China (on the outskirts of the Guilin Hills), seven distinctive mountains along the banks of the Li River. Nearby is the 'Seven Star Park' supposedly named for these seven peaks but according to some writers there is an older astronomical association.[192] To some their geo-physical outline is thought to resemble the celestial pattern of the 'Big Dipper' (the seven prominent stars of Ursa Major), but given that the Pleiades star cluster is said to mimic that constellation in the Northern Hemisphere, there may well be a Pleiadian connection as well. Another frequent, related theme is the reference to the existence of a cave inside the hill or mountain or nearby that enables the girls to take shelter or retreat from the world during their stay here on Earth. Coincidentally, there exists such a cave within the park, although it is simply known as the 'Seven Star Cave'.[193]

The tendency of some researchers to immediately suggest a connection to Ursa Major, as opposed to the stars of the Pleiades, whenever and wherever there is some marking septenary feature — from seven dots painted on an ancient Egyptian calendar to seven hills in the landscape — is somewhat biased. It appears that many researchers (astronomers, archaeologists and the like) are far too eager to dismiss the possibility that the seven stars of the Pleiades could have a direct bearing on ancient artefacts, monuments or locales in favour of the stars of Ursa Major. This is remarkable, considering that the stars of the Pleiades are visible to skywatchers in both hemispheres, whereas the stars of Ursa Major cannot be seen at all in the lower regions of the Southern Hemisphere, and then only partially so in the northern regions above 30° S latitude. That covers most Australian cities, with the exception of Brisbane and Darwin, where observers would only have partial glimpses of the Big Dipper low on the horizon, let alone the full constellation of Ursa Major.

To favour an Ursa Major connection over the Pleiades suggests an inbuilt prejudice that only serves to wipe out any possible traces of the Seven Sisters' Dreaming in areas where Pleiadian memories may be fading from human consciousness such as in Egypt. Here in a country predominantly lacking natural elevation and physical features such as hills or mountains, the presence of pyramids takes on enormous significance as 'cosmic mountains'.[194] From this perspective, Egyptologist Toby Wilkinson of Cambridge University in England argues that pyramids may be seen as symbolic 'stairways to the stars' that act as 'launch pads for the pharaoh's journey to the afterlife among the stars.'[195] This begs the question — if scientists are willing to consider the possibility that pyramids may represent virtual 'launch pads' to the stars (albeit on a spiritual rather than a physical basis) then surely the reverse may be true. In other words, to what extent might they also be regarded as 'arrival platforms'? Could they represent two-way celestial traffic, whether symbolic or real? And if so, what does this make of other pyramids, Devils Tower and other hillocks associated with the Pleiades? These extraterrestrial notions are further explored throughout the book and especially in chapter six, which looks at the ongoing relationship between the Egyptian pyramids and the Pleiades. Beyond the physical, environmental features and the odd monument or building site, there are mythical references to whole towns and cities following the septenary patterns set down by the Seven Sisters. Some authors have linked these legendary Seven Cities to Atlantis and ultimately to the Pleiades.[196]

Honeybees, honeycombs and bee goddesses

The alleged sweet nature of the Pleiades is expressed in the Book of Job, as we have already seen. The clues to this mysterious phrase might be connected with what I have identified as the 'honey theme' in much of the world's mythology on the Seven Sisters of the Pleiades. In the Bundjalung myth of Aboriginal Australia, the Berai Berai men of Orion make repeated attempts to woo the Sisters with honey, which they collect from bees' nests.[197] In the Greek legends at least two of the Sisters — Electra and Merope — are linguistically connected with bees and honey through the meaning of their names; for instance, one of the many interpretations of Merope's name is that of 'bee eater'.[198] Even Electra's name suggests an association with honey; for in Latin this means 'Amber', which the ancient Greeks originally thought to be made from honey because of its golden hue.[199]

A further linguistic connection of bees and honey with Electra is through her son Dardanos who founded Troy. In *The Sacred Bee*, Hilda Ransome says 'that another name for bee in Asia Minor was *darda*, so perhaps the old town of Dardanos, near the modern Dardanelles, was originally the Bee-town.'[200] In *Star Names*, Richard Allen says the star Taygete marked the beginning and end of the two honey harvesting seasons of the Greek calendar[201] but according to Ransome, 'the old writers on apiculture all agree that the bees only collect honey between the rise and setting of the Pleiades', that is all of the Sisters.[202]

The honey theme is vicariously extended in Native American legends where a bear or pack of bears pursues young girls or boys who escape to the skies by becoming the stars of the Pleiades. Bears are renowned for their love of honey and, like other animals in Indigenous cultural traditions, they have lessons to teach human beings. The particular lessons that bears impart are discussed in chapter four on the Seven Little Star Girls of Devils Tower. Why is there this widespread correlation between honey and the Pleiades, and what are we to make of it? It seems that to unravel the mystery of the honey theme is to gain some understanding and insight into this biblical passage and the wisdom it contains. The answer lies somewhere within the mythology surrounding honeybees and honeycombs, and in the meaning ascribed to them by the Ancients.

What do we know about honey? We know it was a prized food in the ancient world, not just for its sweet taste but also for its perceived qualities and benefits, medicinal and otherwise. Exploring the symbolism of honey, Jeanne Cooper says that the imagined 'parthenogenic origin of bees' in the ancient world meant that honey was seen as uncontaminated, which made it an ideal food offering to the gods.[203] Subsequently, as a symbol of purity it was 'poured on the hands and tongues' of initiates in the patriarchal Mithraic cult and used in other chthonic rites.[204] In some cultures honey was considered an aphrodisiac, connected with fertility and even thought to bestow 'poetic genius, eloquence and wisdom.'[205]

But above all, says Barbara Walker, honey represented immortality and rebirth.[206] This explains why the dead were often embalmed in honey and placed in a foetal position in large jars awaiting their rebirth.[207] Apart from its perceived purity, this immortal aspect might explain the common reference to honey as 'the food of the gods.'[208] Certainly Ransome argues the case that honey constituted one of the key ingredients of magical elixirs of immortality; from ambrosia and nectar of the ancient Greeks, to mead of the Celts and Anglo-Saxons, to soma and madhu of the Vedic hymns of India. In her chapter on 'The Food of the Gods' she says, 'The conception of nectar and ambrosia and of honey both included the ideas of sweetness, loveliness, pleasant odour, of healing and life-giving qualities, of the gift of sweet speech and prophetic power.'[209] This explains why children 'were fed with honey immediately after birth among the Greeks, Indians, Germans and Hebrews.'[210]

Beyond honey's spiritual qualities, its medicinal and healing properties were well renowned in the ancient world and celebrated in the Finnish epic the *Kaleva*, which tells how Lemminkainen's mother used honey to bring her son back to life and that Mehilainen the Bee assisted in her ministrations.[211] This connection of honey with the bee goddess in European cultures reveals its sacred nature and its mysteries as they relate to stories of the Seven Sisters. Several goddesses bore the epithet 'Queen Bee', says Walker, including the 'biblical matriarch Deborah.'[212] Like the oracular dove priestesses at Delphi, the term 'bees' or *melissae* was given to priestesses in the servitude of the Goddess. Hence at the famous Temple of Artemis at Ephesus (one of the Seven Wonders of the Ancient World) eunuch priests known as *essenes* or drones, accompanied these bee priestesses.[213] It's entirely feasible these monastic communities were modeled on observations of honeybees in their natural habitat. As Monica Sjoo points out in *The Cosmic Mother*, 'Among honeybees the drone group is produced and regulated by the sterile daughter workers and the fertile queen.'[214] Within this structure, drones exist purely 'to mate with the queen.'[215] Furthermore, 'an average of seven drones per hive' is needed to 'accomplish this act each season, and then the entire male group is destroyed by the workers.'[216] The presence of bees was also visible at the Oracle of Delphi, says Ransome where, according to several ancient historical sources, the original temple was made of bird feathers and bees wax.[217] Ancient coins and tokens from Delphi illustrating bees confirms this.[218]

According to Ransome, the combined presence of birds and bees in ancient Greek mythology suggests the possibility of an ancient religious cult that worshipped both animal and insect together and that 'the remembrance of the bees survived in the name *Melissae*.'[219] This association with bees more adequately explains the biblical reference to the sweet influence of the Pleiades rather than their depiction as doves alone. Curiously, in *The Feminine Monarchie* written in 1609 by Charles Butler, there is a strange reference to bees as the 'birds' of Muses.[220] One can only wonder what this mysterious passage is alluding to, but I suspect it has something to do with this ancient cult. Part of the allure of the bee, says Walker, was that it symbolised 'the feminine potency of nature,' and that it stored its magical elixir within the revered honeycomb.[221]

Pythagoreans were especially drawn to the geometric mysteries contained within the hexagonal shape of the honeycomb, which represented all that was perfect and reflected 'the underlying symmetry of the cosmos.'[222] Of special interest was the numerical value of its six-sided shape with its sixty-degree angles. This represented number six as the special number of the Goddess in her aspect as 'the dual Triple Goddess.'[223] The sacred nature of the honeycomb and its relationship with the Goddess explains why a golden honeycomb was put on display as her representation in Aphrodite's temple at Eryx.[224] Explorations of the mathematical properties of the hexagonal shape of the honeycomb are revealing of the relationship between the numerals six and seven and their role in the stories of the Seven Sisters. Hexagons are not only six-sided, thereby embodying the goddess Aphrodite, but they extend outwards in sixes, creating the mathematical series of 1,6,12 and so on, thus duplicating itself in sixes. Within the structure of the honeycomb you will see that one hexagon provides the base or foundation for its symmetrical reproduction. In the first two inner layers, six others surround one hexagon. In others words, six sisters surround the seventh or missing sister, who completes the mystery of the six and the seven, hence its connection with the Pleiades.

A closer look at the physical features of the honeybee, particularly its wings, reveals some intriguing insights into the legend of the lost or missing sister, especially her clandestine nature. Said to hide her face in a veil or turn her head in shame for marrying a mortal in the Greek legends, these mythological stories attempt to explain why the Lost Pleiad cannot be seen as clearly as her more resplendent Sisters.[225] Their association with honeybees is also interesting because bees are scientifically classified as *hymenoptera,* or 'veil-winged'.[226] This recalls 'the hymen or veil that covered the inner shrine of the Goddess' temple, and the officiating nymph (high priestess) who bore the title of Hymen and ruled over marriage rituals and the honey-moon,' says Walker.[227] The word 'hymen' is of Greek origin, which means veil or membrane and was sometimes used to refer to the covering of the womb, 'the temple of the female body.'[228] Thus, the bride 'who lifts her veil to receive her bridegroom's kiss' still symbolically enacts 'the rending of the veil in a first sexual intercourse,' says Walker.[229] This consecrated notion of the hymen lives on in songs of praise sung in Christian churches today for the word 'hymn' derives from *hymen,* 'for the original *hymnos* was a wedding song.' An example survives in the Song of Solomon in the Bible.[230] With the rise of patriarchy 'their archetypal connection with the 'hymeneal rites' was largely forgotten,' says Walker, and today 'all wedding music is secular.'[231]

From the microcosm to the macrocosm we see this notion of veiled maidens in the night skies where all of the Sisters are enshrouded in blue veils of nebulosity (an interstellar dust cloud otherwise referred to as a nebula).[232] The nebula is more pronounced around the star Merope, which may provide a scientific basis for the myth of the lost sister, in the sense that one of the Sisters is said to have turned her face away in shame or else veiled herself. Taking the individual nomenclature of the star cluster and their individual brightness as proof that science does not reflect mythology, some astronomers point out that shining at 4.1 apparent magnitude,[233] Merope is the fifth brightest star in the cluster and therefore cannot be the Lost Pleiad.

In actual fact this does not necessarily discount mythology, for the astronomical nomenclature may have been wrongly applied in the first place. In other words, the stars may not have been strictly named as the Ancients viewed them. So, for example, the stars Atlas and Pleione may have represented two of the Sisters instead, and so on. Secondly, what matters more is that the story exists in the first place; not whether there is physical, tangible proof in the night skies. In other words, it is the symbolism and meaning attached to these stories that is of significance. And in this regard it is fascinating that a nebula enshrouds the stars in a veiled fashion in the first instance, and even more so that it is especially pronounced around one of them. Astronomically, the star Pleione would be a far better candidate for the Lost Pleiad because not only does she shine at the low magnitude of 5.09, but more importantly, she is a variable star.[234] As their name suggests 'these stars vary in brightness — some dramatically in a matter of days, others more gradually over a period of years.'[235] Whatever the true explanation, certainly the correspondence of the Pleiades with honeybees lends a fascinating analysis to the symbolism of the *veil* and to their role as Goddesses of fate.

Cords, knots, webs and nets of the Pleiades

A less apparent theme is the association of knots, cords and nets with stars in general, and more specifically with the Seven Sisters of the Pleiades. In the epic fantasy novel *The Hobbit*, J. R. R. Tolkien tells us the star cluster was known as *Remmirath*, or 'the Netted Stars', during the ancient time of Middle Earth.[236] Likewise many European peoples including the Finns and Lithuanians referred to the Pleiades as a 'sieve' or a net.[237] We see these representations carved on the omphalos stones of the oracles at Delphi on the Greek mainland and Dodona in northern Africa. In addition to these stone monuments, several Greek, Japanese and Polynesian myths speak of Pleiadian nets in the skies. What are we to make of these images? Apart from their practical application, what do nets symbolise within these cultures? Robert Lawlor provides some interesting insights into net motifs. In *Voices of the First Day*, he says that symbols provide a pathway to spiritual transformation and that images of nets, in particular, indicate 'the capacity of an image to trap the transformative spirit.'[238] This explains why the Rainbow Serpent 'is often represented within a fish trap' in Australian Aboriginal art.[239] The ancient Egyptians also 'used a net for capturing birds in the same way,' says Lawlor.[240] In this way, 'the subtle, the volatile, the ever-moving is captured or contained in a symbol.'[241] These representations of the Pleiades with nets and cords resonate on two levels — the spiritual and the scientific. Spiritually it does this in terms of the relationship between the Seven

Sisters and that of fate and destiny. On the scientific level it does this from the perspective of sacred science and the way in which scientific information may be encoded in mythology. The depiction of celestial nets in stone engravings or other illustrations can be thought of as geodesic markers or as the very web or matrix of creation where time and space are weaved together in intricate but defined patterns. These designs speak to us of fate, destiny and of a Dreamtime where the past, present and future are merged together as One Time.

In some legends the nets are spoken of as gravitational nets and the three-dimensional structure of knots, cords and nets bears a striking similarity to computer-generated graphics depicting the curvature of space and of wormholes. The widespread application of mathematical knot theories such as the Alexander Polynomial, the Hopf Bifurcation Theorem and the Kalman Filter is assisting scientists in the world of genetics to understand the winding and unwinding processes of DNA, assisting astronomers in predicting planetary motions and by clarifying control theory in their understanding of space travel.[242] Lawlor sees within the fish trap or net image a scientific model for explaining the paradoxical nature of light, essentially its wave characteristics versus its particle aspects.[243] 'One image of light is not true,' he says, 'and the other false,' and yet 'each explains some aspect of light.'[244] The fish trap or net, he argues, serves to resolve the apparent contradictions of the inherent nature of light. This occurs through the recognition that 'one image or symbol is effective in one phase of individual or collective development, and another image of the same phenomenon is more potent at a different stage.'[245] In this way symbols 'can be applied to the unresolved questions about the relationship of an object to its image that persist in contemporary thought,' and deepen our understanding of the universe.[246] These sciences are mere reflections of the omniscience of the Creator whose mysteries reveal the workings of the universe and whose knot theories are embodied in many diverse worldly legends, especially those of the Pleiades.

On the spiritual level, in terms of their connection with the goddesses of fate, nets are 'linked with aqua-cosmogony' or the 'water of life' (amniotic fluid and uterus) says Marija Gimbutas.[247] Images of female divinities or their objects bearing the net symbol mean 'that she is the source and container of life-generating and life-sustaining water.'[248] Thus, 'in this aspect, she is the youthful Goddess, the Life-Giver.'[249] Furthermore, nets are made from thread or other similar material and are woven or sewn together, emphasising her fate aspects. But the ultimate thread upon which all human life depends is the umbilical cord, which joins mother and child. The esoteric lore surrounding this sacred object in Australian Aboriginal spirituality is explored in chapter nine, along with all things connected with weaving.

Pursuit of the Pleiades

The singular most widespread theme in Pleiadian mythology is the pursuit of the Sisters by an individual man, or group of men. Often, he or they are identified with the Orion constellation, but sometimes with the orange star Aldebaran in Taurus, or with the planet Venus, who appears in the skies as the evening or morning star. For instance, in Greek legends, the famous hunter Orion is notorious for chasing the Sisters, also Zeus the king of the gods, who in ancient times represented the constellation of Taurus.[250] According to the Ngarrindjeri people of South Australia, their Supreme God Ngurunderi ordered the hunter Orion to chase the Seven Sisters and bring them back to Earth. 'They didn't want to be caught,' explains Ngarrindjeri elder Veronica Brodie, 'so they headed up to the sky, up and up and over the Milky Way and hid and there became the Seven Sisters.'[251]

Despite the fact that all of the Sisters are described as beautiful women and all are chased as a group of maidens, often the single motivation appears to be that their pursuer has fallen in love with one of the Sisters. The majority of Aboriginal legends point to the youngest sister, as in the Greek legend, but sometimes she is said to be the eldest. Where there is a lone pursuer he is usually described as a much older man, as in the Aboriginal legend. In some stories though he is considered a contemporary of the girls, such as Karambil of the Bundjalung myth, who became the star Aldebaran, where he continues to follow the Seven Sisters in the night skies.[252]

In some Aboriginal Australian legends the Moon is said to pursue the Sisters. One western desert myth tells how Wadi Bira (Moon Man) chases after the Seven Sisters but even then he was not their initial main pursuer. This happens only later in the story after the Wadi Gudjarra (Two Men of the western desert mythic landscape) deceived him. Ronald and Catherine Berndt retell the story in *The Speaking Land* as 'Moon's Journey and Death.'[253] Western desert cosmogony maintains that Moon Man continues to pursue the Gunggurunggura across the night sky as they flee from him. This is another example of mythic statements corroborated by scientific fact for, as Robert Burnham Jr points out, 'the Pleiades cluster lies about 4° from the ecliptic, and thus is subject to occasional occultations by the Moon.'[254] When this occurs, the Bundjalung people of northern New South Wales poetically refer to this celestial phenomenon as 'When the Moon kisses the Seven Sisters.'[255]

More captivating is the application of animal imagery, where a single bear or pack of bears or dingos chases the girls, as in some Native American and Aboriginal Australian legends of the Pleiades. The bear motif may reflect the men's totemic relationship, like that of the dingo men of the Aboriginal legend. It may have particular significance to the constellation of Ursa Major the Great Bear, who can be seen hovering above the Pleiades in the Northern Hemisphere night skies. Or it could represent Kali the Black, the Hindu Goddess of death and destruction who destroys the old age to make way for the new. These connections are further explored in the Native American and Hindu chapters respectively.

Gendered skies

In the chase theme section Orion has already been identified as one of the main pursuers in world mythology on the Pleiades. In fact, Orion emerges as the consummate celestial stalker of the Seven Sisters *par excellence*. The question remains, why is there this close alliance between the Pleiades and Orion, more so than Taurus, Aries or some other nearby constellation? Why Orion? And why is there a *gendered* division? In other words, why are the Pleiades predominantly regarded as 'female' and Orion as 'male'? Almost everywhere he is depicted as a male warrior or representative of a legion of male hunters, such as in the Australian Aboriginal legend of the Dog Men of Orion who pursue the Emu Women of the Pleiades. These gendered divisions of the heavens are especially more pronounced in regard to the stars of Orion and the Pleiades than perhaps any other star cluster or constellation.

From the stories a distinct mythical delineation along gender lines becomes apparent, where the Pleiades are designated essentially as 'women's country' or women's domain, and Orion is represented as 'men's country' or men's domain. Is this just another example of gender stereotyping of the stars or is there some deeper, hidden underlying message within this description? What factors might account for this representation? How do so many of the world's cultures see Orion as some sort of hunter or warrior figure? What is it about Orion that conjures up, essentially, male warrior energy in the human psyche? And how is it that the Pleiades are primarily viewed as women? Could the world's peoples and cultures have shared a more common history than we have been led to suppose, and might this explain the overwhelming evidence of the similarities of this particular legend alone? Shirley MacLaine poses similar questions:

> All of this information only scratches the surface of the ancient's references to the Pleiades. What are we to take from these histories? Were there beings from the Pleiades that visited Earth and imparted knowledge that helped to bring us from stone wheels to computers and space flight? Were they arbiters of human destiny? Are they, as believed by the Apibones (of Brazil) and others, our forefathers?[256]

Certainly my grandmother and her people believed that our planet and its occupants have a mutual bond and ancestry with that of the stars and beings of the Pleiades. Pleiadian mythology from around the world also suggests this is a commonly held belief among Indigenous peoples, who not only see themselves as caretakers of our planet but who claim to be either descended from the Pleiades, or at the very least, related to them in some way.

A Pleiadian mystery

As I grew up and began reading Pleiadian legends and stories other than my Aboriginal culture, I began to see enormous similarities. Equally I saw that just as many questions surrounding this famous star cluster remained unanswered. The more I continue to read, investigate and discover all things Pleiadian, the more amazed I become at the profound similarities of stories that are quite literally, if not culturally, worlds apart. Clearly, what has begun to emerge from my research over the years is that the Pleiades remain a great *mystery* akin to the Sirius and Orion 'mysteries' that have been written of in recent years. Their impact upon the human psyche is even more remarkable, given that they are not very bright in comparison to other stars in the night skies, although their combined light makes them more visible than they would otherwise appear.

As an amateur astronomer I am intrigued why this faint group of stars, barely discernible to the naked eye, should feature so prominently in the world's mythologies. Why in all the stories is there a preoccupation with the number seven, when much keener eyesight can observe at least as many as nine or more stars in the Pleiades? Certainly, a pair of binoculars reveals a much larger cluster of stars and, as previously noted, estimates vary as to the exact number in the cluster. But even in those situations where more than seven stars have been sighted, they are still referred to as the Seven Sisters and not as the eleven, twelve or thirteen sisters. How then do we explain the extraordinary degree of commonality between so many of the stories, given the official version of historical contact between the world's nations and continents of people?

Apart from Europe, Asia and Northern Africa, this contact is said to have only occurred in the last few hundred years or so. For instance, in the Great Southern Land of Bandaiyan (Australia), the standard version of history claims that Aboriginal contact with European peoples is relatively recent, compared with other Indigenous cultures. Yet how is it, that the Dreaming of the Seven Sisters is shared by so much of the rest of the world? Does this suggest the possibility of earlier historical contact between the various cultures and civilisations? What does it tell us about our common humanity? Is it possible that genesis occurred in Australia, not in Africa or Mesopotamia as leading archaeological and other alternative theories suggest? The Australian continent is, after all from a geological perspective, the world's oldest continent. This is highly significant. Furthermore, its Indigenous peoples, reputedly the world's oldest living culture, can trace their history back at least a hundred thousand years and more.

Just what exactly is the Pleiadian mystery anyway?

These questions and more are explored throughout this book as I attempt to provide some answers to the timeless and ancient mystery of the Seven Sisters of the Pleiades . . .

Notes

1. See the entire poem in *The Works of Alfred Lord Tennyson*, pp. 149–55.
2. See Richard Allen's chapter on the Pleiades in *Star Names* at pp. 391–413 for references to the individual poems by the various writers. See also William Tyler Olcott's *Starlore of All Ages*, pp. 407–27 and *Burnham's Celestial Handbook*, vol. 3, pp. 1863–88.
3. Allen, *Star Names*, p. 399. See also the relevant footnotes under the individual chapters.
4. Ibid, p. 393.
5. Jobes and Jobes, *Outer Space*, pp. 336–41.
6. Ibid, p. 340.
7. See *Maya Cosmogenesis* by Jenkins.
8. Jobes and Jobes, *Outer Space*, p. 336.
9. Ibid, p. 337, and Watterson, *Gods of Ancient Egypt*, p. 120.
10. Jobes and Jobes, *Outer Space*, p. 87.
11. Graves, *The Greek Myths*, p. 152.
12. Ibid, p. 41.
13. Blavatsky, *The Secret Doctrine*, vol. 2, p. 768.
14. Graves, *The Greek Myths*, pp. 165 and 775.
15. Allen, *Star Names*, pp. 396–97.
16. Robert Hendrickson tells us in *Word and Phrase Origins*, p. 36 that when geographer Gerardus Mercator published his collection of maps in the

sixteenth century he placed a drawing of Atlas bearing the world globe on his shoulders. From that time, all map collections were called an atlas.

17. Ibid.
18. Allen, *Star Names*, p. 402.
19. Ibid, p. 405.
20. Jobes and Jobes, *Outer Space*, p. 392. See also *The Book of Job*, 38:31.
21. See Shirley MacLaine's website at <www.shirleymaclaine.com>.
22. Motz and Nathanson, *The Constellations*, p. 105.
23. Ibid.
24. Ibid.
25. Audouze and Israel (eds), *Cambridge Atlas of Astronomy*, p. 251.
26. Burnham Jr, *Burnham's Celestial Handbook*, vol. 3, p. 1878.
27. Levy, *Skywatching*, p. 215.
28. Ibid, p. 89.
29. Ibid, p. 39.
30. Ibid, p. 215.
31. The research paper, *Aurochs in the Sky* by Frank Edge is available from PO Box 2552, Pinetop, Arizona, USA. Referred to in *Heaven's Mirror* by Graham Hancock at pp. 28–29. The Lascaux Caves are near the village of Montignac in the Dordogne region of France. See also Michael Rappenglueck's article, 'The Pleiades in the "Salle des Taureaux", Grotte de Lascaux,' available on the internet at <www.infis.org/study7b.htm>.
32. Edge, *Aurochs in the Sky*, p. 6, and Hancock, *Heaven's Mirror*, p. 29.
33. Ibid.
34. Hancock, *Heaven's Mirror*, p. 28.
35. Ibid.
36. Levy, *Skywatching*, p. 215.
37. Ibid.
38. These two original books by Katherine Langloh Parker are now combined in a joint publication entitled *Australian Legendary Tales*. See the reference list for further details.
39. Lambert, *Wise Women of the Dreamtime*, p. 5.
40. Ibid, pp. 44–50.
41. Isaacs, *Australian Dreaming*, pp. 152–53.
42. Bell, *Ngarrindjeri Wurruwarrin*, pp. 545–94.
43. See, for example, the Blackfoot myth of the Pleiades in *The Lost Children* by Paul Goble that tells the tale of six little orphan boys. See also the Navajo legend of the Flint Boys (Pleiades) in *Ancient Astronomers* by Anthony Aveni at pp. 131–32.
44. Bell, *Ngarrindjeri Wurruwarrin*, p. 585. Says Bell, at 'several hundred million years, one tenth of the age of the solar system, the Pleiades are considered to be of adolescent age.'
45. Ibid, p. 577.
46. De Santillana and Von Dechend, *Hamlet's Mill*, p. 175.

47. Levy, *Skywatching*, p. 215.
48. Burnham et al., *Astronomy: The Definitive Guide*, p. 366.
49. Ibid, p. 190.
50. Krupp, *Beyond the Blue Horizon*, pp. 241–55.
51. Aveni, *Stairways to the Stars*, p. 153, and *Skywatchers*, p. 310.
52. Simons, *The Meeting of the Waters*, p. 396.
53. Ibid.
54. Ibid.
55. Peat, *Blackfoot Physics*, p. 155.
56. Ibid.
57. Aveni, *Stairways to the Stars*, p. 153, and *Skywatchers*, p. 310.
58. Krupp, *Beyond the Blue Horizon*, p. 249.
59. See Patrick Moore's discussion of the shifting skies according to the seasons and hemispheres in *Stars of the Southern Skies*.
60. Ibid, p. 114.
61. See the discussion by Jobes and Jobes on the symbolism of number seven in *Outer Space*.
62. De Santillana and Von Dechend, *Hamlet's Mill*, p. 415.
63. Blavatsky, *The Secret Doctrine*, vol. 2, pp. 618–19.
64. Ibid.
65. Peat, *Blackfoot Physics*, p. 153.
66. Cooper, *An Illustrated Encyclopaedia of Traditional Symbols*, p. 113.
67. Peat, *Blackfoot Physics*, p. 161.
68. Cooper, *An Illustrated Encyclopaedia of Traditional Symbols*, p. 117.
69. Ibid.
70. Ibid.
71. Ibid.
72. Blavatsky, *The Secret Doctrine*, vol. 2, p. 602.
73. Michell and Rhone, *Twelve-Tribe Nations*, pp. 85–6.
74. Ibid, p. 86.
75. Sitchin, *The Cosmic Code*, p. 174.
76. Ibid.
77. Ibid, p. 175.
78. Jobes and Jobes, *Outer Space*, p. 87.
79. Blavatsky, *The Secret Doctrine*, vol. 2, p. 601.
80. Rees, *Just Six Numbers*, p. 51.
81. Temple, *The Crystal Sun*, p. 362.
82. Ibid, p. 364.
83. For a general introduction to the sacred science of numerology, see *The Mystery of Numbers* by Annemarie Schimmel, and *Numerology: The Power in Numbers* by Ruth Drayer. Schimmel has a separate chapter for each number from one to ten, including number seven at pp. 127–55.
84. Walker, *The Woman's Dictionary of Symbols and Sacred Objects*, p. 76.
85. Sykes, *The Seven Daughters of Eve*, pp. 195–6.

86. Ibid, p. 196.
87. Ibid, p. 8.
88. Ibid, p. 54.
89. Jobes and Jobes, *Outer Space*, p. 87.
90. Doniger O'Flaherty, *The Rig Veda*, pp. 239–40.
91. Peat, *Blackfoot Physics*, p. 156.
92. Ibid.
93. Wilson, *The Oxford Dictionary of English Proverbs*, p. 739.
94. Ibid.
95. Nasr, *An Introduction to Islamic Cosmological Doctrines*, p. 7.
96. Krupp, *Beyond the Blue Horizon*, p. 245.
97. Ibid.
98. Allen, *Star Names*, p. 393.
99. See Von Del Chamberlain's website, Project ASTRO UTAH, at <www.clarkfoundation.org/astro-utah/vondel/dilyehe.html>.
100. Bell, *Ngarrindjeri Wurruwarrin*, p. 586.
101. Ibid.
102. Jobes and Jobes, *Outer Space*, p. 340.
103. Best, *The Astronomical Knowledge of the Maori*, p. 52.
104. Rathbun, *First Encounters: Indian Legends of Devils Tower*, p. 7.
105. Alcyone's magnitude or brightness is said to be 2.86 (apparent magnitude) and −2.6 (absolute magnitude) or nearly 1,000 times more luminous than our Sun. See *Burnham's Celestial Handbook*, p. 1876.
106. Graves, *The White Goddess*, p. 182.
107. Krupp, *Beyond the Blue Horizon*, p. 242.
108. Ibid.
109. Allen, *Star Names*, p. 399. See also *Isis Magic* by Isidora Forrest, p. 132.
110. Jobes and Jobes, *Outer Space*, p. 342.
111. Ibid.
112. Allen, *Star Names*, p. 399.
113. Ibid.
114. As Vivianne, the Lady of the Lake, would often say to Morgaine Le Fey in *The Mists of Avalon* by Marion Bradley, in her interpretation of the Arthurian legend. See p. xi for one of the sayings:

    'For all the Gods are one God,' she said to me then, as she had said many times before, and as I have said to my own novices many times, and as every priestess who comes after me will say again, 'and all the Goddesses are one Goddess, and there is only one Initiator. And to every man his own truth, and the God within.'
115. Gimbutas, *The Language of the Goddess*, p. 316.
116. Ibid.
117. See *40,000 Years of Australian Dreaming* edited by Jennifer Isaacs for a short version of the story, p. 153.

118. Best, *The Astronomical Knowledge of the Maori*, p. 60.
119. Isaacs, *Australian Dreaming*, p. 153.
120. De Santillana and Von Dechend, *Hamlet's Mill*, p. 386.
121. Berndt and Berndt, *The Speaking Land*, pp. 44–45.
122. Ibid, p. 45.
123. Allen, *Star Names*, p. 399.
124. Makemson, *The Morning Star Rises*, p. 79.
125. See the story of the halcyon in Graves, *The Greek Myths*, pp. 163–65, and his commentary in *The White Goddess*, pp. 181–83. The European kingfisher or *Alcedo atthis* is the only bird species of its kind in Europe. It has a brilliant blue plumage with a 'dagger-like' bill. Although referred to as 'European' it is also found in northern Africa and in the Solomon Islands. See the following website for further information: <http://mbgnet.mobot.org/fresh/animals/king.htm>.
126. Allen, *Star Names*, p. 404.
127. Graves, *The Greek Myths*, pp. 164–65.
128. Gimbutas, *The Language of the Goddess*, pp. xv–xxi.
129. Ibid, p. 321.
130. Ibid, p. xxiii.
131. Ibid, pp. 25–29.
132. Ibid, p. 3.
133. Ibid.
134. Ibid, p. xxii.
135. Ibid, p. 1.
136. Ibid, p. xxii.
137. Diane Bell discusses the legal history and political struggle of Ngarrindjeri women to protect their spiritual beliefs and sacred sites in the Hindmarsh affair throughout her book *Ngarrindjeri Wurruwarrin*. For a more straight-forward account of its history unencumbered by social science-speak, see *The Meeting of the Waters* by Margaret Simons.
138. Bell, *Ngarrindjeri Wurruwarrin*, pp. 545–94.
139. Ibid, pp. 205 and 562–4.
140. Ibid, pp. 562 and 569.
141. Ibid, p. 574.
142. Ibid.
143. Mowaljarlai and Malnic, *Yorro Yorro*, p. 136.
144. Ibid.
145. Ibid.
146. Ibid.
147. Gimbutas, *The Language of the Goddess*, p. 265.
148. Ibid.
149. Allen, *Star Names*, p. 380.
150. For a colour photograph of the astronomical ceiling of Set I see *Heaven's Mirror* by Hancock, p. 89.

151. Allen, *Star Names*, p. 399. See also *Isis Magic* by Isidora Forrest, p. 132.
152. Jobes and Jobes, *Outer Space*, p. 254.
153. Eliade, *A History of Religious Ideas*, p. 26.
154. De Santillana and Von Dechend, *Hamlet's Mill*, p. 124.
155. Hand Clow, *The Pleiadian Agenda*, p. 98.
156. De Santillana and Von Dechend, *Hamlet's Mill*, p. 369.
157. Hand Clow, *The Pleiadian Agenda*, p. 109.
158. Ibid.
159. Ibid.
160. Elphick, *The Atlas of Bird Migration*, p. 32.
161. Allen, *Star Names*, pp. 395–6.
162. Graves, *The Greek Myths*, p. 165. See also Allen, *Star Names*, p. 395.
163. Ibid, p. 154.
164. Allen, *Star Names*, p. 395.
165. Ibid.
166. Sitchin, *The Cosmic Code*, p. 175.
167. Allen, *Star Names*, p. 395. Guiseppe Sesti also agrees that the Pleiades are named after their mother, Pleione.
168. Walker, *The Woman's Dictionary of Symbols and Sacred Objects*, p. 399.
169. Allen, *Star Names*, p. 396. See his footnote #1.
170. Ibid.
171. Ibid, p. 395.
172. Ibid.
173. Krupp, *Beyond the Blue Horizon*, p. 251. See also the arguments advanced by American scholar Alexander Krappe in 'Les Péléiades,' *Revue Archéoligique* 36 (Series 5, 1932), pp. 77–93.
174. See *The Power of Compassion* by His Holiness, the Dalai Lama, especially chapter 5 on human compassion, pp. 58–82.
175. Krupp, *Beyond the Blue Horizon*, p. 251.
176. Graves, *The Greek Myths*, pp. 180–1.
177. Ibid, p. 178.
178. Hancock, *Heaven's Mirror*, p. 252.
179. Allen, *Star Names*, p. 399. See also *The Glorious Constellations*, by Giuseppe Sesti at p. 446.
180. Hancock, *Heaven's Mirror*. See the photograph of a copy of the original omphalos stone that clearly shows these engraved mesh patterns.
181. Tompkins, *Secrets of the Great Pyramid*, p. 298.
182. Ibid, p. 349.
183. Cook, *Avian Visual Cognition*, see the website at <www.pigeon.psy.tufts.edu/auc/tvc.htm>.
184. Temple, *The Sirius Mystery*, p. 289 quoting Stecchini.
185. Bauval and Gilbert, *The Orion Mystery*, p. 205.
186. Burnham Jr, *Burnham's Celestial Handbook*, p. 1884.
187. Eliade, *The Myth of the Eternal Return*, p. 16.

188. Mowaljarlai and Malnic, *Yorro Yorro*, pp. 190–92 and 205.
189. Walker, *The Woman's Dictionary of Symbols and Sacred Objects*, p. 217.
190. Personal communication with Sib Bresland, an Aboriginal woman from the Atherton Tablelands region in Queensland, Australia.
191. Gertrude Jobes lists the Seven Hills of Rome in *Mythology, Folklore and Symbols*, Part 2, p. 1422, as Aventine, Caelian, Capitoline, Esquiline, Palatine, Quirinal and Viminal.
192. Buchanan, *Discovering the Wonders of Our World*, p. 185.
193. Ibid.
194. Krupp, *Beyond the Blue Horizon*, p. 290. He says *all* pyramids are cosmic mountains, not just the Egyptian ones.
195. See the article by science editor Tim Radford, in *The Guardian*, 14 May 2001, reporting on a paper presented by Toby Wilkinson, an Egyptologist based at Cambridge University in England. The article was reprinted in *The Sunday Age*, 20 May 2001, p. 20.
196. For a discussion on the Seven Cities of the Atlantic Ocean, see *Gateway to Atlantis* by Andrew Collins. See also Erich Von Däniken who briefly mentions three variants of the theme in Australia, Brazil and the Canary Islands in the Atlantic Ocean, in his work *In Search of Ancient Gods*, pp. 138–39.
197. See the story of 'Where the Frost Comes From,' in *Wise Women of the Dreamtime* by Johanna Lambert, pp. 44–50.
198. Graves, *The Greek Myths*, p. 770.
199. Ibid, pp. 418 and 760.
200. Ransome, *The Sacred Bee*, p. 100. See her footnote at bottom of page.
201. Allen, *Star Names*, p. 407.
202. Ransome, *The Sacred Bee*, p. 136.
203. Cooper, *An Illustrated Encyclopaedia of Traditional Symbols*, p. 84.
204. Ibid.
205. Ibid.
206. Ibid.
207. Walker, *The Woman's Dictionary of Symbols and Sacred Objects*, p. 488.
208. Cooper, *An Illustrated Encyclopaedia of Traditional Symbols*, p. 84.
209. Ransome, *The Sacred Bee*, p. 137.
210. Ibid, p. 136.
211. Walker, *The Woman's Dictionary of Symbols and Sacred Objects*, p. 415.
212. Ibid, pp. 414–5.
213. Ibid, p. 414.
214. Sjoo, pp. 2–3.
215. Ibid, p. 3.
216. Ibid.
217. Ransome, *The Sacred Bee*, pp. 98–9.
218. Ibid, p. 98.
219. Ibid, p. 99.

220. See Ransome's note 'To the Reader' where she reproduces a poem from Butler's book. For a reference, see her bibliography.
221. Walker, *The Woman's Dictionary of Symbols and Sacred Objects*, p. 488.
222. Ibid.
223. Ibid.
224. Ibid.
225. Allen, *Star Names*, p. 406.
226. Walker, *The Woman's Dictionary of Symbols and Sacred Objects*, p. 415.
227. Ibid.
228. Ibid, p. 317.
229. Ibid.
230. Ibid, pp. 317–8.
231. Ibid.
232. See 'The Pleiades Nebulosity' in *Burnham's Celestial Handbook*, vol. 3, pp. 1880–3.
233. Burnham Jr et al., *Space Watching*, p. 204.
234. Burnham Jr, *Burnham's Celestial Handbook*, p. 1876.
235. Burnham Jr et al., *Astronomy: The Definitive Guide*, p. 362.
236. Burnham Jr, *Burnham's Celestial Handbook*, p. 1868.
237. Allen, *Star Names*, p. 397.
238. Lawlor, *Voices of the First Day*, p. 291.
239. Ibid.
240. Ibid, p. 292.
241. Ibid.
242. See *Five More Golden Rules* by John L. Casti for more information on the application of knot theories and its application in science and mathematics.
243. Lawlor, *Voices of the First Day*, p. 292.
244. Ibid.
245. Ibid.
246. Ibid.
247. Gimbutas, *The Language of the Goddess*, p. 81.
248. Ibid, p. 87.
249. Ibid.
250. Graves, *The Greek Myths*, p. 152.
251. Bell, *Ngarrindjeri Wurruwarrin*, p. 578.
252. See Janet Mathews, *The Opal That Turned Into Fire*, pp. 57–8.
253. Berndt and Berndt, *The Speaking Land*, pp. 221–3.
254. Burnham Jr, *Burnham's Celestial Handbook*, p. 1884.
255. Personal communication with local Bundjalung man, Roy Gordon, Jr.
256. See Shirley MacLaine's website at <www.shirleymaclaine.com>.

CHAPTER TWO

Pleiades

Seven Daughters of Atlas and Pleione

> And if longing seizes you for sailing the stormy seas,
> When the Pleiades flee mighty Orion
> and plunge into the misty deep
> and all the gusty winds are raging,
> Then do not keep your ship on the wine-dark sea
> but, as I bid you, remember to work the land
>
> — from *Works and Days* by Greek poet, Hesiod c. 740 BCE.[1]

Once upon a time in ancient Greece the story of the Pleiades was a modern tale. Their story has travelled around the world, from older times and across ancient cultures. Although their light is not that bright, their gentle glow has inspired many dreamers and illuminated the psyche as they sailed across our galaxy during endless starry, starry nights. The Seven Sisters of the Pleiades were celestial Greek celebrities who influenced a broad spectrum of life in those times. Firstly, their status within the family of Greek gods is a fascinating and complex story of power dynamics between men and women told against the background of mythology. Secondly, their official status in our understanding of Greek history is not as simple as it first appears; tales of the Greek gods sit side by side with the knowledge of Indigenous peoples who shared that particular area of southern Europe at the time. Thirdly, these shared stories tell as much about the social and physical environment as they do about the divine in human relationships. Although we tend to think of ancient Greece as the society from which logic and reason grew to influence subsequent civilisations, the symbolism of the Pleiades and their individual names indicate far more. We all exist in spiritual, psychological and physical dimensions and reality is not only the tangible world in which we live. Beyond the temple of our senses the imaginary is just as meaningful. To enter this world of the magical and mysterious land, let's begin with some stories of the gods and goddesses in which the Pleiades played a major role.

As legend tells, the maidens Maia, Alcyone, Asterope (or Sterope), Caelano, Taygete (or Taygeta), Electra and Merope, their youngest sister, were the daughters of the Titan giant Atlas and his consort Pleione, the Oceanid.[2] Some people suggest Atlas, Poseidon and Prometheus were brothers; others say Zeus, Prometheus and Poseidon were siblings and fail to mention Atlas.[3] However this fraternal relationship plays out, its significance rests in the division of the realms by the gods, who drew lots to decide lordship of the sky, sea, and underworld, 'leaving the earth common to all.'[4] This disputed parental lineage makes some sense of the Titan revolt, for Atlas may have been cheated out of drawing lots with Zeus, Poseidon and Hades if indeed he was Poseidon's or Prometheus' brother. Alas, poor Atlas was condemned by Zeus, king of the Greek gods, for participating in the unsuccessful war between the Olympian gods and his fellow Titans. Atlas' punishment was to hold the heavens on his shoulders 'for all eternity',[5] except for one occasion when Heracles 'temporarily relieved him of the task' during his eleventh labour.[6] This was one of twelve such labours that the hero had to undertake as part of a kingship ritual. Two of the labours involved the daughters of Atlas and are part of the story of the Pleiades.

The hunter gives chase

One popular tale tells of the Sisters and their mother, Pleione, walking through the Boeotian countryside near Athens when they were spotted by the giant hunter, Orion. On seeing the beautiful women, this son of the sea god Poseidon (Neptune) set off in hot pursuit.[7] The chase verges on incest for Orion may have been Atlas' nephew (if Poseidon were his brother, and this in turn would have made him a cousin to the Pleiades). Incest aside, Orion was reputed to be of giant stature, so that when he walked along the sea floor his head extended above the waves. Considered 'the most handsome man alive', he was especially enamoured with the youngest sister, Merope, although some accounts say he was really after her mother, Pleione.[8] No matter the real object of his desire, he pursued the Sisters for five (some say seven) years until in desperation they turned to Zeus for protection, whereupon he turned them into doves and placed them in the constellation of Taurus the Bull.

However, Orion's earthly exploits not only involved the Sisters, for the goddess of the hunt, Artemis, once fell for his charms.[9] We cannot be sure how her father Zeus viewed this attraction, but her younger twin brother Apollo disapproved of the liaison, and his deception of Artemis led to Orion's demise. Upon his death, Zeus immortalised Orion by placing him in the sky, although some say it was Artemis who elevated him to his heavenly abode.[10] Unfortunately for the Seven Sisters, there appears to have been little room for escape from the unwanted attention of male admirers and the dubious character of these male gods. For one, their lustful neighbour and eternal suitor, Orion, was now placed in their immediate proximity. Zeus' protection also came at a heavy cost as he seduced four of the Pleiades and fathered children to three of them. This licentious connection comes from the colloquial reference to the central star of Taurus, Aldebaran, as 'the eye of the bull.' The Arabic translation means 'the follower', for it is seen to follow the sisters across the night sky.[11] We now use the colloquial term 'old bull' to refer to a male philanderer, for Zeus gained wide notoriety for his sexual exploits and even once turned into a bull to seduce another goddess, Europa.

The weeping Hyades and the gardening Hesperides

Let's return to the relationship between Zeus and his daughter, Artemis. She was undoubtedly a forthright child for on her third birthday she presented her father with an unusual list of gifts that she desired, including 'sixty young ocean nymphs.'[12] The Seven Sisters of the Pleiades were among this special group of maidens who were placed under the direct tutelage of Artemis, and so fell under extended patronage of the House of Zeus.[13] Although she later became a strong woman of independent means, Goddess of the hunt and of the Moon, this privileged position could not fully protect the Sisters. Like many women who seek sanctuary from the threats and ravages of menfolk in patriarchal times, as the story shows, not even the king of the gods, the heavenly father, could be entrusted with their care. Among the sixty nymphs given to the train of the young goddess were two sets of half-sisters to the Pleiades, the Hyades and the Hesperides, who all shared a common father in Atlas.[14] The individual names of the Hyades sisters were Phaesyia, Ambrosia, Coronis, Eudore and Polyxo.[15] Their only brother was Hyas, from whom the group's collective name derives. Their mother Pleione's sister was Aethra, whose parents were Oceanus and Tethys, which made the whole family Oceanids. They were Titans who ruled the outer seas before Poseidon became lord of the oceans — thus the Pleiades and the Hyades are known as sea, or ocean, nymphs.[16]

Following the death of their brother Hyas, after whom they are named, Zeus placed the Hyades in the sky in the constellation of Taurus where they form the distinctive V-shaped asterism of the bull's horns. The tears they cried for their dead brother earned them their nickname of Weeping Hyades and, like their siblings the Pleiades, both star clusters were seen to herald spring showers in the Northern Hemisphere.[17] The fact that both sets of daughters were Oceanids would have enhanced their perception as water maidens. The other set of half-sisters to the Pleiades, the Hesperides, are immortalised in the tale of Heracles (Hercules) and the eleventh labour. There were twelve labours in all, each one involving different tasks that Heracles was ordered to complete. The Hesperides were the three daughters of Atlas and Hesperis. However, myths are never certain. Some accounts assert they were really the daughters of Ceto and Phorcys; others say they were the daughters of Night, alone.[18] In any event, they are known in Greek mythology for guarding the golden apples of Hera's orchard.[19]

The Garden of the Hesperides was said to be located in the 'Far West' although many writers are uncertain as to its exact location, including our central hero who embarked on a circuitous and difficult route to find the fabled gardens.[20] The most popular view is that it was situated on an island, thereby equating it with Celtic notions of 'western paradises' that 'grew the apples of eternal life,' says Barbara Walker in *The Women's Encyclopedia of Myths and Secrets.*[21] This island was thought to be in the vicinity of Mount Atlas, whose modern day ranges stretch across three northern African countries — Libya, Algeria and Morocco. As a wedding gift on her marriage to Zeus, Gaia (Mother Earth) gave Hera a golden apple tree from which the orchard sprung.[22] Though the three sisters Hespere, Aegle and Erytheis guarded the golden apples with their father Atlas, as added protection, the Titan giant built a wall of stone to enclose the sacred fruit.[23] When Hera discovered that Atlas' daughters had pilfered the apples she set the hundred-headed serpent Ladon who spoke with 'diver's tongues to coil around the tree as its guardian.'[24]

After Zeus' infamous decree following the war in heaven, Atlas was burdened with the unenviable task of holding up the heavens. It was only during Heracles' eleventh labour involving the collection of the golden apples from the garden of the Hesperides that he received a brief respite. This came about because Nereus had advised Heracles not to pick the apples himself but to get the Titan giant to do it instead.[25] It was not difficult to persuade Atlas to undertake the task, for any relief from the cumbersome duty of holding the world aloft was welcome. The only barrier was the serpent Ladon, whom he feared greatly. Remembering the old sea god's advice, Heracles was quick to oblige and fired an arrow over the garden wall, killing the serpent. Bending his knees and leaning over to receive the weight of the globe on his back, Heracles now became like Atlas — the 'Long Suffering One.'[26]

Atlas was now free to collect the three golden apples from his daughters. The sense of freedom as he set about this task was tantalisingly sweet. He offered to take the apples to Heracles' task-master if the strongman could just 'hold up the heavens for a few months longer.'[27] Heeding the warnings of Nereus, Heracles was naturally suspicious. Pretending to agree, he asked Atlas to hold the globe for just a moment so he could place a pad on his head to help balance the weight. The Titan was easily deceived and after laying the golden apples on the ground, he replaced the globe on his shoulders. No sooner had Atlas done so, then Heracles promptly made off with the golden treasures. Returning to Mycenae, Heracles triumphantly presented his precious booty to his taskmaster but Eurystheus remained unimpressed and simply handed the golden apples back to Heracles who, in turn, passed them on to the goddess Athene, who returned them to the nymphs.[28]

It is assumed that Hera never expected to lose control over her property and this is why some say that she, rather than Eurystheus, set the twelve labours. This argument is further supported by the fact that Heracles' name supposedly means 'Glory of Hera', given to one dedicated to her service.[29] In a way, the labours can be seen as a form of initiation, and as in many traditions the initiate is given a new name once they have completed the ordeal. This is certainly true of Heracles, who was originally known as Alceus or Palaemon or Pyrmidion. This idea of Heracles' service to the Goddess explains the source of the modern term 'hero'.[30] Initiation always involves deeper levels of understanding of the journey of consciousness towards enlightenment, and Heracles had to perform his labours in order to qualify for kingship.

Here we can explore the many rich layers of symbolism represented in this hero's story. Apart from the obvious link of his labours with the twelve signs of the zodiac that characterise him as a solar god, the hero's tasks and their completion are part of an ancient kingship ritual tied to various cycles — calendrical, mythological, psychological and political. The quest to obtain the golden apples, in particular, represents the soul's journey into the mysteries of life, death and rebirth, whose cycles are governed by the Goddess to whom apples were sacred. The actual number of sisters and their individual names reveals deeper levels of understanding of her Triple Deity aspect that regulates agricultural and astronomical cycles and practices. The garden of the Hesperides is situated in the far west where the dying sun sets. Its changing hues of red, yellow and green not only reflect the three colours of its fruit, but are commemorated in the Hesperides' names that 'refer to the sunset.'[31] Just moments before his death on the western shores, the sun god makes his final appearance like a red apple cut in half as he sinks below the horizon.[32]

This brings us to the concept of death as liberation, seen as a transformation of the enemy serpent (Ladon) into a spiritual ally in the next dimension: the ego must be conquered for us to transcend the limitations of physical and emotional existence. Only the true hero succeeds on this journey. We must also regard the soul's direction as crucial to successful completion of the journey — the garden sits in the west where the Sun sets, representing adaptation and death. Also, before he could gain his freedom Heracles served a brief term as Atlas, holding up the Earth (or serving Mother). Apart from the apples, a connection between nature and lifecycles is also clearly evident in the time needed for Heracles to complete his labours within a particular time span following a lunar cycle. The Great Year was measured by one hundred lunations[33] and this progression of numbers relates to the meaning of 1 as standing for unity, 10 as completion of the divine cycle, and 100[34] as the alternation[35] of all natural forms from seed to plant by the process of logarithmic growth. This was emphasised by the fact that the serpent-dragon was hundred-headed. The serpent's many tongues signify its prophetic powers, says Graves, and represents the form the hero's 'oracular ghost would assume after he had been sacrificed.'[36]

Maia, the beautiful goddess

The more famous of Atlas' daughters were the Seven Sisters of the Pleiades. Although they were all renowned for their beauty, Maia the eldest was singled out as exceptional.[37] A shy mountain nymph, she lived alone in a cave on Mount Cyllene in Arcadia where Zeus visited her at nights. He seduced the maiden and she gave birth to the messenger of the gods, their son Hermes.[38] A gifted and precocious child, his many notable inventions included the panpipe and lyre.[39] At the age of four, he stole cattle from his older half-brother, Apollo, and used cow-gut to fashion a seven-stringed lyre made from tortoise shell. Their father Zeus demanded that the stolen cattle be returned, but Apollo was so charmed by the beauty of Hermes' lyre-playing that he forgave the theft and rewarded his younger brother with some of the cattle and other gifts.[40] This musical side of Hermes' character reveals a deeper connection between the Pleiades and music on many levels. Firstly, each one of the seven strings of the lyre can be related to the Sisters, whose voices perform in perfect harmony. Secondly, harmony expressed as divine sound is the source of all creation, as Plato immortalised in the *Song of Lachesis*, which celebrates the Harmony of the Spheres, whose seven sounds (scales) form the foundation of cosmic order and time.[41] Thirdly, this harmonic resonance captivates the Greek gods and beautifies the Seven Sisters in the form of Muses. This relationship of the Pleiades with music and time is explored in other chapters.

A popular interpretation of Maia's name in Latin is 'mother' or 'grandmother' and this can be linguistically traced to the older Sanskrit name Maya, the Hindu goddess of creation and mother of the Buddha.[42] By some accounts she is also identified as the eldest of the Seven Sisters in Hindu folklore. In addition to rearing her son Hermes, Maia fostered Arcas after Hera turned his mother, Callisto, into a bear and placed her in the skies as the constellation Ursa Major.[43] Some argue that Zeus cast the spell, which placed their son, Arcas beside his mother Callisto as the Little Bear (Ursa Minor), or as the star Arcturus, meaning 'bear warden', in the constellation of Bootes.[44] Other writers like Hyginus claim that it was Zeus' daughter Artemis. In any event, another interpretation of Maia's name means 'nurse' or 'the Great One.'[45] The first meaning is obvious. The second may possibly reflect a title bestowed by Zeus who was very grateful to Maia for hiding Arcas from the wrath of his jealous wife, Hera.

As the Bona Dea, the 'Great and Fruitful Mother', Maia was known as the 'increaser', and worshipped by the Romans as their Spring

Goddess.[46] It is from Maia that we have our name for the month of May. Her natural and astral aspects cautioned Greek farmers not to sow grain before the time of her setting. Allen says that some earlier writers and astronomers have previously referred to Maia as the brightest star in the cluster.[47] When we look into the complexity of her cosmological past, could it be that Maia tells a cautionary tale? Is she burdened as the older Sister who becomes a faded beauty? Despite her acclaim, Maia's younger sister, Alcyone is the Pleiades' brightest star, known to the Arabs as Al Nair, the 'Bright One', or Al Jauzah, the 'Central One'.[48] Whether this confusion is the result of historical error in the original naming of the stars, or whether Maia may have outshone Alcyone in the distant past is not known, but the distinction between the two Sisters is played out at many levels, not only in astronomy or mythology.

Alcyone of the halcyon days

However this sibling rivalry unfolds, the Greek legends clearly speak of Alcyone, as the group's leader. Following in this tradition Robert Graves attributes her with the 'mystical' power of leadership in his erudite classic *The White Goddess*.[49] Perhaps this mysticism is due, in part, to her relationship in Greek traditions with the legendary halcyon days, a magical time when the Mediterranean becomes 'smooth as a pond', a tranquil and safe haven for sailors.[50] Thomas Bullfinch lends a poetic touch to the tale when describing this interlude, when 'the sea is given up, for the time, to his grandchildren.' [51] Lingering for seven days on each side of the winter solstice, this meant there were fourteen halcyon days in total every year, says Graves.[52] The legend serves to accentuate and highlight the very common practice of depicting the Pleiades as birds in many of the world's mythologies, each having its own hidden symbolism and meaning. Perhaps this zoological imagery is no more inspiring than the poignant tale of the bird's everlasting love, fidelity and devotion to its mate. Not only does the halcyon mate for life, she carries 'her dead mate on her back' over the sea, mourning him with a peculiarly 'plaintive cry.' [53]

The widowed halcyon translates into human experience in the tragic love story of Alcyone and Ceyx, King of Thessaly.[54] Legend has it that Ceyx was the son of the morning star and Alcyone the daughter of Aegiale and Aeolus, 'guardian of the winds.' The lovers were happily married until Alcyone dared to proclaim them to be Hera and Zeus personified.[55] Translating the polite tones of Graves' commentary on the tale, the gods were not impressed by this human masquerade of divinity. Waiting for an opportunity when the earthly lovers were apart from each other in their capital Trachis, Graves explains, 'Zeus . . . let a thunderstorm break over the ship in which Ceyx was sailing to consult an oracle, and drowned him.'[56] Soon after, his ghost appeared to Alcyone who, 'distraught with grief . . . leapt into the sea.'[57] The legend says both were transformed into kingfishers by some 'pitying god', other accounts claim Ceyx 'was turned into a sea mew.'[58]

Bullfinch disagrees with the ghostly apparition as the direct cause of the Pleiad's death and transformation; his ghostly presence forecast the tragic end to this tale.[59] Waiting forlornly on a sea wall for her husband to return from his oracular journey, the haunting apparition became a shocking reality when Alcyone spotted her shipwrecked husband's dead body floating on the water below. Traumatised, she threw herself into the sea. Bullfinch's poetic version says it all:

> There was built out from the shore a mole, constructed to break the assaults of the sea, and stem its violent ingress. She leaped upon this barrier and flew, and striking the air with wings produced on the instant, skimmed along the surface of the water, an unhappy bird. As she flew, her throat poured forth sounds full of grief, and like the voice of one lamenting.[60]

The winged grief of the halcyon pallbearer carrying her dead mate to his watery grave is a haunting image. Through the magical play of nature and her creation of such rich tapestries, the woeful flight of these widowed birds transports us to the deepest realms of the human soul. She has been the muse of many a poet, as in John Milton's hymn *On the Morning of Christ's Nativity*.

But peaceful was the night
Wherein the Prince of Light
His reign of peace upon the earth began;
The winds with wonder whist
Smoothly the waters kist
Whispering new joys to the mild ocean
Who now hath quite forgot to rave
While birds of calm sit brooding on the charmed wave.[61]

In this inner landscape of the mind, John Keats' ode to the young Greek poet Endymion lends a psychological edge to the fable:

O magic sleep! O comfortable bird
That broodest o'er the troubled sea of the mind
Till it is hushed and smooth.[62]

Alcyone's absence on her husband's shipwrecked boat confirms the meaning of her name, 'princess who averts evil.'[63] Although the Aeolians were competent sailors, they still placed their trust in their Sea Goddess, Alcyone, to whom they prayed for protection from rocks and rough weather. Zeus' destruction of Ceyx's ship was an act of defiance against the Goddess, says Graves.[64] Ever since, sailors have carried a talisman of the halcyon's dried body to keep storms at bay and its magical powers protected them from lightning, for they mistakenly believed that lightning only strikes once in the same place.[65]

The Greek historian Pliny reported that the kingfisher is rarely seen and then only during the summer and winter solstices.[66] For his life of service to the Goddess, Robert Graves was granted the privilege of sighting the halcyon, not once but twice, albeit 'with an interval of many years' in between.[67] In *The White Goddess* he describes his mid-summer experiences of the bird skimming over the still waters of a Mediterranean bay, 'Its startlingly bright blue and white plumage made it an unforgettable symbol of the goddess of the calm seas.'[68] Surely it is no coincidence when we look to the stars through our modern binoculars and telescopes and see the Pleiades bathed in a heavenly glow of blue nebulosity? Is this not the same description of the Bird Goddess of the calm *celestial* seas?

Asterope and Celaeno, the middle siblings

Typical of many families, so often the middle child tends to be outshone by the older and younger siblings. The Pleiadian family is no exception to this modern popular psychology truism. Maia was renowned for her beauty and Alcyone for her mysticism, but very little is written or known about the middle sisters, Asterope and Celaeno. In the legend and literature of the night skies they appear to be literally outshone by their sisters and parents, Atlas and Pleione. We do know that both sisters, like their older siblings, mothered kings who founded city-states. Asterope (or Sterope as she is sometimes called) conceived her son Oenomaus during her liaison with Ares (Mars) the god of war.[69] Some writers such as Tripp claim he was not her son but her husband, with whom she had three children.

However the relationship unfolded, we know Oenomaus became King of Pisa. Asterope's name derives from the Greek aster for 'star', which is where we get the terms astrology and astronomy.[70] Graves gives the meaning of her name as 'sun-face', which bears some linguistic resemblance to other star goddesses such as Astarte, Astraea, Esther and Ishtar.[71] Celaeno had two children (including Lycus) to Poseidon, who also fathered Alcyone's four children. Disputed lineage tells us he may also have been uncle to the Sisters. Apart from concerns related to inbreeding that marks present genetic understanding and morality, these relationships may reflect the formation of early societies.

The symbolic and mythological significance of the pairings also indicates many layers of understanding and knowledge obscured by time's passage. In this regard, Celaeno's name becomes ironic for it means 'melon' in Greek.[72] It does not necessarily mean that she was a voluptuous woman or large-breasted, for another meaning of her name 'signifies both a sheep and an apple.'[73] Knowing so little about these two middle daughters we can only guess the connections between sheep, melons and apples that relate one sister to the environment and meteorology. According to Theon the Younger, lightning struck Celaeno and she is therefore reputed to be the Lost Pleiad or missing sister, a controversy surrounding her younger siblings, Electra and Merope.[74] As Electra means 'shining' or 'bright' we can discount her as the lost sister, plus the fact that she is the third brightest star in the cluster.[75]

The majority of Australian Aboriginal legends agree that the lost sister was the youngest. In Greek traditions she is Merope, who shamelessly married a mortal instead of a god and therefore hides her face. Precedence indicates that it is more likely she is the missing sister — even though Asterope is, astronomically speaking, the dimmest Pleiadian star — followed by Celaeno, and then Merope. Legend tells of another Asterope, 'a daughter of the river Cebren' whom Aesacus, one of the many sons of King Priam of Troy, fell in love with.[76] Given the uncertainty of myth and the layering of characters upon one another, we cannot be sure this was the same Pleiadian maiden. Still, it bears a striking resemblance to the story of Alcyone and Ceyx, for after Asterope had died, Graves tells how Aesacus 'tried repeatedly to kill himself by leaping from a sea-cliff until, at last, the gods took pity on his plight.'[77] Like the two vexed lovers of the halcyon days, Aesacus was transformed into 'a diving bird' to enable him to 'indulge his obsession with greater decency.'[78]

Taygete the golden Ceryneian hind

Taygete, like her older sister Maia, lived alone on a mountain west of Sparta that now bears her name. As we know, isolation provided little sanctuary and she too became an object of Zeus' desire. He pursued her relentlessly and she turned in desperation to Artemis her benefactor, for help. To enable her to escape this harassment, the goddess turned Taygete into a doe.[79] Grateful for this respite, she carved Artemis' name into the golden horns of a doe. Unfortunately her safety was short-lived for Zeus was not deceived for long and soon had his wicked way when she was knocked unconscious during another of his lustful chases. Lacedaemon, later to become king of the ill-fated city Sparta, was the product of this rape. Perhaps this tragedy explains some claims that the maiden later hanged herself on 'the summit of Mount Amyclaeus,' subsequently renamed Mount Taygetus to commemorate her spirit and her strength.[80] These days it can be found in the mountain ranges dividing Messenia from Laconia.

As with the Hesperides, Taygete was yet another of Atlas' daughters who all played key roles during Heracles' labours, the third of which involved the capture of the Ceryneian hind, highly valued for her golden horns.[81] Some ancient Greek writers claim the very hind that Heracles captured was the Pleiad in anthropomorphic form. Instructed to capture and bring the golden hind alive to Mycenae, Heracles was compelled to pursue her for well over a year. The chase took him across the whole known world, even as far as Hyperborea in the far region to the north. Exhausted, the hind eventually sought 'refuge on Mount Artemisium' before descending 'into the river Ladon.'[82] And this was where the wily Heracles found her resting. Not wanting to draw blood, he drew his bow and 'shot an arrow through her forelegs.'[83] She could not escape and was carried easily across his shoulders to their destination. Graves' commentary on the third labour suggests its significance differs from all the others, and today its prophetic implications resonate throughout the modern world. Spilling blood becomes a compelling metaphor through the link between Taygete's bloodless capture and the cycle of women's wisdom, which characterises Heracles' triumph over fear. The Hyperboreans were also experts in mystical and secular knowledge, and violence and bloodshed would have been a sacrilege on their land.[84]

Apart from those associations, the hero's initiation on the path to the getting of wisdom symbolises the male possession of shrines to the Mother Goddess.[85] In classic history she was worshipped in deer form and her subjugation represents subsequent patriarchal control of the Western world. Today, in supposedly more enlightened times we know that conquest of the feminine is far from true wisdom. Though Graves suggests Heracles was recognised as wise, this knowledge comes only with death, and in this respect his journey to the garden of the Hesperides 'was really a journey to the Celtic Paradise.'[86] Perhaps this notion of male acumen is double-edged, for another meaning of his name is Melon.[87]

Electrical Electra

This little sister was quite a lively Pleiad, as her name suggests, for her impact endures through her gifts to us in archaeology (Troy), science (electricity) and psychology (Electra complex). Certainly her fertility was pronounced, for she had four children to Zeus in addition to another set of daughters from her marriage to Thaumas; namely the Rainbow Goddess Iris and the Harpies.[88] One of the illegitimate children, her son Dardanus, later founded Troy and this connection to the city marks her reputation. One story claims she is actually the Lost Pleiad who disappeared in grief over destruction of the city and her son's death.[89] Others say she deserted her astral home to witness the city fall and became so distraught that her brilliance diminished. Wandering aimlessly, she eventually settled as Alcor the other half of Mizar, the double star in Ursa Major, the Great Bear.[90] Several other cultures place a Lost Pleiad in that constellation, including India.

Electra of the Pleiades has inspired many thinkers and influenced our understanding of natural phenomena — electricity, electrum and electron owe their meanings to her character. The Latin translation of her name also has a connection with the organic gemstone 'amber'[91] which is not a mineral but was formed from fossilised resin or sap that once flowed from ancient coniferous trees.[92] The Sun's rays were said to lend the 'golden gem of the ages'[93] its beautiful radiance and the Greek word for amber was *elecktra* or *elecktron*; the Sun's energy mythologically captured in the stone gives us the modern word electron.[94] This elementary play of light and energy expressed as the celestial tones of night and day, Moon and Sun, is encapsulated in the real world alloy of silver and gold *electrum* used to make the world's first coins.[95]

In real terms the Greek philosopher Thales of Miletus carried the interrelationship between Electra, amber and energy further. In about 600 BCE he noticed that when pieces of amber were vigorously rubbed together, the friction heated the amber to produce a negative static charge, which attracted bits of lint, paper and straw.[96] In this way the electrical charge became linked with the Goddess and, together with the magical properties of amber, sealed Electra's fate as one of the most influential of the Seven Sisters.[97] As Rice points out, amber sap often contains the remains of plants and insects.[98] As bees were commonly found preserved within the translucent yellowish golden gemstone, the Ancients believed it was made from honey and imbibed with special talismanic qualities.[99] Even polished fatty or turbid amber gave 'the appearance of whipped honey' which added to illustrious associations that speak of its divine origins.[100] Flakes of this malleable stone were used in rituals as incense to add fragrance in temple ceremonies, where its perfume filled the air with the essence of pine.[101] As pine is of the *araucaria* genus, one of the oldest plant families, a connection with the Tree of Knowledge becomes compelling.[102]

We can find further solar connections between plant growth and species ecology in the belief that amber was 'honey melted by the sun and dropped into the sea from the mountains of Ajan, whereupon it became congealed by the water.'[103] Considering the incidence of bee fossilisation in amber, the gemstone's spiritual value in ancient societies literally leads us to the ocean and maritime trade where sailors manually harvested it from the sea 'since it is buoyant enough to float in salt water.'[104] Much of the gemstone supplied to the ancient Greeks was collected from the Baltic Sea. Apart from the fossil's value, it is an intriguing fact that bees navigate through a minute crystal in their brains, which is magnetically attuned to the Sun.[105] This pathway of connections between species, elements, society and the gods guides us to the Amber Route, the very same path along which Heracles travelled during his quest for the golden hind.[106]

Yet another legend refers to amber as being the teardrops of the Heliades, daughters of Helios the Sun who grieved the loss of their reckless brother Phaeton after he drove his father's fiery chariot into the river Eridanus.[107] The Pleiades and the Heliades were cousins through their respective mothers, the sisters Pleione and Clymenne. Their tears remind us of the other set of half-sisters — the Hyades — who mourned the untimely death of their only brother Hyas. Where the Hyades' tears brought rain, those of the weeping Heliades rained down as honey drops. The cycle of life and death is reflected here in these kin relationships between the Hyades, Heliades and Pleiades, and in the striking similarities of their tearful experiences.

Rain and honey play integral roles in the Australian Aboriginal notion of the 'sweet water' of the Wandjina that is the seed from which all life is formed. In scientific terms this phrase can be translated into the sugar-based constituents of nucleotides which, along with phosphates and amino acids, are the essential components of DNA. Could this be what is meant by the biblical reference to the Pleiades as shedding their 'sweet influence'? And what might be its relationship to the fabled elixir of life, the nectar of the gods known to the Ancients as ambrosia and soma to the Hindus? Homer tells us that the Pleiades in their zoological form as doves carried this mysterious drink of immortality to quench Zeus' heavenly thirst.[108] No one seems to know what ingredients comprised these drinks, but the general view is that their purpose was to communicate with the divine. This almost certainly involved the inducement of altered states of consciousness, so was possibly a drug of some kind.

Robert Wasson was one of the first scientists to explore the pharmacological identity of soma in the early seventies.[109] He concluded that soma was the particular species of mushroom *Amanita muscaria*, but Terence McKenna argues otherwise. In *Food of the Gods*, McKenna says that this plant is virtually ineffective as a mood-altering substance and doubts it is the soma of the Vedas.[110] And as we saw in the first chapter, Hilda Ransome linked all these magical elixirs of immortality with honey as their key ingredient in *The Sacred Bee*.[111] Whatever the true identity of these elixirs, certainly there are suggestions of heavenly influences on earth. Rain falls and nurtures plants, which are in turn fertilised by honeybees; the matriarchal cycle repeats mythologically in these seasonally sweet tears, the nectar of the Goddess is an aspect of the one divinity.

Electra and the sacred Palladium of Troy

Besides being the mother of Troy's founder, Electra was commemorated in that city by its Palladium, a holy object allegedly removed to

Rome after the city's fall. Although Barbara Walker maintains 'no one knows exactly what the famous Palladium was',[112] nonetheless Edward Tripp suggests it was a small statue. In *The Handbook of Classical Mythology* he points out that 'palladia' were sacred objects that fell from heaven.[113] During a mock battle in which their tempers flared, Pallas was about to strike her jousting partner Athena. Zeus distracted her by hurling the aegis and Athena accidentally struck a fatal blow.[114] Grief stricken, she promptly carved a small statue and placed this icon in the heavens to honour her childhood companion. The statue, just over a metre high, brandished a spear in one hand, a distaff and spindle in the other, and her feet were joined.[115]

Legend tells that Zeus spied Electra alone on the Greek isle of Samothrace and abducted her. Overcome with desire he carried her to Mount Olympus where he raped her as she vainly clutched the Palladium for protection. Enraged, Zeus hurled the statue out of heaven with Electra clinging precariously to it. Perhaps this clash of celestial energies partially explains his reputation as the god of lightning.[116] Dardanus was the illicit child of that rape who later founded the city of Ilium (Troy). Some accounts say Ilus was really Dardanus, who had prayed to Zeus for a sign to indicate the ideal site for the city. After the Palladium struck the ground beside his tent he built a temple to house the meteoric object. Yet another account says Zeus gave the Palladium 'to Ilus' great-grandfather Dardanus, the Samothracian founder of Dardania.'[117] Despite the confusion over names over time, the statue was closely guarded in the Peramun within the citadel, for Trojans feared its loss would bring down the mighty city.[118]

This faith in the power of revered objects to maintain empires is a universal theme that still survives in contemporary societies. Think of the black ravens that stand watch over the Tower of London, or the Stone of Scone upon which many British monarchs have sat. Certainly, the Palladium occupied a similar status after it was stolen and taken to Rome. Walker claims the hallowed article was placed within the temple of Vesta and guarded by seven vestal virgins that tended the Goddess' sacred hearth.[119] Just as Trojans feared the loss of their Palladium, the eternal flame was never allowed to expire for fear it would lead to the ruin of the Roman Empire. The temple fire was a vestige from long ago when it provided the focus of 'clan and tribal life' and was 'presided over by the ruling matriarch,' says Walker.[120]

As in other cultures, the central fire came to symbolise the warmth of loving relationships of families and other bonded groups, which suggests an etymological link between the hearth fire and the muscular organ that circulates and warms our blood, the heart. This matriarchal power served to uphold the Roman Empire until the vestal flame was extinguished sometime in the fourth century CE[121] when Christianity turned away from love of woman as the spiritual pathway for unification between heaven and earth.[122] Similarly, Hindu traditions depict the Seven Sisters as flames lighting up the night sky and their guardianship of that particular region has endured for millennia.

However we interpret this custodial role of Electra in her transformative responsibility as a matriarchal guardian object, her name takes on chilling significance in the psychological realm immortalised in the works of Aeschylus, Sophocles and Euripides. Those Greek tragedies speak of the vengeful and destructive power of sexual passion experienced by Electra's unfaithful mother Clytaemnestra and her paramour Aegisthus, who killed Electra's father Agamemnon. To avenge her father's murder Electra and her brother Orestes slayed the adulterous lovers.[123] Accounts of the affair differ between dramatists and its psychological dimension has become known as the 'Electra complex.' Although the Electra featured in this ancient Greek drama may not be the fiery astral sister, her primordial influence and sexuality shape our understanding of the importance of tending the sacred fire within us. Unless we honour this divine power, unbridled sexual fire and passion destroys our capacity to respect our families and take responsibility for our carnal desires.

Merope the mortal bee-eater

Finally we come to the youngest of the Sisters, Merope who, unlike her older siblings, married a mortal man and not a god. For this sacrilege she hides her face in shame and is therefore said to be the missing sister or Lost Pleiad.[124] Thus Richard Allen gives the meaning of her name as 'mortal'[125] whereas other writers suggest it means 'bee eater.'[126] She bore three sons, Glaucus, Ornytion and Sinon to her husband, King Sisyphus of Corinth, a notorious womaniser and underworld figure.[127] Some say Merope was ashamed, not because she married a mortal but because she found herself with a criminal husband serving time in Tartarus, the Greek slammer of the underworld. As part of his punishment, Sisyphus was condemned to continuously roll a heavy stone uphill, which persistently rolled back on him before he could topple it to the other side.[128] Robert Graves says the prisoner's 'shameless stone' was 'originally a sun-disk, and the hill up which he rolled it is the vault of Heaven.'[129] In this regard he shares a similar fate with his father-in-law Atlas, who bears the weight of the sun disk on his shoulders. It may be that Sisyphus acts as the proverbial celestial plug, whose removal is said to cause the Deluge. This aspect is looked at in further detail in the final chapter that discusses floods and other catastrophes that occur at the turn of the world's ages.

Despite this unfortunate liaison, Merope was not without suitors. The handsome giant hunter Orion fell madly in love at first sight when he saw the Sisters walking through the Boeotian countryside north of Athens. Others claim he was smitten when visiting Hyria in Chios 'an island off the Ionian coast.'[130] In this tale she is said to be the daughter of King Oenopion, son of Dionysus, and not of the Titan giant, but this apparent discrepancy may simply be a different version of the same tale. It is common practice for mythology to shape its essence to the surrounding environment to reflect and highlight differences in regional characters and localities. There is every reason, therefore, to believe this is the same Merope of the Pleiades whose story has been adapted to suit local conditions.

As legend tells, when her father refused Orion's request for her hand in marriage, Orion raped the hapless maiden. Oenopion took revenge by plying the giant with wine and when he fell asleep the king poked out Orion's eyes and dumped his drunken body by the sea. Upon waking and stumbling blindly, the hunter eventually came upon a soothsayer who told him that if he travelled east, turning his eyes towards the Sun where Helios rises above the ocean, his sight would be regained. Orion immediately set off in a small rowboat and following the sound of a hammer across the ocean, he reached the Isle of Lemnos, abode of Hephaestus the disabled god and blacksmith. Taking pity on the temporarily blind hunter, Hephaestus gave him a young, lame apprentice named Cedalion to act as his eyes. Perched on the giant's shoulders the young guide directed their boat towards the east where they eventually reached the farthest edge of the ocean. There at first light, Eos (the dawn) fell in love with handsome Orion who regained his sight, courtesy of her brother, Helios.

This particular myth of the loss of Orion's eyesight is loaded with astronomical symbolism, say Edwin Krupp and Robert Graves. Orion faces east as the Sun sets in the west and to restore his eyesight he must face the Sun in the east. To do that, he must travel west in order to meet the eastern sunrise. The constellation's last appearance in the western evening skies marks the period of its conjunction with the Sun. Perhaps Orion's heliacal rising in the east lies behind the myth of his regained sight. The hunter's pursuit of the Sisters refers to his appearance just after the Pleiades rise in the Northern Hemisphere's evening skies.[131] This celestial cycle, says Graves, refers to the earthly reluctance of kings to relinquish power at the end of their time in office, a high drama consistently played out in contemporary politics and world affairs.[132] Whether it is the heir apparent or incumbent ruler, a ritualistic sacrifice precedes any bloodletting before a change of regime and, as Graves claims, the new resurrected leader 'remarries the queen and continues his reign.'[133] This electoral sham distracts so many people that the real issues become obscured and nothing really changes.

Although Graves equates this drama with the fabled character Orion, behind the scenes the regal association of the Pleiades with kingship runs deeper, and in political life the struggles and aspirations take on mythical dimensions. For instance, conjuring up Arthurian images of a youthful king at the height of his power with his lovely queen by his side, the United States Kennedy administration of the 1960s was often referred to as the Camelot years. As we now know, real power rests with the kingmakers and electoral spin doctors that fabricate these political mythologies and illusions to beguile a public susceptible to romantic fantasy and the power of their charms. Astonishingly, the election of the American President in November derived from 'the ancient convocation of tribes' whose ceremonies were governed by the rising and setting of these stars.[134] This tradition, says Graves, has its origins in ancient pagan worship of the 'Goddess of Life-in-Death' who, 'in every Great Year, early in November, when the Pleiades set, sent the sacred king his death summons.'[135]

Similarly, Freemasonry holds to a ceremonial cycle connected to the November rising of the Sisters and natural hibernation during winter's vernal equinox.[136] The seven stars of the Pleiades were interpreted as a sign of immortality and the continuum within which death occurred in autumn, rebirth in spring. Ultimately, says Graves, there can be no greater kingship than that bestowed by the Goddess. Undoubtedly this 'marriage' of natural celestial phenomena with its physical manifestation (the *hieros gamos* of the Greeks) is the same karmic aspect of empirical cycles so precisely identified by Blavatsky in *The Secret Doctrine*. She documents those cosmic, physical alliances with the Pleiades who mothered heroes and wed kings, and whose children founded the great nations and city-states of ancient times.[137]

Perseus turns Atlas into stone

Heracles' eleventh labour was not to be the last of Atlas' troubles, for it was prophesied that the time would come when a son of Zeus would one day 'rob him of his Golden Apples.'[138] Perseus was that son. After he killed and beheaded Medusa, he flew with the Gorgon's head over land and sea eventually arriving in the Gardens of the Hesperides.[139] Announcing his parental lineage, Perseus asked Atlas for food and a place to rest but when Atlas remembered Themis' prophecy, he ordered Perseus to leave. Offended by Atlas and his refusal to help, the Gorgon slayer turned his head away, pulled Medusa's head out of a bag and thrust it in the Titan's face. Her grotesque, cadaverous stare turned Atlas to stone and thereafter he was transformed into Mount Atlas where his beard and hair became forests, his upper limbs cliffs, and his bones rocks. His gigantic form reached into the heavens with the stars resting on his shoulders.[140] That part of the world we now know as Libya is where Atlas is said to have lived, and his rule gave us the notions of navigation and astronomy, which occupy a special place in the atlas of world knowledge.

Often referring to the Pleiades as the Atlantides, Madame Blavatsky suggests that Atlas' name is derived from the fabled lost continent of Atlantis. She takes the association further and links the karma and destiny of nations and individuals with the one symbol of the world-bearing Titan who personifies the lost continents of Atlantis and Lemuria.[141] Modern geology tells us that ancient mountains have been dramatically reduced by the ravages of time and that the single primordial landmass they refer to as Pangaea originally split into two continents, Gondwanaland and Laurasia.[142] Lemuria, says Blavatsky, was part of Gondwanaland and Atlantis part of Laurasia.[143] She claims that Atlas — who supports the new landforms and their horizons between his shoulders — represents the lost continent of Atlantis.[144]

Rather than reading this as purely geomythical, we can interpret it as Atlas providing spiritual support to the world and bridging the creative energies of the double helix. Blavatsky's deeper, esoteric meaning is therefore linked to the origins of humanity and her theory of the seven root races from which all peoples were derived. In this matrix there were seven subdivisions, and she asserts that Atlas represents the fourth race and his seven daughters, the Atlantides, the 'seven sub-races.'[145] They, in turn, were descended from the Lemurians, who were of the third root race. Unlike the Lemurians, however, the Atlanteans had reached the pinnacle of human development with their advanced minds and technology.

Before Blavatsky's claims are dismissed as sheer nonsense let's consider Mircea Eliade's explication of symbols and their role in human imagination. He argues that 'symbolic thinking precedes language communication and reasoning by discursive analysis and, containing certain of the deepest aspects of reality, symbols defy rational segmentation.'[146] This holistic form of knowledge transfer occurs within the mindscape,[147] the place between consciousness, phenomenology, and recognition where the psyche brings to light 'hidden modalities of being.'[148] Apart from this symbolic form of communication, we may consider theoretical nuclear cosmology since Einstein, by exploring the possibility that our three-dimensional existence may be part of multidimensional reality. From this radical perspective, the pivot point of time and space is seen as a portal through which we might enter parallel universes and other worlds. Viewed this way, the heroic Atlas and his lost continent cannot be dismissed as psychobabble, but rather as a figurative conduit and realm known to ancient peoples and cultures as a modality hidden from sight of the profane. It is precisely these 'hidden modalities' that are explored throughout this book in an effort to comprehend and understand the mystery of the Pleiades.

Atlantis and the seven root races of the world

Although some people steeped in the notions of rational scientism reject these concepts as theoretical speculation, the spiritual aspects of Blavatsky's work belie her veracity. Arguably, she was one of the most prophetic and visionary thinkers of her time — today's political correctness obscures much insight hidden beneath the extreme racism of language during her era. In the context of world knowledge, her root races theory complements Australian Aboriginal wisdom and is tremendously valuable in gaining an understanding of the Seven Sisters of the Pleiades and their importance in the framework and spirituality of our own psyche. Much of her thesis draws on an ancient text encoded in a secret language known as the *Book of Dyzan*, which she claims Tibetan spirit guides known as the Brotherhood of Mahatmas illuminated for her. Eleanor Van Zandt and Roy Stemman in *Mysteries of the Lost Lands* summarise the essence of her theory:

> Describing the emergences of life on earth, Madame Blavatsky declares that we are the 'Fifth Root Race' to inhabit the earth, and that our

planet is destined to have seen such races, each composed of seven sub races. The First Root Race, invisible beings made of fire-mist, lived on an Imperishable Sacred Land. The Second, who were just visible, inhabited the former Arctic continent of Hyperborea. The Third Root Race was the Lemurians, gigantic, brainless, apelike creatures. The Fourth Root Race was the fully human Atlanteans, who were destroyed through black magic. We are the Fifth and the Sixth will evolve from us and return to live on Lemuria. After the Seventh Root Race, life will leave our planet and start afresh on Mercury.[149]

People may choose to criticise Blavatsky or dismiss her as a crazed individual with 'weird' ideas, but we should remember that she is not alone in her perception. Several well known and respected civilisations — including the ancient Greeks, Persians, Aztecs and Mayans — all spoke of 'Five Ages' of humanity that have already passed.[150] Blavatsky's theory suggests there are two more root races yet to come. Although we cannot be certain whether the ancient Greeks agreed, it is entirely feasible because at the time of their civilisation they acknowledged they were living in the Fifth Age. Or they chose not to mention the other two ages for reasons unknown. Certainly the Aztecs and the Mayans maintain that not only have there been Five Ages or Five Suns but that we are rapidly approaching the dawning of the Sixth Sun.[151] Nor is the idea strange to Christians, for the Book of Revelations in the Bible 'foretold that the world would end after *seven ages* had passed.'[152] Also, there are further cryptic references to the passage of 'seven kings, of whom *five* have gone,' and the 'seven Sabbaths.'[153] In fact, number seven is central to many world creation stories, including the Finnish epic *Kalevala* that tells how the universe was created from seven eggs laid on the terrestrial knee of the Ocean Mother by the teal duck[154] and in Babylonian tradition, the god Ea created seven men and seven women after their Great Flood.[155]

What is enormously interesting about our widespread preoccupation with the number seven in many creation stories is the revelation that genetic analysis of people of European descent traces DNA (deoxyribonucleic acid) carried only in female mitochondrial genes (mtDNA) to seven primordial 'clan mothers.'[156] These are *The Seven Daughters of Eve* who feature in Bryan Sykes' book, which overturned previously held archaeological and anthropological conjecture about early hominids and spontaneous agricultural development. Inspired by scientific revelation, Sykes created his own mythology in naming those universal parental sisters Ursula, Xenia, Helena, Velda, Tara, Katrine and Jasmine, whose initials are cryptic abbreviations for the classification of their genetic codes.[157]

Whether it is Blavatsky's theosophical views, evidence extracted by Sykes' forensics, or the cultural beliefs of diverse societies, it is becoming clearer that we hold cellular genetic memory of our common creative origins. Does this mean our holotropic memory of the Pleiades is not only cultural but that the Dreaming of the Seven Sisters may be literally carried in our blood and our genes? When we consider that our bodies are made of the same stuff of stars (stardust), we can begin to literally feel our atomic bodies as affiliated with the Seven Sisters of the Pleiades. As Sykes himself concedes, further genetic research needs to be undertaken in other parts of the world to shed more light on the subject. Certainly any investigation of this nature is to be encouraged and supported for its potential to enlighten our sense of personal and human identity. That identity can be shared across cultural boundaries and through time, for the Greek philosophers Diodorus and Strabo lend credence to Blavatsky's claim that Atlas and his seven daughters symbolise the seven root races of the world. Those ancient sages reported a strange island in the Atlantic Ocean accidentally discovered by some Phoenician sailors blown off course in a storm. Strabo writes that the island 'was called Meropis, and its people Meropes.'[158] Could this island have been named for Atlas' youngest daughter, Merope? Although this story has not been corroborated elsewhere it raises some intriguing questions about Blavatsky's theory.

A universal theme emerging from reports about Atlantis is of that civilisation's highly advanced technologies, which were destroyed by Atlantean misuse of powers, their moral degeneration, or both. Edgar Cayce's numerous psychic readings indicated that the destruction of

Atlantis was specific to their citizen's misuse of crystals. Known as the 'Sleeping Prophet' for delivering these readings while in deep trance, he declared the Atlanteans invented — among other things — aircraft, television, the laser and 'death rays', atomic energy and many other scientific achievements.[159] Seneca elder, Grandmother Twylah Nitsch recounts Native American legends of Atlantis saying, 'Marvels poured forth from (their) brilliant minds,'[160] and she describes their creative talents, including the discovery of 'cures for all the human diseases.'[161] Blavatsky's Atlanteans were 'learned in magic and high in arts and civilisation' and they possessed extraordinary psychic skills.[162] One of these skills described by Herodotus was their ability to exist in a dreamless state. The theosophist's physiological perspective suggests the earliest Atlanteans' brains were not sufficiently evolved 'to permit the nervous centres to act during sleep,' whereas Atlanteans of the later subdivisions were able to dream.[163]

Paula Gunn-Allen's more empowering explanation says the earlier dreamless state was made possible by their use of crystals to rejuvenate.[164] It may be that dreams were made redundant as a means of mental housework that is necessary for our health and spiritual wellbeing due to the healing energies of crystals. Another claim made by Herodotus was that the Atlanteans cursed the scorching heat of the Sun, which tormented them every day.[165] Blavatsky says they did not do so because of its infernal temperature but for their own 'moral degeneration.'[166] However, Grandmother Twylah refutes this. She claims instead that they prayed to Grandfather Sun for his blessing light to energise and heal their physical bodies.[167]

The universal symbolism of the Deluge

However it is discussed, Madame Blavatsky believes the Deluge is full of occult meaning related to the spiritual and physical realms. Although she acknowledges the universalism of this theme running through stories from many different cultures, she distinguishes between the *first* 'Cosmic Flood' of primordial creation and the Atlantean deluge. In the Cosmic Flood a mortal woman becomes the receptacle of human seed fertilised by the Sun or male principle, giving physical form to the Fourth Race.[168] The *navis* or ship is a metaphor for her womb as vessel on which humanity's seven ancient mariners sailed to salvation, an event commemorated as the Ark of the Covenant.[169] In *The Women's Encyclopedia of Myths and Secrets*, Barbara Walker tells us that the Semitic word for ark derived from the Hindu *Argha*, which means 'great ship.'[170] This was the boat on which the Hindu progenitor Manu and his six sons sailed.[171] Metaphorically it symbolised 'the Great Yoni — a female-sexual vessel bearing seeds of life through the sea of chaos between destruction of one cosmos and creation of the next.'[172] Or, as Marija Gimbutas refers to it, the 'Ship of Renewal', whose symbolism is retained by the maritime custom of naming ships after women, and the prows that in ancient times bore the faces or forms of women and goddesses.[173]

The Atlantean flood, on the other hand, is one of a series of periodic deluges that mark the end of the subdivisions of the root races. Blavatsky says the uninitiated cannot distinguish between the various deluges and consequently see only one flood and one hero.[174] She claims the biblical Noah is a purely fictitious rendering of older traditions such as that of the Chaldean Xisuthrus, the Hindu Vaivasvata Manu, the Greek Deucalion, the Chinese Peirun and the Swedish Belgamer.[175] In this regard it is interesting that the meaning of Atlas' daughter's name, Merope, not only refers to her marrying the mortal Corinthian king, but it may also mean 'mortal' itself. Could this mortal woman who seeded the human race be Merope of the Pleiades in another incarnation as a sevenfold deity? Might this explain the strange race of people called the Meropes on the Isle of Meropis that Phoenician sailors allegedly stumbled across in their travels? What is the symbolism behind Merope's daughter, Niobe, being the daughter of a seventh daughter, who in turn gives birth to seven daughters and seven sons?

In a similar vein, Mircea Eliade cites the *Gilgamesh* myth as indicative of the archaic theme of human redemption in which spiritual cleansing ushers in a new age to atone for the 'sins of men and decrepitude of the world.'[176] In Eliade's view this cycle is integral to human relationships with nature, driven by the cosmic rhythm of aquatic origin and life forces. He argues that this celestial pattern of decay and revival underpins our continuing celebration of the New Year when resolutions are made to resolve the residue of past imperfections.[177] It is a modern tragedy that we are so removed from the realisation that the rising of the Pleiades traditionally heralded a time of deep reflection and aspiration for renewal in many cultures.

The cycle of ignorance that led to Atlantean destruction continues today in the spiritual and physical corruption of many individuals and institutions including church and state, where the Cardinal is pederast and the President is non-elect. This archetypical patriarchy and its diabolical heritage are reflected in the tale of Niobe, daughter to one of the Pleiades and granddaughter of Atlas. As Queen of Thebes and the first mortal woman to be raped by Zeus,[178] her earthly vanity and homophobic criticism of Latona's 'masculine daughter' and 'womanly son' compelled the Goddess to beseech her twins Artemis and Apollo to avenge this arrogant sleight.[179] Driven to distraction by Niobe's tears over the slaughter of her husband, seven sons and seven daughters, Zeus turned the bereft widow into a mountain of stone from which a torrent flowed that can still be seen to this day.[180] Blavatsky sees this atrocity as an allegory for the fall of Atlantis and argues that it stands for the seven subdivisions of the fourth race.[181]

As the astronomic manifestation of the Sun and Moon, Artemis and Apollo are also cosmological energies, which influence the earth's axis and magnetic polarity.[182] As we now know, variation of that polarity has precipitated the onset of Ice Ages and other catastrophes throughout geological history. Several writers, including Blavatsky, suggest that survivors of the Atlantean cataclysm emigrated to other regions including Europe, Asia, Egypt and the Americas. Mayan shaman and author Hunbatz Men claims his ancestors came from Atlantis and brought with them the mysterious crystal skulls to be used for the spiritual benefit of humanity.[183] Many believe the power of those sacred artefacts will assist our evolutionary growth and their wisdom will be revealed sometime early in the coming Aquarian Age. However we interpret the Atlantean legend, we are left with some graphic symbols and images, which inspire our continuing search for common understanding. Foremost among these is the well known image of an altruistic, solitary individual — a heroic figure bearing the weight of the world on our endless journey beyond time and space.

Atlas shoulders the heavens

Besides Blavatsky, Barbara Hand Clow is particularly stirred by his archetypal symbolic value in *The Pleiadian Agenda*. Before you dismiss the following information as too 'New Age', it is worth remembering that some of our greatest scientific and technological achievements have been dreamed into being or emerged from sheer creative human imagination. German chemist Friedrich Kekulé, for example, came upon the chemical structure and molecular shape of benzene while dreaming of a snake biting its own tail.[184] Such stories may once have been dismissed as science fiction, but even that genre has given us Jonathan Swift's *Gulliver's Travels*, Jules Verne's *20,000 Leagues under the Sea* and *Journey to the Center of the Earth*, Arthur C. Clarke's *2001: A Space Odyssey* and the work of Gene Roddenberry, who created the popular television sci-fi series *Star Trek*. In turn, each of those far-fetched tales has spawned its own inventions and contributed to the body of scientific knowledge which once saw our world as flat. Perhaps we should consider that sometimes the wackiest, craziest ideas and theories are those that prove the norm or exception, and it may be those that form the general matrix of our understanding of the universe and the world in which we live.

Steeped in this understanding, Hand Clow builds upon Blavatsky's belief that Atlas and his daughters literally signified the fourth of the seven root races of humanity. She maintains we share a common affinity and destiny with the Pleiades, even though the cluster is hundreds of light-years away. The idea that we might have something in common with these stars is perhaps not as controversial as some of her other theories. In short, she asserts that our Sun and Sirius in the constellation of Canis Major are twin stars and part of the Pleiades star cluster.[185] Together, these celestial bodies are linked to the Pleiades by a 'photon band' radiating from the galactic centre of the Milky Way.[186] This 'five-dimensional' band of energy spirals through Alcyone to the other stars of the Pleiades, penetrating our solar system and that of Sirius. This spiralling pattern, she claims, is personified in legends and in images of Atlas holding the world upon his shoulders.[187] This symbol, in particular, is coded astronomic information and proof that we belong to the Pleiades star cluster. Although scientists have rejected Hand Clow's theories, there is something intriguing about her interpretation of the Atlas symbol, in light of ancient Greek mythology and the claims of many diverse Indigenous peoples that we are either descendants or relatives of the Pleiades. And although her ideas are considered reckless and contentious by conventional science, I believe that some of her more esoteric ideas are worth exploring from Indigenous perspectives of spirituality and science, especially Indigenous cosmologies.

First, let's examine the seemingly bold suggestion that our Sun is a sister star or twin to Sirius. The orthodox scientific view teaches that our solar system is exceptional because we have only one Sun, unlike the majority of stars in the universe which are at least binary and often multiple.[188] By contrast, several Dreamtime legends speak of a time past when our Sun was once a binary star.[189] Australian Aboriginal cosmology speaks of *two* suns — Mother Sun and Daughter Sun — who once lived in the sky world. Legend tells that their combined heat made it difficult for life to flourish on Earth and so they held council to discuss the problem. Daughter Sun's painful decision to leave her mother by going to live in another solar system was her altruistic and ultimate sacrifice. These mother–daughter or twin suns become multiple in the Ngarrindjeri legend of South Australia.[190]

Here where the mighty Murray River flows into the Great Southern Ocean, this epic tells of Mother Sun staying behind to bring life to the Coorong and the confluence of fresh and saltwaters of its many tributaries, while her Seven Daughters return to their heavenly home in the Pleiades.[191] So the suggestion that our Sun may be — or once was — a double star is not such a strange concept to Aboriginal people. These stories raise all sorts of intriguing questions and even though they don't state where Daughter Sun went, it is possible that she could have gone to live in the Canis Major constellation as its brightest star Sirius. Whether scientists are willing to explore this hypothesis remains to be seen, but once again the idea may not be as fanciful as it first appears. After all, the concept of runaway stars is already known to modern astronomy.

However, even supposing Daughter Sun *is* Sirius, there still remains the matter of the distance that our Sun and Sirius are from the Pleiades, which seems to determine the question of whether we 'belong' to a particular constellation or stellar system. At a mere distance of 8.6 light-years away, Sirius is one of the closest stars to Earth,[192] whereas the Pleiades are much further away at some 400 light-years distance. Astronomers have therefore rejected Hand Clow's suggestion that Sirius and our solar system are part of the Pleiades star cluster because of the enormous distances involved. This raises the question, to what extent does distance determine our sense of belonging in the universe? Do we have to revolve around a specific Sun to be part of an astral domain? By whose definition do we create our sense of belonging or feeling of kindredness?

Take Virgo for instance. This super cluster of galaxies is some 50 million light-years away from us[193] and yet the Local Group — the smaller cluster of galaxies of which we are members — is, in turn, considered to be part of this much larger conglomerate. And, unlike the Pleiades, Virgo is right outside our own galaxy, and millions of times further away. Science tells us the centre of the universe is a matter of perspective; it is only decided from a position relative to the observer. Therefore, if the centre is arbitrary, why can't we simply draw a straight axis line from the Pleiades to our Sun and give ourselves that astral address? Ultimately, whether or not we belong to the Pleiades is not the real issue, and need not diminish our sense of affiliation with these stars. What matters is that we feel a holistic sense of kinship with our entire universe, with all its stars, moons and planets.

Now to the most controversial aspect of Hand Clow's theory, that which has caused great consternation among scientists: her claim of a band of light emanating from the centre of our galaxy, passing through the Pleiades, on to our solar system and that of Sirius. Once again, the concept is a curious one but let's break the analysis down to explore its assertion. Firstly, the issue of the photon band — what exactly is it and does it exist or not? Secondly, the matter of its alleged spiral motion. Although Hand Clow describes this photon band (for want of a better term) as a form of spiritual energy that impacts on human consciousness in a positive manner, scientists have taken her to task because it cannot be seen or measured, therefore it does not exist. But I believe they have simply missed the point; essentially it is a band of light that is tied up with the notion of spiritual *enlightenment*. Within this context, light is used either as a metaphor for religious knowledge and wisdom, or else its physical and tangible personification. This explains why Jesus Christ says in the Bible, 'I am the Light and the Way.'

Conservative theologians would argue this is nothing more than a metaphorical statement; yet Albert Einstein, arguably one of the twentieth century's greatest scientists who saw the hand of the Creator in the cosmos, sought to understand its physical dimensions along with its sacred meaning. To be fair, Hand Clow clearly states its principal purpose is to mark the Age of Aquarius, otherwise known as the Age of Light.[194] This forthcoming era is generally perceived and hailed as a time of greater spiritual growth and evolution. Furthermore, this band of light only comes into play during two specific zodiac ages, principally that of Aquarius and its opposite sign of Leo. During these enlightened ages we are more spiritually awakened. The rest of the time we are considered to be travelling in darkness and therefore spiritually asleep at the wheel, as it were.[195] This concept of cycles of light and darkness tied to our evolving spiritual growth and conscious-ness that is associated with specific time periods is part of the great Eastern philosophies and religions such as Hinduism and Buddhism. How they relate to the stars of the Pleiades in one of these traditions is looked at in more detail in the chapter on the Krittika.

Apart from its symbolic nature, we should consider the possibilities of its physical properties. Light is composed of small particles or quanta known as photons.[196] What we commonly understand as 'light' is really just one portion 'of the electromagnetic spectrum visible to the human eye, but can also refer to other forms of electromagnetic radiation.'[197] Effectively, this means that light can be visible and invisible, but where it is invisible it must be measurable in order for it to exist. Insisting there is no evidence of this photon phenomenon, scientists have rejected the notion, but given the existence of elementary particles called *neutrinos* her suggestion may be valid.[198] There are many sources of neutrinos in our universe, including our planet's natural background radiation, atmospheric neutrinos, and those that are human generated by nuclear power stations. The other two sources include solar neutrinos that originate from the nuclear fusion powering the Sun, and other stars that daily bombard the Earth and that left over from the Big Bang as cosmic background radiation.[199]

Although previously undetected because its mass is so low, scientists agree on their existence. If this so-called 'photon band' turns out to be no more than a collection of these minute particles — which cannot be perceived by the human eye and which has only been recently detected by scientific means — then scientists may one day regret the ferocity of their attacks on Hand Clow. If, however, this explanation is not scientifically plausible, then the simple fact remains. There is a photon band streaming down from the Pleiades towards our planet, only it is better known in popular parlance as *starlight*. Admittedly, the light we see tonight left that star cluster some 400 years ago.[200] Does this starlight influence our human and planetary consciousness? Certainly our fascination for these stars and the way in which they have stimulated human creativity demonstrates their power as Muses. And given this realisation, who can say for certain that starlight is not another form of cosmic intelligence?

Finally, does starlight travel in spirals? Vincent Van Gogh often painted stars in this fashion, says Hand Clow. The renowned artist was an open, creative spirit, so in tune with this energetic environment that he naturally painted them this way. Although freedom of artistic expression took Van Gogh beyond the prison of normal cognition, his insight came at a heavy cost for he was labelled 'mad'. Similarly, Hand Clow's outsider perception deserves more credit from some members of the scientific community, who have dismissed and devalued her thesis. Outside conventional boundaries, her arcane knowledge reveals ancient wisdom shared across Native American and other diverse Indigenous peoples. Among some of these societies the concept of spiralling stars in motion is already familiar. In *The Gift of Power*, Lakota medicine man Archie Fire Lame Deer explains that heated rocks twirled by the fire-keeper's pitchfork in sweat lodge ceremonies imitate this pattern of moving starlight.[201] This spiral pattern, he says, reminds us of the movement of heavenly bodies, the Moon, Sun and stars.[202]

Although we cannot be sure of the exact aerodynamic motion or flight path of the Palladium as Zeus flung the sacred icon out of heaven with Electra still clinging to it, this mythic image suggests an unidentified flying object or some form of energy spiralling from the Pleiades towards Earth aeons ago. The notion of spiralling starlight is not altogether strange because we know that galaxies, which are made up of the light from billions and billions of suns, rotate in spiral fashion. Like gigantic cosmic catherine-wheels, they spin through space. This spiralling pattern is personified in the symbol of Atlas holding the world upon his shoulders.[203] His physical shape is twisted or coiled by cosmic energy snaking through his body; the orb represents the unification of those forces passing through him. This human Kundalini helix speaks across the esoteric and empiric fields and appears in many images, especially the caduceus and its entwined serpents whose third head signifies the planetary and solar orb, as well as the bindu or circle of creation.[204]

Whether a photon band exists or not, Hand Clow's reading of the Atlas symbol is enormously exciting. Her ideas inspire endless possibilities of interconnectedness and kindredness between our solar system and other celestial beings. Her 'Vision of the Future' (as in Tennyson's poem) is far more powerful and offers hope beyond the empty, meaninglessness and bankrupt spirituality of Nietzschean Nihilism, such as that portrayed in Clarke's Space Odysseys. Her views, and those of many Indigenous peoples, reassure us that we are not alone. For we are children of the universe and need only look into the night skies among the Star People to see our beloved heavenly mother, father, brothers and sisters and all our relations. It's just a matter of perspective, after all.

Notes

1. There are many interpretations of Hesiod's classic poem *Works and Days*, and of this particular passage. See, for example, Richmond Lattimore's translation in *Hesiod* pp. 91 and 93 similar to this.
2. As named by Aratos in Tyler Olcott's *Starlore of All Ages*, p. 419. See also Gertrude Jobes, *Dictionary of Mythology, Folklore and Symbols*, vol. 2, p. 1279.
3. Graves says Atlas was the eldest of his two younger siblings Epimetheus and Menoetius at p. 143 of *The Greek Myths*, but earlier at p. 43 he refers to the 'brotherhood' of Hades, Poseidon and Zeus. At p. 143, he says Prometheus and Atlas were brothers, along with the two siblings named above.
4. Graves, *The Greek Myths*, p. 59.
5. Ibid.
6. Ibid, p. 144.
7. Jobes and Jobes, *Outer Space*, p. 338, and Graves, *The Greek Myths*, p. 152.
8. Graves, *The Greek Myths*, p. 151.
9. Bullfinch, *Myth and Legend*, p. 254.
10. Ibid.
11. Allen, *Star Names*, p. 383.
12. Graves, *The Greek Myths*, p. 83.
13. Ibid.
14. Ibid, p. 144.
15. The individual names of the Hyades vary among ancient Greek writers. Hesiod gives their names as Eudora, Koronis, Phaeo, Kleea and Phaesula, but legend says there were originally seven sisters. These were given as Diona, Ambrosia, Thyrene, Aesula, Polyxo, Koronis and Eudora, by Pherecydes. For further information see *Star Names* by Richard Allen at p. 387, *The Greek Myths* by Robert Graves, p. 108, and Robert Burnham

Jr's, *Burnham's Celestial Handbook* at p. 1819.

16. Allen, *Star Names*, p. 387.
17. Ibid.
18. Graves, *The Greek Myths*, p. 129.
19. Ibid, p. 50.
20. Ibid, p. 127.
21. Walker, *The Women's Encylopedia of Myths and Secrets*, p. 48.
22. Graves, *The Greek Myths*, p. 507.
23. Ibid.
24. Ibid, p. 507.
25. Ibid, p. 508.
26. Ibid, p. 513.
27. Ibid.
28. Ibid.
29. Graves, *The Greek Myths*, p. 764 and *The White Goddess*, p. 119.
30. Graves, *The Greek Myths*, p. 52.
31. Ibid, pp. 129 and 513.
32. Ibid, p. 130.
33. Ibid, p. 513.
34. Nasr, *An Introduction to Islamic Cosmological Doctrines*, p. 50 and Shah, *The Sufis*, p. 174.
35. Lawlor, *Sacred Geometry*, p. 38.
36. Graves, *The Greek Myths*, p. 513
37. Allen, *Star Names*, p. 405.
38. Graves, *The Greek Myths*, p. 55.
39. Ibid, p. 65.
40. Ibid, p. 64.
41. De Santillana and Von Dechend, *Hamlet's Mill*, p. 334.
42. Walker, *The Woman's Dictionary of Symbols and Sacred Objects*, pp. 205, 349, 372 and 465.
43. Ibid, p. 84.
44. Ibid.
45. *The American Heritage Dictionary of the English Language* 2000, 4th edn, (online).
46. Allen, *Star Names*, p. 405.
47. Ibid, p. 403.
48. Ibid.
49. Graves, *The White Goddess*, p. 182.
50. Ibid, p. 181.
51. Bullfinch, *Bullfinch's Mythology*, p. 92.
52. Graves, *The White Goddess*, p. 181.
53. Graves, *The Greek Myths*, p. 163.
54. Ibid. See the story of the halcyon at pp. 163–64.

55. Graves, *The White Goddess*, p. 182.
56. Graves, *The Greek Myths*, p. 164.
57. Ibid.
58. Ibid.
59. See Bullfinch's account in *Myth and Legend*, pp. 85–93.
60. Bullfinch, p. 92.
61. See Verse 5, *On the Morning of Christ's Nativity* by John Milton, p. 14.
62. Keats, *Endymion: A Poetic Romance*, pp. 24–25.
63. Graves, *The White Goddess*, p. 182.
64. Graves, *The Greek Myths*, p. 165.
65. Graves, *The Greek Myths* and *The White Goddess*, p. 182.
66. As reported by Graves, *The Greek Myths*, p. 165 and in *The White Goddess*, p. 182.
67. Graves, *The White Goddess*, p. 182.
68. Ibid.
69. Graves, *The Greek Myths*, p. 394.
70. Jobes, *Dictionary of Mythology, Folklore and Symbols*, p. 144.
71. Graves, *The Greek Myths*, p. 753.
72. Spence, *History of Atlantis*, p. 28.
73. Ibid.
74. Allen, *Star Names*, p. 407.
75. Jobes, *Dictionary of Mythology, Folklore and Symbols*, p. 144.
76. Allen, *Star Names*, p. 625.
77. Ibid.
78. Ibid.
79. Graves, *The Greek Myths*, p. 473.
80. Ibid.
81. See the tale of Heracles's Third Labour in Graves, *The Greek Myths*, pp. 472–73.
82. Ibid, p. 473.
83. Ibid.
84. Ibid, p. 474.
85. Ibid.
86. Ibid.
87. Graves, *The Greek Myths*, p. 474. Graves says Heracles was called 'melon' as in 'of apples' because these were offered to him, 'presumably in recognition of his wisdom; but such wisdom came only with death.'
88. Jobes and Jobes, *Outer Space*, pp. 499–500.
89. Allen, *Star Names*, p. 406.
90. Ibid.
91. Graves, *The Greek Myths*, pp. 418 and 760.
92. Rice, *Amber: The Golden Gem of the Ages*, pp. 4 and 15.
93. Ibid, p. 3.

94. Ibid, pp. 7 and 41.
95. Kraay, *Greek Coins*, p. 383.
96. Rice, *Amber: The Golden Gem of the Ages*, p. 41.
97. Ibid, p. 7.
98. Ibid, p. 17.
99. Ibid, p. 3.
100. Ibid, p. 12.
101. Ibid, p. 7.
102. *Websters New World Encyclopedia*, p. 58.
103. Rice, *Amber: The Golden Gem of the Ages*, p. 112.
104. Ibid, p. 4.
105. Elphick, *The Atlas of Bird Migration*, p. 32.
106. Graves, *The Greek Myths*, p. 474.
107. Rice, *Amber: The Golden Gem of the Ages*, p. 41.
108. Allen, *Star Names*, p. 395.
109. Wasson, *Soma: Divine Mushroom of Immortality*.
110. McKenna, *Food of the Gods*, p. 110. See also ch. 7, pp. 97–120.
111. Ransome, *The Sacred Bee*, p. 137.
112. Walker, *The Woman's Dictionary of Symbols and Sacred Objects*, p. 102.
113. Tripp, *The Handbook of Classical Mythology*, p. 440.
114. Ibid, p. 441.
115. Ibid.
116. See *Ancient Meteorology* by Liba Taub and her discussion of how meteorological phenomena were perceived and discussed in ancient Greece through the works of ancient writers, poets and historians. See especially her references to Zeus and lightning at pp. 5–7 and thunderbolts at pp. 184–85.
117. Tripp, *The Handbook of Classical Mythology*, p. 440.
118. Ibid.
119. Walker, *The Woman's Dictionary of Symbols and Sacred Objects*, p. 102.
120. Ibid, p. 137.
121. Ibid, p. 227.
122. Nasr, *An Introduction to Islamic Cosmological Doctrines*, pp. 6–7. See his footnote #11.
123. See Robert Graves' interpretation of this Greek tragedy in *The Greek Myths*, pp. 418–26.
124. Allen, *Star Names*, p. 406.
125. Ibid.
126. Graves, *The Greek Myths*, p. 770.
127. Ibid, p. 216.
128. Ibid, p. 218.
129. Ibid, p. 219.
130. Ibid, p. 151.

131. Krupp, *Beyond the Blue Horizon*, p. 214.
132. Graves, *The Greek Myths*, p. 153.
133. Ibid.
134. Jobes and Jobes, *Outer Space*, p. 345.
135. Graves, *The Greek Myths*, p. 165.
136. Jobes and Jobes, *Outer Space*, p. 345. For a detailed discussion on Freemasonry, see Steven Sora, *Secret Societies of America's Elite*, ch. 17.
137. Blavatsky, *The Secret Doctrine*, vol. 2, p. 769.
138. Graves, *The Greek Myths*, p. 507.
139. Bullfinch, *Myth and Legend*, p. 143.
140. Ibid.
141. Blavatsky, *The Secret Doctrine*, vol. 2, p. 762.
142. Cotterell, *The Supergods*, p. 5.
143. Blavatsky, *The Secret Doctrine*, vol. 2, p. 493.
144. Ibid, p. 762.
145. Ibid, p. 493.
146. Eliade, *Images and Symbols*, p. 12.
147. Rucker, *Infinity and the Mind*, pp. 62–63.
148. Eliade, *Images and Symbols*, p. 12.
149. Van Zandt and Stemman, *Myteries of the Lost Lands*, pp. 198–99.
150. Graves, *The Greek Myths*, pp. 35–37. See also Hesiod's famous poem *Works and Days* for the Greek account, and West's commentary on 'The Myth of Ages', pp. 172–205 in Hesiod, *Works and Days*.
151. Hancock, *Heaven's Mirror*, p. 15.
152. Lippincott et al., *The Story of Time*, p. 258.
153. Blavatsky, *The Secret Doctrine*, vol. 2, p. 565. See verses 2 and 10 in 'St John's Revelation' and verses 15–18 in 'Leviticus'.
154. Husain, *The Goddess: Power, Sexuality and the Feminine Divine*, p. 53.
155. Eliade, *A History of Religious Ideas*, vol. 1, p. 63. See his footnote #17.
156. Sykes, *The Seven Daughters of Eve*, pp. 196–99.
157. Ibid, p. 201.
158. Spence, *History of Atlantis*, p. 30.
159. Cayce, *Edgar Cayce on Atlantis*, pp. 76–81.
160. Nitsch, *Other Council Fires Were Here Before Ours*, pp. 68–69. Also quoted in Morton and Thomas, *The Mystery of the Crystal Skulls*, p. 355.
161. Ibid.
162. Blavatsky, *The Secret Doctrine*, vol. 2, p. 761.
163. Ibid.
164. Quoted in Murray Hope, *Atlantis: Myth or Reality?*, p. 298 and in Morton and Thomas, *The Mystery of the Crystal Skulls*, p. 354.
165. Blavatsky, *The Secret Doctrine*, vol. 2, p. 761.
166. Ibid, p. 762.
167. Nitsch, *Other Council Fires Were Here Before Ours*, pp. 68–69. Also

quoted in Morton and Thomas, *The Mystery of the Crystal Skulls*, p. 355.

168. Blavatsky, *The Secret Doctrine*, vol. 2, p. 140.
169. Ibid, pp. 139 and 313.
170. Walker, *The Women's Encyclopedia of Myths and Secrets*, p. 57.
171. Jobes and Jobes lists the crew of seven in *Outer Space* at p. 262 as Manu or Vashishtha, a form of Brahma (Zeta Ursae Majoris) and his six sons — Kratu (Alpha Ursae Majoris), Pulaha (Beta Ursa Majoris), Pulastya (Gamma Ursae Majoris), Atri (Delta Ursae Majoris), Angiras (Epsilon Ursae Majoris) and Marici (Eta Ursae Majoris).
172. Walker, *The Women's Encyclopedia of Myths and Secrets*, p. 57.
173. Gimbutas, *The Language of the Goddess*, p. 249.
174. Blavatsky, *The Secret Doctrine*, vol. 2, p. 141.
175. Ibid, p. 774.
176. Eliade, *A History of Religious Ideas*, vol. 1, p. 63.
177. Ibid.
178. Graves, *The Greek Myths*, p. 260.
179. Ibid, p. 259.
180. Ibid.
181. Blavatsky, *The Secret Doctrine*, vol. 2, p. 771.
182. Ibid.
183. Morton and Thomas, *The Mystery of the Crystal Skulls*, p. 336.
184. Jung, *Man and His Symbols*, pp. 25–26.
185. Hand Clow, *The Pleiadian Agenda*, p. 91.
186. Ibid, p. 32.
187. Ibid.
188. Burnham et al., *Astronomy: The Definitive Guide*, pp. 180–81.
189. Roberts and Roberts, *Dreamtime Heritage*, p. 54.
190. See Bell's treatment of the Ngarrindjeri story of the Pleiades in *Ngarrindjeri Wurruwarrin*, pp. 545–94.
191. Ibid, p. 577.
192. Burnham et al., *Astronomy: The Definitive Guide*, p. 348.
193. The estimates of the distance of the Virgo super cluster of galaxies varies from anywhere between 50 to 65 million light-years. For instance, Patrick Moore estimates this being 50 million light-years away in *Stars of the Southern Skies*, p. 104; whereas David Levy claims it is 65 million light-years in *Skywatching*. The real problem is that this super cluster is made up of hundreds of galaxies, but they are all individually at different distances. For example, spiral galaxy M 99 is 50 million light-years away from us, whereas M 87 is 70 million light-years distance. See Burnham et al., *Astronomy: The Definitive Guide*, at pp. 325 and 402.
194. Hand Clow, *The Pleiadian Agenda*, p. 48.
195. Ibid, p. 27.
196. *Wikipedia Encyclopedia* (online) <http://en.wikipedia.org/wiki/Light>

197. Ibid.
198. Segal, *The Amazing Space Almanac*, p. 42.
199. *Wikipedia Encyclopedia* (online).
200. Burnham et al., *Astronomy: The Definitive Guide*, p. 366.
201. Lame Deer and Erdoes, *The Gift of Power: The Life and Teachings of a Lakota Medicine Man*, p. 179.
202. Ibid.
203. Hand Clow, *The Pleiadian Agenda*, p. 32.
204. Lawlor, *Sacred Geometry*, p. 22.

CHAPTER THREE

Maimai

Seven Sisters of the Dreamtime

During late in winter when (the Pleiades) first appear in the eastern sky, the icicles falling from their bodies cover the Earth beneath with a layer of frost. It is at these times that the older boys and girls go to the nearest spring, and, scraping the frost from the ground, rub it over their naked bodies.

The ritual makes the boys grow into strong and successful hunters and the girls into beautiful women with large breasts.

But they have to be most careful at these times for, should a beam of sunlight strike their bodies when covered with frost, they will always be weak and puny.

— from *The Dawn of Time* by Charles Mountford.[1]

In the world's oldest continent the creative epoch known as the Dreamtime stretches back into a remote era in history when the creator ancestors known as the First Peoples travelled across the great southern land of Bandaiyan (Australia), creating and naming as they went. Their historic deeds are recorded in stories, ritual, song and dance by the various tribes or nations in epic *songlines*, which detail these events. The diverse Indigenous cultures were orally based, and so these traditions were committed to human memory and passed down through the generations. Legends tell of numerous floods which covered the lands, of mountains of fire that shook, of strange lights and strange beings and giant animals that once shared the land with the black people. Against this backdrop of 'mythology', the Dreamtime records the lives of real men, women and children, their relationships with one another and their fellow creatures. Among the many creator ancestors were the famous Seven Sisters of the Pleiades who came down from the heavens long, long ago in the Dreamtime and who continue to visit and commune with their kantrimin. Their names are as many and diverse as the numerous Aboriginal nations, but their stories remain essentially the same. To the Adnyamathanha people of South Australia the Sisters are known as the Magara,[2] whereas the Bundjalung people of northern New South Wales refer to them as the Meamai or the Maimai.[3] The Walmadjeri of Western Australia call the Sisters Gungaguranggara,[4] while the Kulin of Victoria tell of the Karagurk, seven women who 'carried fire at the end of their digging sticks.'[5] Like the supreme creator being, the feathered Rainbow Serpent, the Seven Sisters play an important role in Aboriginal 'women's business' — especially in initiation ceremonies involving young women — but the lessons they impart are for all humanity.[6]

Water girls and ice maidens

In many Aboriginal legends the Seven Sisters are often associated with water in all its manifestations — rain, hail, frost, ice or snow. Consequently, they are often referred to as the 'Water Girls' or 'Ice Maidens'.[7] In this respect they share a similarity with ancient Greek legends that speak of the Pleiades as 'Oceanids.'[8] This aquatic theme, as previously noted, has direct links with the Great Mother as Bird Goddess through her life-giving moisture.[9] In *Ngarrindjeri Wurruwarrin*, Diane Bell documents how the Seven Sisters invest the waters surrounding Kumarangk (Hindmarsh Island) with life.[10] In a story from the Northern Territory the adaptability of the Sisters is pronounced, for in the wet tropics they are known as the Yunggamurra

Water Girls that live in lagoons and as the Munga Munga who live in the dry country.[11] As renowned beauties, both sets of sisters are regarded as being responsible for djarada or love magic,[12] and whose songs, like the Greek Sirens,[13] entrap men. In the Yunggamurra tale, their pursuer is not Aldebaran, Orion or some other strange man but their father Dunia, who is later turned into a crocodile for his incestuous ways.[14] Astronomically he translates to the other prominent orange star, Antares, in the constellation of Scorpio. Much emphasis is placed on Dunia's fall from the heavens as he calls out three times, 'I am falling, I am falling, I am falling.'[15] This Aboriginal Lucifer bears a striking similarity to the biblical tale of humanity's fall from grace and to Milton's epic poem *Paradise Lost*.[16] Madame Blavatsky argues this symbolic event represents the creation of the third root race and nothing more than the manifestation of spirit entities into the physical realm of the flesh.[17]

Here we have a vital link between the stars of the Pleiades in Taurus and those of the opposite constellation of Scorpio, which ancient Mayan and Aboriginal people viewed as a crocodile. Its significance lies in the unhinging of the Crocodile Tree in Mayan cosmology as we enter the New Age of the Sixth Sun. Like the Mayans, Aboriginal people placed a tree in the heavens that the Pleiades climbed to escape their father and other pursuers. This tree represents the cosmic axis, whose unhinging causes the various cataclysms that inflict destruction on the world to create a new world age. In this story, Manbuk (Orion) falls in love with Milajun, one of the Seven Sisters,[18] and after lighting a fire to 'smoke' (or smudge) her, she sheds her slimy, grey skin like the sealskin or 'soul skin' of the Celtic Selkie.[19] Curiously, many Aboriginal legends refer to the bodies of star beings as being covered with crystals or some other shiny material, hence their reference as the Shining Ones. The Magara in the Adnyamathanha story are described as ravishing beauties with long hair reaching down to their waists and with bodies covered in glistening icicles.[20] Is this their *soul skin*? Might they have been *spacesuits* instead? Whatever your views may be on the subject, Aboriginal people do not perceive extraterrestrial beings as 'aliens' or as hallucinatory imaginings of the crazed, but as very real entities whose comings and goings are as natural as the rising and setting Sun, Moon and stars. And so it is that the Water Girls come and go about their normal business and shed their celestial influence with those who welcome and value their presence.

The Seven Emu Sisters

In their zoological form, the Magara who bring the early morning frost are the Seven Emu Sisters who are chased by the Dingo men of Orion, known as the Wanjin.[21] The Greeks placed the giant's hunting dogs in the constellations of Canis Major and Canis Minor, the Greater and Lesser Dog in the sky beside him.[22] One story tells how the Dingo men chased the Magara, who hid under a rocky outcrop of tumbled boulders, but 'with their keen sense of smell,' the Wanjin soon discovered 'where the women were hiding.'[23] They lit a fire to lure the women out of their rocky abode, hoping that their wings would get burnt and enable an easy capture. However, the Dingo men only partly succeeded in their attempt, for while the fire did scorch their wings — which prevented them from flying — 'the strenuous efforts of the Magara to step over the burning grass and bushes made their legs grow so long that even to this day the Emu can outdistance almost every creature.'[24] The bird women 'ran to the ends of the earth' to escape the fire but the dog men continued to chase them.[25] In a final act of desperation they ran up into the heavens where they became the stars of the Seven Sisters.

Not to be outdone by the Sisters, the men followed them into the skies where they changed into the constellation of Orion. But the Sisters 'are always the first to reach the western horizon, where, for a time, they are safe from the unwelcome attentions of the Dingo men.'[26] In another version of the story, the Magara and Wanjin are married to each other, and when the women disappear in the western horizon about an hour before the men, it is because they are making camp and lighting fires to cook food for their husbands. The portrayal of the Magara as emus and the Wanjin as dingos more than likely indicates the 'skin' group to which they belong. What this means is that they are in the socially approved marriageable set to one another, and therefore it may only be a classificatory relationship rather than an actual partnership.[27] In any event, the significance of dingo and bird dreamings as they relate to men's and women's business is explored in the following sections; suffice to say that the stars marked an important time for the young people of the tribe, a *rite de passage* into manhood and womanhood.

Dingo Dreaming or Dog medicine

Compared to other Indigenous animals in Australia, the dingo is a relative newcomer, but is nonetheless a favourite companion of many Aboriginal people. Misunderstood by the invading Europeans, the animal is still considered a pariah by some sections of the community who fail to acknowledge its special dreaming qualities or medicine. Dingo or Dog Dreaming is of particular importance in the interior regions of Bandaiyan, where Dingo Dreaming tracks or songlines traverse large portions of the continent from north to south, east to west. Anthropologist Deborah Bird Rose, who lived in a remote Aboriginal community in the Northern Territory, writes of its significance among the Yarralin in *Dingo Makes Us Human*. According to their traditional beliefs, Dingo created human beings and gave us the gift of life, including our distinctive human features and our 'upright stance.'[28] Unlike other creatures, Dingo was the only animal in the Dreamtime who 'walked then as he does now.'[29] What is more, the dingo and human beings were one in the Dreamtime, say the Yarralin, and that is why they are considered very close in their nature. As Bird Rose puts it, 'They (dingos) are what we would be if we were not what we are.'[30]

As the dingo has only been in Australia some four or five thousand years, it might appear as though there is some time discrepancy, given the antiquity of the Dreamtime and the occupation of Aboriginal people, which is conservatively estimated to be at least forty thousand years and possibly more than a hundred thousand years.[31] In *Voices of the First Day*, however, Robert Lawlor suggests that the dog in the Dreamtime stories was probably the now extinct Tasmanian tiger.[32]

Whatever the true situation, some interesting parallels can be drawn between the spiritual beliefs of the ancient Egyptians, Native Americans and Aboriginal Australians in terms of dog dreaming. Like Dingo, Anubis the jackal-headed god of ancient Egypt and Coyote of Native American traditions, are said to have created the first human beings. Coyote, in particular, is said to have done so 'by scratching open the hide of Mother Earth to release the first people from her womb.'[33] In addition to their procreative powers, Dingo, Coyote and Jackal officiate over other related tasks and responsibilities, including funerary rites and other 'sorry business.'[34] Anubis is credited in particular with having invented mummification of the dead following the death of his father Osiris — whose bandaging he presided over — hence his title 'Lord of the Mummy Wrappings.'[35] There were in fact two jackal gods in ancient Egypt — Anubis and Wewawet (or Upuaut) — whose name meant the 'opener of the ways.'[36] Of the two, Anubis is often described as a black jackal. Once again, this is analogous to Aboriginal Australian beliefs where the black dog is a central Dreaming figure, particularly in some areas of Queensland. Quite clearly they were 'different aspects of the same divine archetype.'[37] Like Anubis, Coyote's actions of scratching open Mother Earth's hide to birth humanity may deserve a similar epithet.

In Native American traditions, where there are more than one species of dog — native dog, wolf and coyote — there is more delineation of specific tasks or individual medicines and characteristics that are peculiar to that specific animal. Despite this diversity, these canines all share common dog traits such as loyalty and companion-ship, as well as their own individual medicine or dreaming. The creators of the Native American *Medicine Cards*, Jamie Sams and David Carson, describe these individual medicines and their healing properties in the accompanying book. They point out that through-out history and in many cultures, the dog has been 'the servant of humanity.'[38] So people with Dog Medicine or Dog Dreaming usually serve others and their community in some way — they may be a charity worker, philanthropist, nurse, counsellor, minister or solidier.[39] As a servant, Dog's other roles are to guard and protect those whom they serve, as well as to teach what it means to be a true and loyal friend. Coyote's special brand of medicine is in the realm of trickery, where he is portrayed in Native American traditions as the great trickster. Although gifted with 'many magical powers,' they don't always work in his favour, say Sams and Carson, for his trickery fools even himself.[40] 'No one is more astonished than Coyote at the outcome of his own tricks':[41]

> As Coyote moves from one disaster to the next, he refines the art of self-sabotage to sheer perfection. No one can blindly do themselves or others in with more grace and ease than this holy trickster. Coyote takes himself so seriously at times that he cannot see the obvious; for example, the steamroller that is about to run over him. That is why, when it hits him, he cannot believe it. 'Was that really a steamroller? I better go look,' he says. And he is run over once more.[42]

Although Coyote keeps falling into his own trap, somehow he 'manages to survive' through it all.[43] Battered and bruised he continues to err, not learning from his mistakes, but although he appears to have 'lost the battle,' Coyote is never truly beaten.[44] There is something hauntingly familiar about Coyote's remarkable resilience and innocence, and his trickery reminds us that it is human to make mistakes. Whether we learn from them or not, it is imperative that we get up and start again, no matter what the odds. This message has particular meaning for Indigenous peoples — especially in the face of colonisation — that we can survive and that ultimately, like Coyote, we can never be beaten. It is an empowering message to the oppressed, disenfranchised and downtrodden alike. Above all, Coyote reminds us not to forget to laugh at our foibles, our imperfections, and our faults. Within 'the folly of his acts we see our own foolishness,' our own innocence and our own vulnerabilities.[45] Coyote medicine therefore teaches us humility.

In many ways his antics mirror that of the American cartoon figure, Wylie E. Coyote in the Warner Brothers *Road Runner* cartoons. The cartoon is analogous to the Dreamtime story of the Dingo Men of

Orion who continually chase the Emu Women of the Pleiades, for it is about a bird (Roadrunner) that is constantly chased by a dog (Wylie E. Coyote), thus playing out an old and timeless tale with two very ancient, cosmic arche-types. Of all three canines, Wolf is considered the teacher, for she always returns to the tribe 'to *teach* and *share* medicine', say Sams and Carson.[46] As with Dingo, Native Americans also emphasise the human characteristics of Wolf. Thus they say, 'If you were to keep company with wolves, you would find an enormous sense of family within the pack, as well as a strong individualistic urge.'[47] These qualities 'make wolf very much like the human race.'[48] Wolf teaches us that we can be an individual with our own sense of self and yet remain part of society with shared dreams and aspirations. It is precisely this feeling of sharing, caring and community that Dog Dreaming imparts, and why it is that Aboriginal people at Yarralin claim that dingo makes us human.

Moon Dog howling, Dingo brings death to the world

If Dingo teaches us how to be human, then that includes all the aspects of the life cycle, including living and dying. Like Anubis and Coyote, Dingo not only brought the gift of life to humanity but the gift of death. There are two aspects to Dingo Law, say the Yarralin. The first rule is that 'human beings finish (die) and stay finished forever.'[49] The second rule is that our spirit is eternal and while one half returns to our country, the other half returns to the stars.[50] Within this framework, says Bird Rose, humans for the most part are considered inseparable and intact during life, but upon death the parts separate, and it is these parts that 'endure forever.'[51] In *Wise Women of the Dreamtime*, Johanna Lambert draws parallels between the spiritual beliefs of Aboriginal Australians and those of ancient Egyptians, who believed that when they died their spirits journeyed to the stars along 'energy paths' that pass through the Pleiades and the Dog constellations of Canis Major and Canis Minor.[52] Among these constellations, the bright star Sirius, nicknamed the 'Dog Star', is designated the 'gateway' to the Land of the Dead, says Lambert.[53] This notion of dogs guarding the entrance to the underworld or as psychopomps who lead the dead to heaven and rebirth, is shared by many other traditions, including Hinduism.[54] The *Rig Veda*, for instance, refers to the two dogs of Yama, the first mortal and King of the Dead, as 'keepers of the path.'[55] This may explain the symbolism of two dogs standing between two towers and howling at the Moon in the major arcana card of the tarot. In many cultural traditions the Moon is associated with dog in all its manifestations. Its phases (in particular the full, half and new Moon) provide the perfect imagery of the cycle of life, death and rebirth, and therefore the Triple Goddess. Unlike Native American traditions, however, the Moon is principally male, not female, in Australian Aboriginal and ancient Egyptian cosmology; thus Moon Man is intimately connected to the dingo in Bandaiyan, as Anubis is with Thoth.

Because the Moon does not generate its own light but reflects instead the light of the Sun, it therefore signifies hidden unconscious, psychic energies that hold 'secret knowledge and wisdom.'[56] Hence in the foreground of the Moon tarot card lies the pool of subconscious memories from which a scorpion emerges. Like its zodiac namesake, the creature epitomises secrets, or that which is hidden. 'Baying at the moon' may indicate Dog's 'desire to connect with new ideas which are just below the surface of the consciousness,' say Sams and Carson.[57] The two towers represent a gateway through which the golden path leads from the pool of memories to the spiritual mountain (heaven). Quite apart from their relationship with the Moon, dogs — like cats — are renowned for their ability 'to see ghosts and other spirits invisible to humans.'[58] Their keen sense of smell especially contributed to their association with death, says Barbara Walker. As guardians of the underworld they were able to 'detect the odour of sanctity and decide whether the soul could be admitted to the company of the gods.'[59] Dingo, in all her many manifestations of dog, wolf and coyote, ensures the keeping of our death rituals by ensuring the proper procedures and ceremonies need to be installed and followed. Thus Anubis, the black-headed jackal god of the Egyptian underworld, presided over the dead and funerary rites. Moon Dog's howls, in particular, teach us to how to mourn for our dead in a sorrowful and respectful way. It is no coincidence that our mourning songs and cries imitate their howling.

Men's and Women's Dreaming

Among the Yarralin, as in other regions of Bandaiyan, Dingo Dreaming is largely considered to be men's business. However, as Bird Rose points out, 'Yarralin men, and other men throughout the Victoria River District, recognise that much of their secret ritual and law ultimately derive from women's dreamings, just as all life originates in mother earth, and as they themselves are born of women.'[60] Although women 'gave or lost some of their power to men,' nonetheless, 'women today still hold their own Law and ritual which have been handed to them from Dreaming Women.'[61] These Dreaming Women would no doubt have included the Seven Sisters of the Pleiades.

In the West Kimberley region of Australia, where I come from, there is a similar belief that long ago in the Dreamtime it was women who created and maintained Aboriginal Law, all the songs and all the dances. They were the Wongai women who came down from the heavens, and whose metamorphosis into stones can still be seen today on the Mitchell Plateau.[62] Even though their power was stolen from them, women's law is still recognised as far more powerful than men's. In this part of Bandaiyan, Aboriginal men are taught to work hard to attain a divine equality, unlike women who are sacred by virtue of their sex alone. It is men who have to continually 'tend the spiritual aspects and maintain the connection with the cosmic regions where the law of creation is manifest.'[63] Salvation is not automatically guaranteed to a man, he must earn it. For as David Mowaljarlai and Jutta Malnic explain in *Yorro Yorro*, 'Without these constant services, a man cannot be assured of his passage through death into a new life cycle.'[64] Thus, while Dingo Dreaming may be predominantly a male Dreaming in some regions, there is no doubt that it was once the domain of women, if not still today.

Barbara Walker argues the same in *The Woman's Dictionary of Symbols and Sacred Objects*. 'Women,' she says, 'were the first to domesticate the dog because the dogs were companions of the Goddess in many cultures, long before gods or men appeared with canine companions.'[65] Acknowledging the associative qualities of Dingo Dreaming to women's experiences, Bird Rose notes the Yarralin comparison between women and dingo as creative beings. Like the dingo 'who made people human in the first place, who gave us our characteristic human shape, human brain, and human culture,' so too have women.[66] This is precisely 'what women do', they give birth to human beings. She says, 'It is women sitting in the camps, giving birth and cooking food, allocating food and people who make human life specifically and uniquely human.'[67] Noting the former holiness of dog through its connection with the Mother Goddess, Walker says the animal became demonised 'with the decline of goddess worship' that led to the development of disparaging terms in our lexicon like 'bitch' and 'son of a bitch' to refer to women and their children.[68] She points out that even though Muslims 'still consider dog a mortal insult, there are indications that this prejudice is based on the dog's former sacredness.'[69]

Likewise in Australian slang, a woman is often referred to in derogatory terms as 'a dog'. In any event, Dog Dreaming remains very significant in Native American teachings, where women such as Grandmother Kitty Rasmussen and Grandmother Twylah Nitsch played a pivotal role. Unfortunately, Grandmother Kitty is no longer with us and is now in Spirit world. It was she who initiated me into the Wolf Clan Lodge of the Lakota Sioux, and who named me 'Earth Law Woman' in the early nineties.[70] Thus Dingo Dreaming is a very *big* Dreaming for men and women alike. One of my many Aboriginal teachers, Violet Newman, a Wongai woman from the western desert area near Kalgoorlie in Western Australia, explained in very simple linguistic terms why this is so. 'Munya,' she asked, 'what does *dog* spell backwards?'

The Maimai of the Bundjalung

The Bundjalung people's tribal lands in northern New South Wales extends from Grafton in the south to as far north as Tweed Heads near the Queensland border, east to Byron Bay and as far west as Tabulam.

Situated in an area known as the Northern Rivers, it covers an extensive range of environmental habitats from lush, tropical rainforests and mountainous terrain in the northern regions to low lying plains in the south and a scenic coastline stretching along its eastern shores. Against the backdrop of this panoramic vista the story of the Seven Sisters of the Pleiades is played out in the night skies above and was re-enacted in corroboree around the campfires of long ago. Like the seven daughters of Atlas and Pleione in the Greek legends, the Maimai enjoyed a close, loving relationship with their parents. And as typical parents of teenage children, they were content to stay near their home 'on some distant mountain' and not wander about as their daughters.[71]

Although the legend does not indicate which mountain was their family home, the most likely candidate would have to be the prominent peak Wollumbin (Mount Warning), visible in some places as far as a hundred kilometres away. As befitting a holy icon, the summit of this extinct volcano is the first place to receive sunlight each day as dawn rises over the east coast of Australia. Many people flocked to its peak on the first day of January in the year 2000 to embrace the First Light of the New Millennium. Wherever the true location of their earthly home within this region, the enduring presence of the Sisters in the night skies is felt at many levels, culturally, spiritually and otherwise.

Today their memory is evoked by the early morning frost brought by the Ice Maidens in their earthly visitations.[72] True to their aquatic nature, before they left the Earth the Sisters 'went into the mountains and made springs of water to feed the rivers, so that there would be water for men and women for all time.'[73] The sound of thunder in the distance is caused by the splashing sounds of the young women bathing in the celestial waters above. People know from these sounds that rain will soon fall and the waters of the Maimai will splash down on them.[74] In times past, parents would scoop up the ice particles and hold them against the septum of children's noses to numb them before piercing. As strange as this custom may appear, it was supposedly done to enable the children to sing as beautifully as the Pleiades.[75] The remarkable thing about this ritual practice — and the depiction of the

Girls as emus in some Aboriginal legends — is that pierced septums appear to imitate the prominent holes in birds' beaks, particularly emus. Equally renowned for their cleverness as their beauty, the Maimai possessed secret knowledge and magic, including the use of fire that was previously unknown to humans.[76] Several suitors can lay claim to pursuit of the Sisters in this legend including the Berai Berai (young men from Orion), the young warrior Karambil (Aldebaran) or the old fiery ancestor, Wurrunah (Venus).

In this version of the tale,[77] Wurrunah wanders into a strange country where he accidentally comes across the Seven Sisters and their campfire. Being generous of nature, the Sisters welcomed the old man to their camp, gave him food to eat and invited him to stay and camp there for the night. He stayed overnight and in the morning made as if to leave, but instead hid in bushes nearby where he could spy on the young women, for he was lonely and desired a wife. He watched them dig for honey ants with their digging sticks and while they feasted on this delicacy, he crept up and stole two of their sticks. After the Sisters had finished eating, they decided to head back to camp but noticed that two of their sticks were missing. Thinking they might find them nearby, two of the young women told their other siblings to go on ahead, reassuring them they would catch up quickly. No sooner had they departed, then the two remaining Sisters set about searching for their sticks, but couldn't find them. When their backs were turned, the old man came out of the bushes and planted the two digging sticks together in the ground before hiding himself once again. When they turned around, the girls were delighted to see their sticks standing upright and tried to pull them out of the ground, but as they were struggling to do so, the crafty old man grabbed both of them around the waist and held them firmly together to prevent their escape. The two Sisters struggled and screamed for help, but they soon realised there was no one to help them. The old man explained why he had captured them and threatened violence if they would not be quiet. Wurrunah told the girls he would take good care of them provided they remained quiet and did not attempt to escape. As resistance was futile, they agreed to travel quietly with the old man but made it clear to him that their Sisters would return to find them. Wurrunah was aware of this possibility and travelled quickly and further away from the camp to escape capture.

Several weeks passed and when their relatives had not found them, the Sisters grew increasingly despondent. One day when they were making camp, the old man ordered them to collect some bark from two nearby pine trees, but knowing what would happen if they cut the pine bark, the Sisters objected to this task, telling Wurrunah that they would be forever lost to him if they did so. But the old man ignored their warning and insisted they do as he ordered. So the Sisters took an axe each and proceeded to hack at the two trees, but with each strike, the trees began to rise into the air, bearing the Sisters on their branches. Higher and higher they grew, lifting the two Sisters into the heavens until they could be seen no more. In a desperate bid to reach them, Wurrunah climbed on one of the trees to chase after them but they rose still higher, until at last they reached the top of the sky. The remaining Sisters grabbed the two girls and pulled them safely into their enclave and away from the clutches of the old man. And there they remain in the skies to this day, although the two Sisters who were left behind do not shine as brightly as the other five because their brightness had been dulled by Wurrunah's embraces when he tried to melt the icicles off them.[78]

Besides the old man, a group of young Aboriginal men known as the Berai Berai were hopelessly infatuated with the Sisters and would follow them everywhere.[79] They watched where the Maimai would set up camp, and would leave an assortment of gifts for them and serenade the young women with songs of love. The Berai Berai were especially skilled in finding bees' nests, and honey was a particular delicacy among their offerings, which they placed in containers made of bark. Even though the Sisters would eat the honey and enjoy the other gifts, the efforts and solemn declarations of the young men were futile, for their love went unrequited. When they heard that the Sisters had gone to live in the stars after the kidnapping ordeal, the Berai Berai were heartbroken. They refused to eat, pined away and died. The Old People (spirits) felt sorry for the young men and placed them in the constellation of Orion, not far away from the Maimai so their songs could still be heard.

Another version tells how the young hunter, Karambil fell in love with one of the Sisters, but because they were of the wrong skin group, tribal law forbade them from having a sexual relationship and so he abducted her.[80] The Maimai were upset by what had happened and concerned for their sister's safety, and immediately set about developing a plan to ensure her return. To pressure Karambil they decided to embark 'on a long journey to the west, where they found winter'[81] and sent it back to the young hunter 'as frost, icicles and cold weather.' When he agreed to release the captive sister, the Maimai travelled towards the east 'to commandeer summer to disperse the ice.'[82] The astronomical symbolism in this story is obvious, says Edwin Krupp; the actions of the Sisters mimic what the stars of the Pleiades do in the Australian night skies. Here in Bandaiyan, like other countries in the Southern Hemisphere, the Pleiades disappear in the early evening around April with the approach of winter and 'return to the east in the early evening sky in October' when summer is underway.[83] He can be seen pursuing the Sisters still as the giant orange star Aldebaran in the constellation of Taurus.

The Seven Sisters empowering all women

Astronomical symbolisms aside, a wide range of associations can be made between the Maimai and their mythological counterparts in other traditions. Many of the Aboriginal legends refer to the Sisters as possessing fire, the knowledge of which was much desired by humans and other creatures. In one legend from Victoria, for instance, Wahn the Crow steals fire from the Sisters, who kept it in their digging sticks.[84] In many cultures, fire signifies wisdom, which the Sisters possessed, and this makes them female counterparts of Prometheus in ancient Greek legends. Also, women's digging sticks are esoterically coded in many Aboriginal cultures. An obvious parallel can be drawn with witches and their broomsticks that enable them to journey between worlds. The growing pine trees are reminiscent of the fairytale of Jack and the Beanstalk, and the trees stand for not one but two polar axes: the ecliptic axis and the polar axis. The star-crossed lovers' theme of Karambil and one of the seven maidens is a cultural reminder of the marriage laws that are strictly adhered to and whose violations are costly.

In *Wise Women of the Dreamtime*, Johanna Lambert draws some notable parallels between the Bundjalung legend and that of ancient Greek and Hindu mythologies of the Pleiades. Apart from the obvious connection of the Berai Berai with the ancient Greek character Orion and the constellation named after him — and the association of the Maimai with honey — the strength of her analysis lies in the comparisons between the Maimai and the Indian Krittika in their aspect as Ice Maidens and judges of men.[85] Lambert equates the Maimai's 'cold aloofness' and their 'psychological castration' and rejection of the young men of Orion with the actions of the Krittika, who would sometimes critically wound men that were being judged.[86] Perhaps their perceived demeanour is really more about being indifferent or discerning, as hot-headedness and the warmth of passionate emotions can sway one's better judgment. Within this context their portrayal as Ice Maidens is symbolic of the sword energy of the tarot, whose mental clarity enhances discernment in decision making; hence the correlation of icicles with sharp objects such as knives, razors and swords.

The other aspect that she touches upon is the depiction of the Maimai as strong, powerful young women. Likening their personal characteristics and traits to the hunting goddess Artemis, Lambert describes them as the personification of the 'independent feminine spirit, free from men and male domination.'[87] The Maimai, she says, 'were also aloof and cold, disdaining lovers, maintaining at all times a strong sisterhood.'[88] Lambert sees within them many reflections of the Greek goddess and the practice of these qualities in Aboriginal women's business. She points out that in many Aboriginal cultures 'it was women who were responsible for mitigating the violence of tribal conflicts to ensure that justice took place.'[89] Thus, she concludes, 'In the same way that Artemis embodied certain qualities of the female psyche, the Maimai shone as an example for all women.'[90]

That Lambert equates the Aboriginal Sisters with the Greek goddess Artemis is not surprising, given that in the Greek legends, Artemis is benefactor to the Pleiades who are in her train.[91] As patron therefore, and as Goddess of the Hunt, Artemis was also perceived as the special protector of young women, says Lambert. Likewise, in many Aboriginal traditions the Pleiades are often regarded as strong warrior women like Artemis. As positive and empowering role models they bring strength and wisdom to those who follow their Dreamings, and throughout Aboriginal Australia the Pleiades play a major role in the initiation, spiritual awareness and protection of young women.[92] Among the Ngarrindjeri of South Australia, the Seven Sisters are responsible for issuing the 'sacred orders' of women's business — what has been described as holy 'directives about life, preparation for womanhood, marriage and child-bearing.'[93] In *Legendary Tales of the Australian Aborigines*, Indigenous author David Unaipon provides some insight into rituals surrounding the Seven Sisters, although — as Diane Bell points out in *Ngarrindjeri Wurruwarrin* — his account is somewhat limited in that 'he knows the structure but has not directly observed the women's rites.'[94]

Nonetheless this demonstrates that Aboriginal men are aware of the Dreaming of the Seven Sisters, just as women have some knowledge of male ceremonies. The essential difference between this gendered division of esoteric knowledge lies, as Bell suggests, in 'being present at certain times and coming into contact with certain places, fluids, and persons that is taboo.'[95] In his story of the Mungingee (Pleiades), Unaipon describes how the Sisters undergo a series of trials and ordeals to prove that they could conquer fear. Rivalling the twelve labours of the Greek hero Heracles, the tests included control of their appetites, tooth avulsion, ritual scarring of the body, perforation of the nasal septum and sleeping on a bed of ants.[96] Apart from Unaipon's narrative, much of the content of these rituals remains very much a part of secret women's business. The clash of a text-oriented society and that of an oral *moving stories* culture — together with the reluctance of Aboriginal women to speak publicly about the women's business side of the Seven Sisters' Dreaming — is painfully illustrated by the Hindmarsh Island debacle, forever tainting Australia's legal history.[97] Yet despite the controversy surrounding the struggle of some Aboriginal women to protect and dignify their spiritual beliefs, the Seven Sisters of the Pleiades continue to empower all women through their sheer presence in the night skies, in the stories that remain throughout many cultures and in their celestial memory that lies in the hearts and minds of all humanity.

The birdsong of women's initiation

An elemental component of young Aboriginal women's initiation is what Barbara Hand Clow refers to as birdsong in *The Pleiadian Agenda*. From an Indigenous perspective birds — like all the Earth's creatures — can teach us spiritual things as well as the stars. Their role in the initiation of young Aboriginal women in particular reminds them of the Great Mother in her aspect as the Bird Goddess, and also of our celestial origins. One of my dearest Aboriginal teachers now in Spirit world, Aunty Lorraine Mafi-Williams of the Gittabul clan of the Bundjalung people, taught me many things in the two short years that I had known and lived with her. She was my teacher, my guiding light and my friend, and I loved her dearly. I was honoured and privileged to have shared some of my journey with her. In my eulogy at her funeral I told the assembly how she shared with me a Bundjalung story of how human beings were once taught to speak by the bird people. She would often say that if you listen with open ears, you can hear what the birds are saying and the special messages they have for you. The *willy-wagtail* was of special totemic significance to Aunty. It was her own personal messenger who would bring news of her loved ones. Often she would interrupt our conversations to point out that the bird was calling out to her. 'Listen Munya,' she would say. 'You can hear them calling to me, Mrs Williams. Mrs Williams. Mrs Williams.' I would listen and after a while I began to hear them clearly and I also learned how to *talk* with the birds. A great Aboriginal woman shaman, Aunty Lorraine continues to teach and instruct me from the Spirit world. I swear there is a bird flying around me now that keeps calling out,

'Munya, Munya, Munya' that almost drives me to distraction.

Many Indigenous peoples have an instinctive and common understanding of the kinship that exists between humans and animals because of our cultural traditions. Lakota medicine man Archie Fire Lame Deer tells a fantastic tale of an extraordinary encounter between a Native American elder and a magpie in *The Gift of Power*. He tells how he 'came across a magpie with a broken wing' when he was a young boy and took it home to his grandfather to help it heal. He soon taught the bird to talk and say '*Hau, witko?*' or, 'How are you doing, crazy man?'[98] Then one day an elder friend paid them a visit but was not aware of the new pet. As soon as the bird started talking, the elder chastised the young boy for saying 'bad things' and not respecting his elders.[99] When the magpie flew down from the rafters onto the boy's outstretched finger, he realised that the bird had called him crazy, not young Archie, and the old man howled with laughter. 'The strange thing about this,' says Lame Deer, 'was that Adam was stone deaf,' yet 'he had understood the magpie's high-pitched, piercing sound perfectly.'[100]

In *The Knot of Time*, a book on astrology and the female experience, Lindsay River thanks the woman who taught her the 'language of the birds' in her acknowledgements. River and co-author, Sally Gillespie, tell a beautifully creative fictional tale of how the Mother Goddess dreamed her twelve star children (the zodiac signs) into being and the healing songs of birds:

> They sang of those who would express their anger rather than give up, and of those who would hear it rather than run away. They sang of women who discovered their own strength at the bottom of a pit of oppression and despair. They sang of people in pain who discovered kinship and respect for each other, of prisoners whose songs could not be kept behind bars, of soldiers who chose to desert or die rather than torture or kill. They sang of survivors.[101]

Lambert touches on the special role that birds play in the initiation of young Aboriginal women. Like the vision quests of Native Americans, young Aboriginal boys and girls are taken out into the bush to be instructed in religious matters and to test their mental and physical endurance. One of the objectives of isolation is to develop their sensibilities and powers of concentration to make them aware of the 'living and symbolic interrelatedness of the natural world.'[102] Young girls are taught to 'listen to the first note that any bird sings throughout the day,' and respond accordingly.[103] These are lessons in communication with nature and spirit ancestors. The teachings enhance her listening skills 'so that she is made aware of every sound made by members of her tribe in their distant camp.'[104] This kind of listening develops survival skills so that she can hear and recognise the sounds of danger that alert her to any potential threats to herself, family and kin.

The association of the Pleiades (in their zoological form as birds) with creation through sound and harmonics is common to many other world traditions and philosophies. That birds play a key role in our spiritual enlightenment is a widespread theme in the great religions and eastern philosophies such as Buddhism, Confucianism and Taoism. Joseph Campbell recounts the proverbial story of the Zen Buddhist master who was about to deliver a sermon and 'just as he was about to open his mouth, a bird sang,' whereupon the master stated, 'The sermon has been delivered.'[105] Barbara Hand Clow maintains that our connection to the Pleiades is basically made 'through sound and vibration'; of which birds play a key role.[106] Equating light with sound (a concept not unfamiliar to science), she says that the spiralling light emanating out of the Pleiades 'circulate this non-physical intelligence into sound.'[107] Classifying these sounds into a variety of numeric dimensions, she identifies 'seventh dimensional sound' (7D) as 'birdsong'.[108] Birds 'resonate with 7D sound coding on Earth,' says Hand Clow, hence their connection with the Pleiades.[109] She believes it is possible to 'tune' into the seventh dimension by studying the intonations of birds.[110] Could this be one of the reasons Aboriginal elders taught our young people to listen to birds? Is this the real meaning behind the involvement of the stars of the Pleiades in young Aboriginal 'women's business'? Although seventh-dimensional sound is higher 'than the sound coding of human languages,' Hand Clow maintains it gave rise to human speech.[111]

As fantastic as this claim may sound, there can be no denying it mirrors some Aboriginal beliefs that birds once taught humans how to speak in the Dreamtime. And as we know from the Greek legends of the Seven Sisters of the Pleiades, birds have taught us much of the science of navigation and continue to do so in modern times. It makes you wonder, just how much the Ancients once knew about our world in the past, and how much information may have been lost along the way? Where this information remains, whether intact or fragmented, such as in the Aboriginal Dreamtime or other Indigenous teachings, this ancient knowledge must be preserved for the benefit of all humanity to guide and instruct us as we journey through life.

Drumming up the Pleiades

Hand Clow's assertion that we are able to communicate with the Pleiades through the vibrations of sound is supported by evidence from a diverse range of cultures around the world where music, in particular women's drumming, is linked to the Pleiades. In Aboriginal Australia, for instance, women traditionally played the drums as opposed to men. Pleiadian legends of the Wotjobaluk people of Victoria confirm this, for they tell how the Larangurgg (Pleiades) beat upon drums made from rolled-up possum skins to signal 'corroboree time for the Kulkunbullas, the stars in Orion's Belt.'[112] Neither should we forget that it was Hermes, the son of Maia, the eldest and most beautiful of the Seven Sisters, who reputedly made the first lyre out of a tortoise shell.[113] More startling is the lesser-known fact that the Orphics called the Pleiades the 'Lyre of the Muses' and that it existed 'side by side with Lyra' (the Harp constellation).[114] Our modern term 'music' is derived from the creative influences and inspirations of numerous Greek goddesses known collectively as the Muses, which obviously included the Pleiades.[115] The ancient Mayans also noted this connection between these stars and music, for they depicted the cluster as a rattle — and a deadly one too — the appendage of the rattlesnake.[116]

Besides the use of psychotropic drugs, drums are used in many cultures to induce shamanic experiences, whose rhythms create altered states of consciousness. In some cultures an emphasis is placed on the actual material the drums are made from, or the symbols inscribed on them. It is these actions of consecration that are supposed to give the instruments their supernatural powers. As Clarissa Pinkola Estes points out in *Women Who Run With the Wolves*, 'the skin or body of a drum determines who and what will be called into being.'[117] As she explains:

> Drums made of human bone call the dead. Drums made of the hide of certain animals are good for calling the animal spirits. Drums that are particularly beautiful call Beauty. Drums with bells attached call child-spirits and weather. Drums that are low or high in voice call the spirits who can hear that tone and so on.[118]

In ancient Mesopotamia drums were clearly associated with the Pleiades, for the most revered were those made from the 'hide of a black bull with seven circles inscribed on the drum face.'[119] Within this context, the cryptic comments made by the authors of *Hamlet's Mill* regarding the comparison between the Vedic horse sacrifice and the black bull of Mesopotamia become more comprehensible. Referring to the inscription of the Pleiades on the forehead of Ashvamedha and the bull hide, Georgio De Santillana and Hertha Von Dechend say this observation alone 'should be enough to indicate the level of phenomena brought into play.'[120] In other words, the fact that so many diverse cultures placed their supreme creator in this region of the sky (Anu, Zeus and Shiva, to name a few) clearly indicates who is being summoned and invoked. But of all the gods, say the authors, the Babylonian Anu was 'a far more exact entity' for he gave that civilisation the sexagesimal system from which we have inherited the minute comprised of sixty seconds, and the hour comprised of sixty minutes.[121] The system derives from the cuneiform mark for Anu that was 'written with one wedge, which stands for the numbers 1 and for 60.'[122] Pythagoreans interpreted this to mean that Anu represented 'the One and the Decad,' or 1 and 10 respectively.[123]

From a numerological perspective on a purely symbolic level, the digital expression of ten or the Decad (an ancient term for God) as 1 and 0, which forms the basis of the binary system (their exact placement determining numeric value; for example 15,500 or 5,000) may be seen to represent the Godhead. Binary numbers lie at the heart of computer programming language, without which our computerised age could not survive. Another interpretation would be to view these digits as male (1) and female (0) and therefore creation. As Joseph Campbell says, all things speak of God, including the computer, to which we can add drums. In *The Power of Myth* he describes a revelatory experience he has with computers, describing their elemental components, silicon chips, as 'miracles.'[124]

The Babylonian godhead ultimately stood for the 'fundamental time measure of celestial events,' say Von Dechend and De Santillana.[125] Therefore, to strike Anu's drum 'was to involve the essential Time and Place in heaven.'[126] This exegesis of the Babylonian cuneiform script ties in with Madame Blavatsky's claims that it is the Pleiades that control these cycles of time and, ultimately, our destiny. Numerically, Anu's representation as 1 and 60 also alludes to the lost or missing sister, in that 6 plus 1 equals the mystical seven. These numeric themes of six and one or 666 or 777 are all symbolic of the great cycles of time, as will be discussed in later chapters.

Fishing in the Milky Way, images of the Goddess in Australian Aboriginal art

As we have seen from the Aboriginal legend of the Pleiades where the father is turned into a crocodile, Jennifer Isaacs' claim that 'the theme of pursuit does not exist in Arnhem Land,' is not entirely correct.[127] Having said that, it is still worthwhile to consider some of the fundamental concepts and universal themes of this Pleiadian legend from Groote Eylandt in the Northern Territory of Australia as expressed in the traditional bark painting 'Orion and the Pleiades' by Minimini Mamarika, also from that area. Here, the Pleiades are called Wutarinja and their husbands, the Burumburumrunja fishermen. In this painting the artist has clearly depicted thirteen stars in the Pleiades — not seven

— either with extraordinary eyesight, or else the night skies are exceptionally clear in the northern regions of Bandaiyan. Curiously, the artist has drawn the Pleiades in a womb-like enclosure, but Orion sits within a T-shaped frame that looks like a tool or weapon, or some sort of implement such as a hammer, or it may represent a hammer shark, given the men are fishing from their canoe.

This raises the question — why has the artist painted these stars in this manner? The simple answer is that he has engaged in a gendered demarcation of the cultural landscape into men's and women's country according to Aboriginal traditions through the use of particular cultural icons to mark this division. Clearly the circular, womb-like shape of the Pleiades suggests an obvious correlation with femaleness, whereas the T-shaped figure of Orion resembles Thor's hammer, and therefore conjures up a male warrior image. While this gendered imagery may be explained in simple terms of the shape of male and female bodies and their genitals, their translation to star patterns in the sky is altogether a different matter. With the exception of Scorpius (Scorpio) there are very few constellations in the sky that resemble their actual namesake. So how is it that so many diverse human cultures and civilisations essentially perceive the Pleiades as female and Orion as male? Even more intriguing, why are the Pleiades connected with culture, political stability and the arts, while Orion is allied with conflict, combat, warfare and other destructive energies? Is there something inherent or intrinsic about the energies emanating from these two stellar systems that results in their assigned gender?

A possible clue may lie in the association of the Pleiades with the knowledge of fire that they bring to humanity. Apart from its practical application, fire represents divine wisdom as well as the fires of sexual passion that lead to our procreation. If we look to traditional Aboriginal art icons, campfires are often drawn either as kidney-shaped or womb-like shapes such as this one. As in ancient Greece, campfires are primarily of women's domain in Aboriginal cultures, for they are largely responsible for tending the sacred hearth of Aboriginal family life. Applying the 'language of the Goddess' and the theories of Marija Gimbutas to these Aboriginal icons, we can start to comprehend the more deeper, underlying symbolism of these images to arrive at what Mircea Eliade describes as the 'hidden modalities of being.'[128]

To begin with, the enclosed circular structure housing the Wutarinja women of the Pleiades contains other identifiable Goddess symbols. Firstly, its very shape combines two separate but related images — that of the uterus and vulva, which Gimbutas says represent regeneration. This explains why many Neolithic 'graves and temples assumed the shape of the egg, vagina, and uterus of the Goddess or of her complete body.'[129] Secondly, this circular corral looks like a curled-up serpent as well as an egg. Both are very ancient signs for the creation of our planet and the creation of the universe itself. It combines the Supreme Creator and deity of Aboriginal beliefs, the Rainbow Serpent, with that of the cosmic egg — out of which hatched 'the sun and moon' and all creation, says Barbara Walker.[130] 'It used to be a common idea,' she says, 'that the primeval universe, or the Great Mother who created it, took the form of an egg.'[131] The snake, too, embodies the notion of regeneration because it sheds its skin on a regular basis. 'The symbolism of the bird is interwoven with that of the snake,' says Gimbutas, because the necks of birds appear to be shaped like snakes.[132] Combined with 'their periodic renewal each spring after they have spent the winter months in the south,' this suggests a close relationship.[133] Within this context, the serpent may be considered another aspect of the Bird Goddess, and therefore ultimately relates back to the Bird Goddesses of the Pleiades.

This association of snakes with birds is not merely symbolic but reaches far back into the evolutionary genetic history of our planet's creation and the biodiversification of its many species. Dreamtime legends that tell how birds were once snakes long ago now reveal a scientific truth. Scientists have in recent years accepted this evolutionary theory, whereas the stories date back almost a hundred thousand years or more. That the depiction of the Wutarinja women of the Pleiades should encompass several Goddess symbols in the one image merely reflects her omnipotent presence. 'It seems more appropriate to view all of these Goddess images as aspects of the one Great

Goddess with her core functions — life-giving, death-wielding, regeneration and renewal,' says Gimbutas.[134]

Thirdly, the images themselves and the stories of Orion and the Pleiades in Aboriginal mythology emphasise the structured and bound nature of men's and women's business. Both domains are culturally regulated, clear boundaries established and made visible in the physical environment as in mythology and its enactment in ritual and ceremonies. To this extent, both realms are to be respected in their own regions of the universe. Even where a man, or group of men, pursues or stalks the Sisters, they still manage to elude or fight him off, thereby reasserting their sovereignty. It is not so much a tale of abduction or attempted rape as it is about women rediscovering their own sense of personal power and inner strength, the sacredness of their sex and having a place of refuge in which to replenish, re-nurture and re-sustain themselves.

It is interesting to note that not only Aboriginal people depicted stars in this fashion, for the Dogon people of Mali in West Africa did the same.[135] These people are renowned for their amazing astronomical knowledge, which they say was given to them by extraterrestrial beings collectively known as the 'Nommo'.[136] Astronomers, in particular, have been astounded by the wealth and accuracy of scientific information the Dogon possess about the brightest star in our skies, Sirius. Not only were they aware of the existence of a small white dwarf star orbiting around Sirius, but they identified a *third* member of the stellar system whose existence could not be confirmed by modern astronomy until 1995. This star, simply known as Sirius C, is referred to as Emme ya or the 'Sun of Women' by the Dogon, for they believe women only inhabit its planet.[137] In a drawing of the Sirius star system, the Dogon have depicted Sirius B and the planet of Sirius C as kidney-shaped 'campfires' or womb-like structures. What is more, the Dogon refer to the stellar system as 'Nommo's placenta.'[138]

Following in the footsteps of the Dogon and in a similar vein, Mamarika's traditional painting of Orion and the Pleiades reveals a remarkably accurate and detailed astronomical map of the stellar region comprising Orion's Belt and Sword. Here we see the three stars, which comprise the giant's 'belt', representing the Burumburumrunja fishermen located in the top frame of the T-shape or Tau Cross. The official names for the belt stars in Western astronomy are derived from the Arabic language: Al Nitak (Zeta Orionis), Al Nilam the middle star (Epsilon Orionis) and Mintaka (Delta Orionis).[139] Remarkably, Mintaka's name simply means, 'the belt.'[140] The other stars of Orion as depicted in this bark painting are located inside the long stem of the T-shape figure. Although these particular stars signify the fish that the men caught — which they placed inside their canoe Julpan — when compared to modern star charts of the constellation, they clearly represent the hunter's sword. Other cultures also describe the constellation of Orion as a canoe, including the Maori people of Aotearoa (New Zealand), for whom it represents the Tainui waka intimately connected with the Pleiades in Maori starlore.

The essential thing to bear in mind when viewing Orion's pattern of stars in the night skies is that the constellation appears 'upside down' in the Southern Hemisphere in comparison to star maps drawn from a Northern Hemisphere perspective.[141] What this means effectively is that Orion's sword will appear to Northern Hemisphere viewers as though it were hanging down from the hunter's belt, whereas it is the other way around in the opposite hemisphere. In other words, the sword will appear to point in an upward direction as a general rule. Taking into account the changing sky during an entire year which affects the changing positions of stars, Orion will seem to gradually fall on his side during the autumn months before eventually falling well below the horizon with the onset of winter. The ancient Greeks explained Orion's annual demise in the night skies in their mythology, which tells how the hunter was bitten by a scorpion that led to his death (as opposed to the deadly arrow shot by Artemis). So as Orion begins to sink from view as winter approaches, his nemesis the constellation Scorpius starts to rise in the opposite region.

Let's look specifically then at the depiction of Orion's Sword in Mamarika's traditional bark painting. Among the so-called sword stars, the artist has clearly differentiated between stars (with their rays extending outward) and two other unknown celestial objects (drawn as circles without rays). What might these objects be? Amateur and professional astronomers would know that there are three major nebulae (interstellar dust and clouds) within the constellation of Orion. They include the Great Nebula or Orion Nebula (M 42), a smaller patch of nebulosity nearby (M 43) and the Horsehead Nebula (IC 434).[142] Both M 42 and M 43 lie within the area identified as Orion's Sword, but the Horsehead Nebula is situated in Orion's Belt beside the star Al Nitak or Zeta Orionis. The significance of this particular nebula will be revealed in just a moment. Clearly, the two other celestial objects identified in Mamarika's painting are therefore the Orion Nebula or M 42 and its smaller companion nebula M 43. Now here comes the really exciting part: whereas the Orion Nebula is easily visible to the naked eye and appears as a blurred object, M 43 can only be seen through binoculars or a telescope.

What this traditional painting reveals is that Aboriginal people have long since known of the existence of the smaller nebula without the aid of any modern optical instruments. How is this possible? Some scientists would argue the possibility that the nebula was brighter in the past and the memory of it was passed down through history as it gradually diminished in luminosity. Others would say, as they did with the Dogon, that the artist came across a modern photograph or Western astronomer with knowledge of these stars. I sometimes wonder if scientists realise how racist this suggestion is, for even if this were so, it still does not explain the astounding revelation that Aboriginal people in the Kimberley region have known of the existence of the Horsehead Nebula — only visible through a high-powered telescope of at least 200 millimetre, or eight inch diameter — before European colonisation. To them this was not a 'horse's head' but an image of the Supreme Creator, the feathered Rainbow Serpent whose abode lies within the region of Orion's Belt.[143]

On looking at photographs of the nebula, it is hard to imagine how anyone — let alone scientists — could perceive this formation as a horse's head. A closer inspection suggests it looks more like a dragon or serpent. Of course, most astronomers would argue that we are only talking about celestial cloud formations and therefore what we see is a matter of individual interpretation, but to Aboriginal people this is the face of God. The ancient Egyptians, who are recognised for their great civilisation and scientific knowledge, also revered this constellation above all others for it represented the heavenly home of their god Osiris or Sahu.[144] What is more, they too, like Aboriginal Australians, regarded the stellar region of Orion's Belt to be of enormous spiritual significance.

That Aboriginal people should know of the existence of the Horsehead Nebula, let alone its shape and appearance, is truly remarkable and reveals only a minute detail of the complex body of knowledge or science that Indigenous peoples possess about the stars and the universe in which we live. Some people would argue this knowledge comes from walking between the worlds and recognising that 'all of the world is alive with spirit,' and is therefore ultimately shamanic.[145] However that may be, what is abundantly clear is that this information is constantly renewed and renegotiated with the universe in our waking and dreaming lives. Within this context, the source of this wisdom is available to each and every one of us, so long as we approach it with reverence for nature and respect for one another. So when we gaze into the night skies at the beautiful, gentle celestial glow of the stars of the Pleiades, we might bear in mind the eloquent words of Johanna Lambert who reminds us that we can all 'experience the timeless sense in which the great Dreaming has unfolded and will continue to unfold.'[146] Furthermore, 'With this vision, our fearful and hungry imaginations fill with a mythic dimension, and the seeds of change are born anew.'[147] The Seven Sisters of the Pleiades have planted the seed. It is up to us to engage in dialogue with them on every level and in the deepest recesses of our being, for the Dreaming of the Seven Sisters belongs to everyone. It symbolises our common humanity. Perhaps we are not so different after all.

Notes

1. Mountford, *The Dawn of Time*, p. 38.
2. Ibid.
3. Katherine Langloh Parker refers to them as *Meamai* in *Australian Legendary Tales* and *More Australian Legendary Tales.* Johanna Lambert does the same in *Wise Women of the Dreamtime.*
4. Berndt and Berndt, *The Speaking Land*, pp. 281–82.
5. Massola, *Bunjil's Cave*, p. 108.
6. 'Business' is an Aboriginal English word for the spiritual aspects of community life, specifically rituals and ceremonies. Because much of the practical aspect of Aboriginal spirituality is gendered, there is a clear demarcation between 'men's business' and 'women's business'. See Bell's discussion at pp. 528–42 in *Ngarrindjeri Wurruwarrin.*
7. See, for instance, the story of the Magara in *The Dawn of Time* by Charles Mountford, p. 38.
8. Oceanids were water nymphs that lived in the ocean as opposed to rivers and streams and other watery abodes. All nymphs were young women 'considered independent of men because of her unmarried state' says Walker in *The Woman's Dictionary of Symbols and Sacred Objects*, p. 267.
9. Gimbutas, *The Language of the Goddess*, p. 3.
10. See Diane Bell's discussion of the Ngarrindjeri legend of the Pleiades in *Ngarrindjeri Wurruwarrin*, pp. 545–94.
11. Harney, *Tales from the Aborigines*, p. 39.
12. Ibid, p. 40. Although Kunapipi the Old Crone Goddess principally rules love magic, this legend suggests the Pleiades are also involved.
13. Graves, *The Greek Myths*, p. 611. Like the Pleiades, the Sirens were described as 'bird-women' who lived on an island in the Mediterranean Sea. They were said to lure sailors to their death with their enchanting songs and Jason and the Argonauts were especially warned against listening to the Sirens or docking on their island. They were rumoured to have lost a singing contest to the Muses — see p. 607.
14. Harney, *Tales from the Aborigines*, p. 44.
15. Ibid, p. 45.
16. See Krupp's discussion on Paradise Lost in *Beyond the Blue Horizon* at pp. 197–98.
17. Blavatsky, *The Secret Doctrine*, vol. 2, p. 62.
18. Harney, *Tales from the Aborigines*, p. 42.
19. Pinkola Estes, *Women Who Run With the Wolves*, pp. 257–64. See also the novel *Song of the Selkie* by Cathie Dunsford.
20. Mountford, *The Dawn of Time*, p. 38.
21. Mountford, *The Dreamtime*, p. 68.
22. Levy, *Skywatching*, p. 146.
23. Mountford, *The Dreamtime*, p. 68.

24. Ibid.
25. Ibid.
26. Ibid.
27. 'Skin' is another of those Aboriginal English words that have a specific meaning. In this context it is used to refer to the kin system of moieties and other subdivisions in Aboriginal society that determine all kin relationships on a social basis. See, for example, the chapter on skin in *Dingo Makes Us Human* by Deborah Bird Rose at pp. 74–89.
28. Bird Rose, *Dingo Makes Us Human*, p. 47.
29. Ibid.
30. Ibid.
31. Lawlor, *Voices of the First Day*, p. 102.
32. Ibid.
33. Walker, *The Woman's Dictionary of Symbols and Sacred Objects*, p. 369.
34. 'Sorry business' is an Aboriginal English term for rituals such as funerals and other things associated with death and dying.
35. Storm, *Egyptian Mythology*, p. 17.
36. Bauval and Gilbert, *The Orion Mystery*, p. 58.
37. Ibid.
38. Sams and Carson, *Medicine Cards*, p. 93.
39. Ibid.
40. Ibid, p. 89.
41. Ibid.
42. Ibid.
43. Ibid.
44. Ibid.
45. Ibid.
46. Ibid, p. 97.
47. Ibid.
48. Ibid.
49. Bird Rose, *Dingo Makes Us Human*, p. 69.
50. Ibid.
51. Ibid.
52. Lambert, *Wise Women of the Dreamtime*, p. 48.
53. Ibid.
54. Doniger O'Flaherty, *The Rig Veda*, p. 43.
55. Ibid, p. 44.
56. Sams and Carson, *Medicine Cards*, p. 97.
57. Ibid.
58. Walker, *The Woman's Dictionary of Symbols and Sacred Objects*, p. 371.
59. Ibid.
60. Bird Rose, *Dingo Makes Us Human*, p. 51.
61. Ibid.

62. Mowaljarlai and Malnic, *Yorro Yorro*, pp. 8 and 12–13.
63. Ibid, p. 144.
64. Ibid.
65. Walker, *The Woman's Dictionary of Symbols and Sacred Objects*, p. 370.
66. Bird Rose, *Dingo Makes Us Human*, p. 177.
67. Ibid.
68. Walker, *The Woman's Dictionary of Symbols and Sacred Objects*, p. 371.
69. Ibid.
70. Grandmother Kitty (Catherine Rasmussen) was a Lakota Sioux elder. Her native name was *Dark Sun She Walks*, which is a reference to the solar eclipse.
71. Langloh Parker, *Australian Legendary Tales*, p. 164.
72. Ibid, p. 165.
73. Reed, *Aboriginal Stories*, p. 83.
74. Langloh Parker, *Australian Legendary Tales*, p. 165.
75. Ibid.
76. Reed, *Aboriginal Stories*, p. 81.
77. Langloh Parker, *Australian Legendary Tales*, pp. 43–47.
78. Lambert, *Wise Women of the Dreamtime*, p. 47.
79. See 'Where the Frost Comes From' in *Australian Legendary Tales* by Katherine Langloh Parker at pp. 164–65.
80. Reed, *Aboriginal Stories*, pp. 82–83.
81. Krupp, *Beyond the Blue Horizon*, p. 252.
82. Ibid.
83. Ibid.
84. Cahir, *Mythology*, pp. 27–28.
85. Lambert, *Wise Women of the Dreamtime*, p. 47.
86. Ibid.
87. Ibid, p. 48.
88. Ibid.
89. Ibid.
90. Ibid.
91. Graves, *The Greek Myths*, p. 83.
92. Lambert, *Wise Women of the Dreamtime*, p. 47.
93. Bell, *Ngarrindjeri Wurruwarrin*, p. 574.
94. Ibid, p. 580.
95. Ibid, p. 584.
96. Unaipon, *Legendary Tales of the Australian Aborigines*, pp. 145–49.
97. Diane Bell discusses the legal history and political struggle of Ngarrindjeri women to protect their spiritual beliefs and sacred sites in the Hindmarsh affair throughout her book *Ngarrindjeri Wurruwarrin*. For a more straightforward account of its history, unencumbered by social science-speak, see *The Meeting of the Waters* by Margaret Simons.

98. Lame Deer and Erdoes, *Gift of Power*, p. 43.
99. Ibid, p. 44.
100. Ibid.
101. River and Gillespie, *The Knot of Time*, p. 272.
102. Lambert, *Wise Women of the Dreamtime*, p. 78.
103. Ibid.
104. Ibid.
105. Campbell, *The Power of Myth*, p. 22.
106. Hand Clow, *The Pleiadian Agenda*, pp. 98–99.
107. Ibid, pp. 97–98.
108. Ibid, p. 98.
109. Ibid, p. 97.
110. Ibid, p. 98.
111. Ibid.
112. Massola, *Bunjil's Cave*, p. 108.
113. Graves, *The Greek Myths*, p. 65.
114. De Santillana and Von Dechend, *Hamlet's Mill*, p. 369.
115. Walker, *The Woman's Dictionary of Symbols and Sacred Objects*, pp. 264–65.
116. Aveni, *Skywatchers*, p. 34 and Milbrath, *Star Gods of the Maya*, p. 38.
117. Pinkola Estes, *Women Who Run With the Wolves*, p. 158.
118. Ibid, pp. 158–59.
119. Lambert, *Wise Women of the Dreamtime*, p. 48.
120. De Santillana and Von Dechend, *Hamlet's Mill*, p. 125.
121. Ibid.
122. Ibid.
123. Ibid.
124. Campbell, *The Power of Myth*, p. 20.
125. De Santillana and Von Dechend, *Hamlet's Mill*, p. 125.
126. Ibid.
127. Isaacs (ed.), *Australian Dreaming*, p. 153.
128. Eliade, *Images and Symbols*, p. 12.
129. Gimbutas, *The Language of the Goddess*, p. xxiii.
130. Walker, *The Woman's Dictionary of Symbols and Sacred Objects*, p. 5.
131. Ibid.
132. Gimbutas, *The Language of the Goddess*, p. 317.
133. Ibid.
134. Ibid, p. 316.
135. See the Dogon illustrations of the Sirius star system in *The Sirius Mystery* by Robert Temple, p. 94.
136. Ibid, p. 111.
137. Ibid, p. 103.
138. Ibid, p. 111.
139. Allen, *Star Names*, p. 314.

140. Ibid.
141. To understand the changing night skies from a seasonal and hemisphere perspective, see *Stars of the Southern Skies* by Patrick Moore. Although written from a Southern Hemisphere point of view, nonetheless the astronomical principles remain the same for the Northern Hemisphere.
142. Levy, *Skywatching*, p. 195. For a more in-depth discussion of the stars and nebulae surrounding Orion and the region of the Hunter's Belt, see chapter 4 of *The Constellations* by Lloyd Motz and Carol Nathanson at pp. 124–31.
143. Mowaljarlai and Malnic, *Yorro Yorro*, p. 132.
144. See *The Orion Mystery* by Robert Bauval and Adrian Gilbert.
145. Lambert, *Wise Women of the Dreamtime*, p. 49.
146. Ibid.
147. Ibid.

CHAPTER FOUR

Mateo Tipi

Seven Star Girls of Devils Tower

The Indians called Devils Tower, 'Bear Lodge', because so many bears lived there. They believed it was put there by Great Spirit for a special reason, because it was different from the other rocks, rising high up in the air, instead of being on the ground. For this reason it was looked upon as a 'Holy Place', and the Indians went there to worship and fast. Dream houses were built there and the ruins of these should be there yet, as they were built out of stone.

— Chief Max-Big-Man[1]

Rising high above the plains surrounding the Belle Fourche River in the northern state of Wyoming in the United States of America lies the spectacular, natural volcanic formation of Mateo Tipi, or Devils Tower. Its extraterrestrial links with the Pleiades in so many Native American legends fired the imagination of film director Steven Spielberg, who made the classic sci-fi movie *Close Encounters of the Third Kind* in 1977. Devils Tower was not only used as an aesthetic backdrop to the movie, but its imposing, majestic nature provided powerful and evocative imagery of things not quite of this world. The unworldly look and feel of Devils Tower was central to the film's overall mystical theme of spaceships and alien landings that lent support to the cinematic suggestion that we are not alone. Devils Tower is known by many different names by the various Indian tribal nations of those who lived near, visited or heard about it from the stories of other Native peoples. Bear's House, Bear's Tipi and Bear's Tower are the traditional names given by the Arapaho, Cheyenne, Crow and Sioux nations.[2] The Crow name for the Tower is Dabicha Asow, which means Bear's Tipi or Bear's House. Mateo Tipi or Mateo Tipilla is the Sioux name that means Bear's Lodge. The Kiowa name for the Tower is Tso-aa, which means 'tree rock' because 'it grew tall like a tree,'[3] as in the legend of the Seven Star Girls.

The tower is located in a national park that was established in 1906 and given the official name of Devils Tower National Monument. Colonel Richard Dodge is credited with giving the formation its present name. In his book *The Black Hills* published in 1876, Dodge referred to the geological formation as 'Devils Tower', an adaptation of the local Indian people's name of 'Bad God's Tower'.[4] This name was subsequently adopted by surveyors and appears in one of the earliest maps of the region. The name choice clearly reveals the colonel's Christian bias by turning Bad God's Tower into *Devils* Tower. Anyone who has ever visited there would know that there is nothing diabolical about the place at all, in fact, quite the opposite. Or as one Native American elder remarked, 'If there had been such a thing as a Devil living there when the Indians were there, all the Indians would be dead.'[5]

The centre of the universe

To see the tower rise some 1,200 feet high above the prairies reaching up into the wide, blue Wyoming skies from a distance is a mystical experience, not unlike gazing on that other prominent big red rock Uluru (Ayer's Rock) in the central Australian desert. Its sheer majesty and spiritual presence is breathtaking. The feeling of grandeur, mystery, magic and awe that these gigantic monoliths engender is not surprising, given both rocks are considered sacred sites by the Indigenous peoples of the respective lands. In fact, Devils Tower is not only a major place of pilgrimage but is *so* sacred that many

Native Americans refer to it as the 'Center of the Universe'.

But what exactly does this mean? Several writers including Mircea Eliade, Joseph Campbell and Edwin Krupp have contemplated this concept in their various writings and each has identified its chief characteristics. In *Images and Symbols*, Eliade, the renowned historian of religion, says that all cultures possess some notion of a centre — 'a place that is sacred above all.'[6] This is not to be confused with a geographic centre, he adds, for there are many so-called 'centres' of the world. The centre in this context is determined by 'sacred, mythic geography.'[7] Places are sanctified either because holy objects are buried or kept there, or through epiphanies such as the manifestation of divinity and or revelations. The central place, says Eliade, is where 'the sacred manifests itself in its totality.'[8] This 'same archaic image' of the centre is repeatedly represented as 'the Cosmic Mountain, the World Tree or the Central Pillar that sustains the planes of the cosmos.'[9] Devils Tower embodies all of these characteristics as a cosmic mountain rising above the plains, as a tree rock to the Kiowa and as a pillar above which the cosmos revolves.

There is another facet to the central place, says Joseph Campbell. In *The Power of Myth*, he recounts a vision that the great Sioux Indian chief and seer, Black Elk, experienced as a young boy growing up in the Black Hills of South Dakota. In his vision, Black Elk saw himself standing 'on the sacred mountain of the world.'[10] At first he thought it was a familiar peak in his homelands, but soon came to the realisation that 'the central mountain is everywhere.'[11] Says Campbell:

> That is a real mythological realization. It distinguishes between the local cult image, Harney Peak, and its connotation as the center of the world. The center of the world is the *axis mundi*, the central point, the pole around which all revolves. The central point of the world is the point where stillness and movement are together. Movement is time, but stillness is eternity. Realizing how this moment of your life is actually a moment of eternity, and experiencing the eternal aspect of what you're doing in the temporal experience — this is the mythological experience.[12]

In *Skywatchers, Shamans and Kings*, American astronomer Edwin Krupp states that the centre in all cosmologies marked the place 'where creation began, where order and life first emerged.'[13] This explains why many world centres are often referred to as 'the navel of the Earth, the point at which creation began,' says Mircea Eliade.[14] Stones or other objects such as golden rods in Cuzco, the ancient Inca capital, or meteorites in the ancient Egyptian city of Heliopolis (now an outer suburb of modern Cairo) marked many of these so-called 'centres' of the world. The ancient Greek name for these marker stones is especially revealing, for *omphalos* means 'navel'.[15] Given this link with creation, it would be interesting to discover how many of these centres have an association either directly or indirectly with the stars of the Pleiades. The omphalos at Delphi may be one of these.

We know, for example from Greek mythology that the Seven Pleiadian Sisters were turned into doves, and that a single dove was responsible for locating the foundation spot of the Delphic oracle known to the Ancients as their 'navel of the world'. This is not to suggest that this dove was one of the Pleiadians, but that there may be some indirect link, perhaps through their relationship with navigation. Whatever the situation, the Pleiades are without a doubt intimately connected with this particular Centre of the Universe, as attested by various Native American legends of Devils Tower, some of which will be examined later in the chapter. Mateo Tipi, therefore, is not just the place of bears — it is the earthly abode of the Seven Sisters and the keeping place of sacred objects.

Finally, the central place is 'where heaven communicated with earth, and where the world was renewed through celestial power,' says Krupp.[16] This communication is multifaceted and takes many forms, from personal prayer and meditation through to elaborate ritual and ceremonies. Like other sacred sites around the world, traditional Native ceremonies are still conducted there including vision quests, sweat lodges and the Sun Dance. A rawhide shirt made by Native American artist Jo Esther Parshall on display at the Edinburgh Museum in Scotland appears to suggest that the Pleiades may have played a central role in this sacred dance.[17] The shirt depicts the celebrated ceremony alongside the stars of the Pleiades, together with our Sun. The stars are drawn exactly as they appear in the night skies and in relative spatial distance from one another, although curiously our Sun is placed *within* the cluster beside Maia.

This is intriguing in light of claims made by Barbara Hand Clow in *The Pleiadian Agenda* that our Sun is part of the Pleiades star cluster and a twin to Maia, one of its more brighter stars (that is, out of the entire stellar system, not of the Sisters themselves).[18] Astronomers have rejected this assertion but the persistence of Indigenous perceptions emphasising an earthly connection with the Pleiades cannot be overlooked or easily dismissed without a more thorough examination. It may be that this connection is spiritual or symbolic rather than physical, or it may be both. Sacred sites do more than recharge our batteries; they take us to a place of inner knowing and understanding that all of creation and the universe are one. Within these places, says Hand Clow in *Catastrophobia*, 'another world opens that is unfolded right in the middle of mundane existence.'[19] Little wonder that Devils Tower remains a place of holy pilgrimage to Native Americans and other kindred souls in search of the sacred.

Pilgrims, prophets, visionaries and seers

Many great Native American leaders are said to have visited the tower to seek counsel from the Great Spirit, including Crazy Horse and Sitting Bull.[20] 'Both warriors were looked upon as holy men by their people, as a result of having experienced powerful visions that influenced their future actions,' says Tom Lowenstein in *Mother Earth, Father Sky*.[21] Sitting Bull's wife gave birth to a son at Mateo Tipi but unfortunately he died soon after his birth.[22] It is not known whether the child is buried there but it remains a possibility, given that there were other known burials at the site, thereby enhancing its sacredness. Chief White Bull, a nephew of Sitting Bull and a member of the Minkowoju (Minneconjui) band of the Sioux, recalls wintering there once when he was fourteen in 1864, and again four years later as an eighteen-year-old.[23] Bear's Tipi, he said, was well known to the Sioux as a place of high significance and only the medicine men and women of the tribe would go there to pray. They would sleep on beds of sagebrush and fast for days at a time. It comes as no surprise that the longest anyone could stay at Devils Tower for prayer and meditation is 'four days and four nights'[24] — as the chief claims — for I know through personal experience that Mateo Tipi is a very sacred and powerful place. So powerful in fact, that it simply overwhelms. On my last visit to the tower in the fall of 2000, I recall sitting on a large, flat rock slab among the hundreds of rocks that are strewn around the base, reflecting on the significance of the place. I immediately began to feel its awesome presence and found myself being drawn into the rock into a meditative state, an inexplicable sense of communion with it and with all the other spirits, human and otherwise, that surround the tower.

As I meditated on the rock slab, I began to think of the profound similarities between many of the Native American and Australian Aboriginal stories of the Seven Sisters. Here before me was this huge, towering rock, rising up into the sky. Then it suddenly struck me. Wasn't this just like the legendary hill of Aboriginal mythologies where the Sisters are said to have landed and to have used as their 'launching platform' to return to the Pleiades? The realisation and recognition of this familiar theme had such an impact on me that I recall commenting to friends that Devils Tower had to be the *biggest* Dreaming place of the Seven Sisters that I had ever seen! Unfortunately I could not stay as long as I wanted as I was on a bus tour that was only making a two-hour stop. Had I stayed longer, I may very well have found myself on top of the tower, like the Sioux warrior who had gone there with his buffalo skull to 'fast and worship the Great Spirit in solitude.'[25] The following account describes the warrior's experience:

> Standing at the base of Mateo Tipi, after he had worshipped for two days, he suddenly found himself on top of this high rock. He was very much frightened, as he did not know how he would get down. After appealing to the Great Spirit, he went to sleep. When he awoke he was very glad to find that he was again at the base of this high rock.

He saw that he was standing at the door of a big bear's lodge. There were footprints of a very big bear, there. He could tell that the cracks in the big rock were made by the big bear's claws. So he knew that all the time he had been on top of the big rock, he had been standing on a big bear's lodge.[26]

When Short Bull told this story to Dick Stone in 1932 he mentioned that the buffalo skull the warrior had taken along with him to meditate was still on top of the tower.[27] Whether it is still there or not is unknown, as there appears to be no reference to it in the official literature. Perhaps the only way of knowing is to ask any of the climbers who have reached its summit whether they have sighted a buffalo skull, or the remains of one. If it did exist, it may have been destroyed or taken by some zealous amateur collector unaware of its scientific and spiritual significance as a sacred archaeological artefact. On the other hand, it may be that the skull never existed in physical form at all but only as a spiritual entity of symbolic import.

What this may signify is discussed further in the chapter, but it is worth mentioning here the ongoing controversy at Devils Tower on the issue of climbing. As we have already seen, Mateo Tipi is an important sacred site and many Native peoples regularly visit throughout the year to participate in religious services, especially in June when the Sun Dance ritual is observed during the Northern Hemisphere summer solstice. In recent years, however, the number of climbers has dramatically increased to such an extent that it has intruded on the observance of these ceremonies. The competing interests of both parties resulted in the matter going all the way to the US Supreme Court to resolve the conflict and the final outcome resulted in a minor victory for Native Americans in regard to their religious practices at the tower.[28] Unfortunately these cultural, spiritual and political clashes remain an ongoing struggle for many Indigenous peoples around the world trying desperately to protect their sacred sites and their way of life.

Similar issues have been raised with climbing Uluru, although they have not led to litigation. Local Aboriginal people have posted signs asking prospective climbers to voluntarily refrain from climbing out of respect for Aboriginal spiritual beliefs. As with the Devils Tower experience, some people choose to abide by the request but others continue to ignore it. The choice is quite simple, really. We can choose to have a powerful, uplifting and enlightening spiritual encounter by approaching these sacred sites with an open heart, open mind and with reverence, or else we can choose to forego the experience. As Hand Clow points out, 'the more people visit these places, the more the mysteries deepen for them, unless they travel like common tourists.'[29] The question is which one are you? Are you a 'common tourist' or are you a Seeker? There is a profound difference. Whatever our choice, the fact remains that we cannot take away from the sacredness or numinous nature of these special places, and that is an empowering realisation.

Of the many pilgrims to have visited Mateo Tipi, perhaps the most famous and influential of all were Sweet Medicine and White Buffalo Calf Woman, hence the significance of the buffalo skull on the tower. The great Cheyenne hero Sweet Medicine is renowned for bringing his people their sacred bundle containing the Four Arrows, which form the basis of their 'religion, religious rituals, and social laws.'[30] These laws laid down the foundation for 'the religious and political organization the Cheyenne would adopt, the proper marriage ritual, the correct way to trap eagles to obtain the emblem feathers the chiefs wore, and many other things,' says Paula Gunn Allen in *The Sacred Hoop*. Sweet Medicine claimed that the 'Sacred Ones' who live on Devils Tower handed these laws to him. A complex and troubled character, this Cheyenne lawman exhibited shamanistic skills from early childhood, but his orphaned upbringing may have led to intense feelings of isolation and alienation, which resulted in 'anti-social' behaviour.[31]

As a consequence Sweet Medicine grew further and further apart from his community, which led to his eventual withdrawal and retreat into the Spirit world for four years. While there he caused a great famine by persuading the animal spirits to withhold their availability to hunters. Upon his return from the other side, Sweet Medicine came across some hungry children. Their desperate plight caused him to feel pity and so he immediately set about finding food for them. This act of generosity extended to the rest of the tribe, with Sweet Medicine instructing them in four days how to remove the animal restrictions by performing certain rituals.[32] Hailed as a great prophet, he foretold the coming of the white man with dire consequences for the Cheyenne, including the 'extermination of the buffalo and the introduction of horses and cattle.'[33] Blessed with 'the gift of longevity from the spirits meant that he could grow old and regain his youth after several times,' says Tom Lowenstein in *Mother Earth, Father Sky*.[34] Outlasting four generations of his tribe, Sweet Medicine eventually succumbed. His body, however, reputedly disappeared and was never found, so a stone cairn was placed to mark the spot where he died in his tipi.[35]

Devils Tower is a place of special significance to the seven tribes of the Sioux nation for reasons similar to the Cheyenne, in that it marks the location of an extraordinary encounter between their ancestors and the Supreme Being Ptesan Win or White Buffalo Calf Woman, as she came to be known.[36] An envoy of Wakan Tanka (the Great Mystery), she appeared out of nowhere during midsummer, a beautiful woman draped in a white robe, bearing gifts in a holy bundle containing the Sacred Pipe.[37] Among the many gifts she brought were corn kernels, and in this regard she bears a striking similarity to the Egyptian goddess Isis and other Corn Maidens.[38] On greeting the Sioux as relatives, this strange woman announced she was of the Buffalo tribe and told them they were all related as one family and as one people.

Ever since that fateful event, all Sioux ceremonies now end with the blessing Mitakuye Oyasin, which means 'All My Relations,' says Archie Fire Lame Deer and Richard Erdoes in *Gift of Power*.[39] Like the Cheyenne hero, White Buffalo Calf Woman imparted spiritual laws, wisdom and prophecies to the Sioux and laid the foundation for their Seven Ceremonies.[40] These include the Keeping of the Soul ceremony, puberty rites for young girls called Preparing for Womanhood, the purification ritual of the Sweat Lodge, the Vision Quest, the Naming of Relations, Throwing the Ball ritual and the sacred Sun Dance.[41] Not surprisingly, the buffalo plays an important role in these ceremonies and remains a central theme.[42] In *Animals of the Soul,* Joseph Epes Brown discusses the symbolism of the buffalo in the spiritual beliefs of the Oglala Sioux — especially its relationship to White Buffalo Calf Woman and its application in certain rituals and ceremonies.

To understand its spiritual significance, says Brown, is to have a deep appreciation of its esoteric symbolism. Like the sacred cow of ancient Greece, Egypt and India that was associated with plenitude, the bison represented similar virtues and characteristics to the Sioux where it stood for Mother Earth and the nourishment that she provides.[43] And because it epitomised the female principle, the animal embodied values of virtue and 'ideal womanhood,' which made it the perfect role model in the initiation of young women.[44] Every aspect of the animal's features, says Brown, was endowed with special meaning and applied to Sioux ritual. For instance, buffalo skulls with horns intact were left outside sweat lodges facing east, while their robes were used to seal sweat lodges, thereby enhancing their dome shapes which symbolised the womb of the buffalo or Mother Earth.[45] These skulls played a major role in prayer and rituals, especially those surrounding abundance, and buffalo dances were central to many of these ceremonies.[46]

The most revered of religious objects, however, was without doubt the Sacred Pipe. Essentially, it consisted of two parts, a round bowl and a long stem.[47] Some say the bowl was made of stone and the stem of wood; others suggest the other way around or else both pieces were composed of either wood or pipestone. Whichever the case, there are as many different interpretations as to what these pieces represented individually as when combined together. Tom Lowenstein, for instance, says the circular stone container signified 'the Earth and all its creatures,' and the wooden stem indicates 'a direct link between the Earth and the sky.'[48] Jamie Sams and David Carson, the authors of *Medicine Cards,* describe the pipe as the union of male and female principle that seeds life.[49] Thus:

The bowl of the pipe was the receptacle that held tobacco, an herb with male and female medicine. The stem of the pipe represented the male entering the female and seeding life. In the coming together of male and female, the connection to the divine energy of the Great Spirit was made. As the pipe was loaded with tobacco, every family in nature was asked to enter into the pipe and share its medicine as prayer and praise the heavens. The smoke was considered to be visual prayer, and was very sacred and cleansing.[50]

Quite apart from its symbolic value and its use in native ceremonies, the pipe performs another very important purpose, says Paula Gunn Allen. The pipe 'is analogous in function to the ear of corn left with the people by Iyatiku, Corn Woman, the mother goddess of the Keres.'[51] The corn's presence in ritual and ceremony ensures a direct link between the two worlds and therefore serves as 'the heart of the people.'[52] Without its presence, or that of the Corn Mother, 'no ceremony can produce the power it is designed to create or release.'[53] These ceremonial objects therefore act as a 'ritual or sacred center of the community,' says Gunn Allen, without which there would be no community.[54] In *Gift of Power*, Archie Fire Lame Deer elaborates on the perception of the pipe as 'the heart of the people.' He recalls how his father would often remind him of this fact by saying to him, 'The pipe is the Indian's heart. The bowl of red pipestone is his blood and flesh. The stem is his spine or body, and the smoke rising from it is Wakan Tanka's breath.'[55] But, as the author points out, 'The pipe itself is not sacred. It is the way in which we use it and the prayers we say when smoking it that make it holy.'[56]

White Buffalo Calf Woman taught the Sioux all of these things in just four days and told the people that she would return after four ages had passed. After she had done all that she needed to do she changed 'into a white buffalo calf,' turned toward the four directions and disappeared.[57] Like those awaiting the promised return of the Messiah there has been much anticipation of her arrival. Toward the end of the twentieth century a truly momentous event took place, which led to much speculation as to whether the 'White Goddess' had returned as she foretold. On 22 September 1994, a white buffalo calf was born on a farm in southern Wisconsin.[58] As word began to spread of this amazing event, many Native peoples and other believers made special pilgrimages to visit the calf because it symbolised the return of White Buffalo Calf Woman. The calf's birth date was especially significant because it was born on the autumn equinox in the Northern Hemisphere.

Predictably, many scientists played down this fact including Edwin Krupp, who dismissed the event as mere coincidence.[59] A Jungian interpretation of this event would suggest otherwise; that it was perfect synchronicity and a timely event, bearing in mind that we are fast approaching the end of one age and about to embark on a new one. Even more startling is the revelation by Standing Elk that White Buffalo Calf Woman is the youngest of the Seven Sisters of the Pleiades.[60] Once again this confirms two other prevalent Pleiadian themes identified in the first chapter — the connection between these stars and young women's initiation, and their association with the cycles of time. Initiation was discussed in the Australian Aboriginal chapter, and their relationship with time is explored in the Hindu chapter on the Krittika.

With so many visitors leaving offerings at the farm, the owner sought advice from Sioux elder Arvol Looking Horse on what he should do with all the gifts people were leaving on the property. The farmer was told to collect these items 'every four days' and burn them, for the smoke from the fire 'would carry them to the spirits.'[61] This advice, says Krupp, is based on 'ancient cosmological' ideas about the ordering of the world, particularly the symbolism of the four directions so common to many Native American peoples and other world cultures.[62] This aspect of quartering the environment and its spiritual significance is looked at in more detail in the concluding chapter on the Pleiades calendar. Suffice to say that the emphasis placed on the fourfold division is borne out by the reference to the four days of teaching by White Buffalo Calf Woman, her turning to face the four directions, the passing of four ages before her promised return and the four legs of the buffalo. This ties in with Sweet Medicine's leave of absence from the waking world for four years and his four-day ceremonial instructions to the Cheyenne on how to lift the long famine that had visited them since his departure. The reference to four such ages is particularly intriguing in light of Hindu perceptions of the cyclical nature of time that is made up essentially of four ages or 'yugas' as they are known in that tradition.

Even more startling is the revelation by Gordon Brotherston in his *Book of the Fourth World*, that 'Buffalo Woman's body incorporates four previous world ages as legs whose hairs are their years,' as recorded in the Oglala Sioux history of High Hawk.[63] Furthermore, the Sioux counted time in 'buffalo units', huge periods of time which 'correspond to the very beginnings of human culture.'[64] This realisation alone suggests there are some strikingly apparent commonalities between Native American concepts of time and that of Hindu philosophy. Additional parallels may be drawn between the Medicine Wheel of Native Americans and the Hindu wheel of life, the Kalacakra.[65] Brotherston also notes the fourfold nature of corn, which may implicate this quaternary division of time.[66]

The Seven Star Girls

Given that Mateo Tipi is a significant site to several Native American nations, there are many diverse stories of its origins and of events that transpired there. Even so, bears feature quite prominently, and among these stories one of the central themes is the pursuit of one or more girls by an individual bear or pack of bears. The most well known and popular story of the origin of Devils Tower is that of the Kiowa legend of the Seven Star Girls. The following version was told by I-See-Many-Camp-Fire-Places, a Kiowa soldier stationed at Fort Sill in the Oklahoma Territory in 1897:

> Before the Kiowa came south they were camped on a stream in the far north where there were a great many bears, many of them.
>
> One day seven little girls were playing at a distance from the village and were chased by some bears. The girls ran toward the village and the bears were just about to catch up to them when the girls jumped on a low rock, about three feet high.

'Rock take pity on us, rock save us.'

The rock heard them and began to grow upwards, pushing the girls higher and higher. When the bears jumped to reach the girls they scratched the rock, broke their claws and fell on the ground.

The rock rose higher and higher, the bears still jumped at the girls until they were pushed up into the sky, where they now are, seven little stars in a group. (The Pleiades)

In the winter, in the middle of the night, the seven stars are right over this rock. When the people came to look they found the bear's claws, turned to stone, all around the base.

No Kiowa living has ever seen this rock, but the old men have told about it — it is very far north where the Kiowa used to live. It is a single rock with scratched sides, the marks of the bear's claws are there yet, rising straight up, very high. There is no other like it in the whole country, there are no trees on it, only grass on top.

The Kiowa call this rock 'Tso-aa', a tree rock, possibly because it grew tall like a tree.[67]

The old man's comment that 'no living Kiowa' has ever seen the tower is not just a statement of the times before the advent of modern travel, but more an historical reference to the great migrations of the various Indian nations across the Americas prior to their eventual settlement in their present tribal lands. As Vic Paddleknee, another tribesman explains, the Kiowa were once a South Canadian Indian tribe. 'During their migration from Canada, and after having crossed the area of what is now Minnesota and Wisconsin,' he says, 'they camped for a time on the shores of the Little Missouri River in the eastern part of Wyoming.'[68] In similar fashion, the Cheyenne, another Plains tribe, migrated from their 'ancestral homelands near the Great Lakes' in the American mid-west.[69] Devils Tower was originally not part of Sioux country, says Chief-Max-Big-Man. It once 'belonged to the Crows . . . but they were pushed back.'[70] Despite the Plains becoming 'the scene of fierce rivalry between competing tribes', Native peoples still managed to share a common affinity with one another, says Lowenstein.[71] The two most influential factors he identifies for this sense of kindredness were a shared sign language and 'a profound belief in a supreme creator, the *Great Spirit.*'[72]

The migratory history of Native Americans has been the subject of ongoing debate, and in some instances has created bitter contention among and between scientists and Native peoples. In *Red Earth, White Lies*, Vine Deloria Jr questions two common assumptions and stereotypes about Native Americans. The first assumption is that Native peoples do not possess an *exact* science, and secondly, that they crossed the Bering Strait to migrate to the new lands in various waves of migration within specific time periods. It is not that he questions whether immigration occurred or not; rather he questions the time periods formulated and suggests that Native peoples may have been on Turtle Island (America) for much longer than previously thought by Western scientists, especially archaeologists. Using the geological history of Devils Tower to support his argument, it raises some interesting questions about Native peoples' scientific observations of their environment. Not only does this implicate the existence of Indigenous sciences but it questions the orthodox archaeological time frame of the migration of Native Americans. Most people, he says, would accept the Kiowa story of the Seven Little Star Girls as a cultural explanation for the tower's unusual geological features and formation, but even so a curiosity remains.[73] He writes:

> If you have ever been to Devils Tower, you will see fluted columns of volcanic material composing the giant structure and will be told by the Park Service guide that you are seeing a volcanic plug from which the surrounding strata have been eroded. The curious thing about Devils Tower is that the rock appears to be badly eroded about a quarter of the way down and the remainder of the structure seems to have rock of a much cleaner and fresher look. It does look like the greater percentage of the rock has recently been pushed upward and that the original top part had been subjected to erosion for a considerable period of time. There are numerous volcanic cones

southeast of Devils Tower, so some correlation could be made between the two sites. But it has always seemed to me that small part of the tribal traditions, that the rock suddenly was raised straight out of the ground to new heights, may reflect some historical observation by Indians.[74]

I have no doubt that many Aboriginal people's Dreamtime stories are in fact based on actual reality, either by direct observation or of some deeper process (scientific and spiritual) of understanding the world and the universe in which we live. To this end Hand Clow argues that shamanic ways of knowing have been devalued by Western academia to such an extent that it denies the ingenuity of Indigenous peoples, their cultures and civilisations.[75]

Alcyone, the Broken Chest Star

Another Plains tribe, the Arapaho — who once camped at Devils Tower but who now live further south — tell a fascinating story of how Alcyone of the Pleiades came to be called the 'Broken Chest Star.' Although this particular legend talks about five young brothers and their two sisters instead of seven sisters, the focus remains on the two young girls who are central to the story.

> An Arapaho lodge was camped at 'Bears Tipi'. The father of this lodge was a head lodge and had seven children, five boys and two girls. The two girls had made an arrangement between themselves that the one who found the end bone (end rib) of a buffalo should receive the most favours from the brothers. The boys often made trips to other tribes. After a long search one of the girls found an end bone of a buffalo and on picking it up she turned into a bear and made some big scratches on her sister's back. The 'Bear-girl' told her sister, 'If you tell the dogs will howl and this will be a signal so I will know that you have told.' The sister did tell her brothers and when they heard the dogs howl and give the signal they were scared and started to run.
>
> The Bear-girl heard the signal and ran after them. The girl who had told was carrying a ball in her hand, which she dropped and accidentally kicked. The ball bounded up on the big, high rock. The

Bear-girl reached over her sister's shoulder to grab the ball, slipped and made very big scratches on the big rock and fell on her sister and broke the sister's chest. The Bear-girl climbed to the top of the big, high rock and told her family that there would be seven stars in the shape of a diamond appear in the east and the first star out would be off to one side and would be brighter than the other stars. The first star would be called 'Broken Chest Star.'[76]

This story raises a number of intriguing cosmological and anthropological questions. Firstly, the common preoccupation or special emphasis on one of the Sisters; whether it is the youngest or the eldest this invariably turns out to be Alcyone. We saw, for instance, in the Greek chapter that Robert Graves refers to her as the 'mystical' leader of the Seven Sisters in *The White Goddess*.[77] And in her channelled *Pleiadian Agenda*, Barbara Hand Clow singles out Alcyone as the repository of earthly knowledge and 'stellar intelligence.'[78] How do we explain this accentuation and what does it mean? One interpretation might be that she serves as role model to her siblings and to humans who hear of her special qualities and attributes in stories such as this. Hence this Sister who stands apart from the others represents the strength and foundation of the group as a whole and often symbolises the embodiment of certain principles and virtues they should aspire to.

Among the Hindus this is none other than Arundhati, the virtuous wife of one of the Seven Sages, for reasons which are explored in that chapter. Secondly, this story raises interesting questions about cosmology, the science of understanding how our universe came to be. Was there a Big Bang as Alcyone's starburst suggests? Is this how the star cluster was born? Certainly it indicates a high level of sophistication of Indigenous knowledge of cosmology thousands of years before the Big Bang theory was even formulated. We see this knowledge shared among Polynesian peoples in the Pacific legend of the Pleiades that tells how a single star gave birth to other members of the cluster. Astronomically Alcyone is identified as Eta Tauri, the brightest star of the Seven Sisters that shines at magnitude 2.87.[79] Whether her leadership and prominent status is due to astronomy or spirituality is not known for certain, but that she might have appeared as the White Buffalo Calf Woman ensures that she remains an enigma in legend and in the night skies.

Shamanism and power animals

Sage, an Arapaho elder, was 81 years old when he told this story to Dick Stone in 1932. It was 'very old' even then, he says, and had been handed down for many generations.[80] Sage's grandfather, Drying-Up Hide, had passed the story on to Sage's father, who in turn told the young Sage when he was nine years old. As with many Indigenous storytellers from orally based cultures, the old man emphasised to Stone the importance and responsibility of telling it 'straight'; that is to say 'true'. Sage's grandfather was buried near Devils Tower and, like many shamans, he had possessed a 'strange power over animals.'[81] He recalls an occasion when his grandfather built a trap near Mateo Tipi and on its completion he then proceeded to sing 'to the animals, buffalo, deer and antelope.'[82] Says Sage:

> The song was sung four times and each time it was pitched a higher key. The first time he sang the song the animals stopped eating, the second time they raised their heads and looked around. After it was sung the third time they started to walk towards the trap and at the finish of the fourth time they walked into the trap.[83]

Sage's reference to his grandfather's powers is not only testimony to the shamanic powers of medicine men and women but it speaks to the particular relationship that exists between humans and animals in Native American and other Indigenous cultures. Not only did animals provide practical commodities like food and clothing, but they 'also played a vital role in the spiritual life of Native American communities.'[84] The idea of animals as guardian spirits and bringers of wisdom is familiar to many Indigenous cultures. In Native American spirituality, as in the Aboriginal Dreamtime, animals are much revered because of the healing *medicine* or *dreaming* that they possess. Jamie Sams and David Carson elaborate as follows: 'When you call upon the power of an animal, you are asking to be drawn into complete harmony with the strength of that creature's essence.'[85] To gain an understanding from 'these brothers and sisters of the animal kingdom is a healing process, and must be approached with humanity and intuitiveness.'[86] Animals are 'a pathway to power' and the power 'lies in the wisdom and understanding of one's role in the Great Mystery, and in honouring every living thing as a teacher.'[87] Within this context, Sage's grandfather's actions of singing to the animals reflect an honouring act as well as an act of power, which is derived from engaging in relationship and communication with these animals that is ultimately shamanic. As we have seen from Native American mythology surrounding the tower, bears are intimately connected with this location. Therefore it remains a special place where human beings can make contact, not only with the stars of the Pleiades, but with the medicinal and healing powers of bears.

Bear medicine and the wisdom of bears

Bears are 'perhaps the most feared and honoured' of all animals, says Tom Lowenstein in *Mother Earth, Father Sky*, not only for their obvious physical strength but more importantly for their wisdom.[88] He recounts a traditional Pawnee story of 'The Man who Lived with Bears', which tells of a hunter who was killed in an enemy attack and brutally dismembered. Two bears later found the wounded warrior and restored him back to life. After living among them for many years he eventually decided to return home to his people.

> As he took his leave, the male bear embraced him; pressed its mouth to the man's lips, rubbed with its paws and fur. The touch of the fur gave the man power, while the kiss gave him wisdom. He became a great warrior and established the Bear Dance among his people.[89]

This was a gift to reward an action taken many years before the hunter was born. The man's father had been out hunting when he came across an orphaned bear cub. Rather than kill it he chose instead to tie 'an offering of tobacco around its neck' and blessed the bear cub by asking the Great Mystery to protect it.[90] Upon his return to camp he told his pregnant wife about the encounter. Soon after a son was born to them who 'grew up feeling a powerful sense of kinship with the bears.'[91] This story, says Lowenstein, demonstrates the 'intense feeling of kinship' that exists between animals and human beings.[92] For the Kiowa this sense of kinship has extended to placing a cultural taboo on the eating of bear meat.[93] The prohibition against eating certain foods is a widespread cultural practice among many Indigenous cultures. Many Aboriginal Australian peoples, for example, are forbidden to eat those foods that are associated with their respective totems or dreamings. Apart from their religious value these spiritual beliefs and practices help to ensure the continual supply of certain animal and plant species through cultural and self-regulation. It provides a perfect example of how environmental knowledge and awareness can be interwoven into daily life and ritual through spirituality and other cultural practices.

In *The Gift of Power*, Lakota medicine man, Archie Fire Lame Deer recalls how many of his family members had bear names because they carried that medicine or dreaming. He says that although the animal can be fierce and dangerous it is also known as 'the medicine man among animals' because of its knowledge of herbs, whose roots it is able to dig up with its strong claws.[94] Thus if someone has this medicine then that person 'will acquire its knowledge of secret herbs to use in doctoring.'[95] Joseph Epes Brown has observed a gendered polarity between the buffalo and bear among the Sioux. As noted earlier, the buffalo symbolises feminine attributes whereas the bear represents the expression of masculine energies.[96] This explains why the animal was the special totem of the Sioux visionary Black Elk, says Brown. The holy man had received the blessings of its powers 'through dreams and in the vision quest.'[97] When troubled or ill, Black Elk was known to emit 'bear-like sounds' and when questioned why he did this, he replied, 'I have the bear's power, and when in need of it this gives me strength.'[98]

The authors of *Medicine Cards* provide further insight into the particular wisdom of bears by explaining what their medicine is all about and what it can teach us. Essentially, bear medicine is all about introspection, say Sams and Carsons.[99] It is about entering the 'Great Void of Silence' to reach the place of inner knowing within each and every one of us. In order to do this, we must become like the bear that 'seeks honey, or the sweetness of truth', to 'enter the safety of the womb-cave', and to 'quiet the mind, enter the silence, and *know*.'[100] This place of inner knowing or meditative practice is what Native Americans refer to as the dream lodge, 'where the death of the illusion of physical reality overlays the expansiveness of eternity.'[101] During meditation in the dream lodge, a council of spiritual elders will sit and meet with the Seeker to guide them on their pathway.[102] The ability of bears to hibernate during the winter months is a reflection of its powers of introspection, and caves in particular are considered inducible to the mystic experience.

But there is another side to this equation, say the authors, and that is to engage in an exercise of sacred geometry and enter the world of symbolism. The cave in Hindu philosophy, for instance, represents Brahma's Cave where the pineal gland 'sits in the center of the four lobes of the brain.'[103] Seen from an overhead position, the right brain, the intuitive side, would sit in the west, the exact location where bear medicine sits on the Medicine Wheel, the wheel of life. Endlessly turning and forever evolving this wheel brings 'new lessons and truths' to assist us on our path.[104] It is interesting to note that not only is Devils Tower associated with bear dreaming that sits in the west of the medicine wheel, but it is physically and geographically located in the American west. The ultimate invitation of bear medicine, according to Sams and Carsons, is to 'journey with bear to the quietness of your cave and hibernate in silence' so as to 'dream your dreams and own them.'[105] After you have done so, 'Then in strength you will be ready to discover the honey waiting in the Tree of Life.'[106]

So what does this mean exactly? What is the significance of honey and what or where is the Tree of Life? The Tree, as we have seen from Eliade's interpretation, is none other than the Centre, the Source of all existence, and the Eye of the Storm where stillness and calmness prevail, where eternity is experienced. So where does honey fit into the equation? Bear provides a clue. Remembering the correlation between the microcosm and the macrocosm or the 'As above, so below' principle, the solution is to widen one's perspective. The connection of bears with Devils Tower speaks to the special affiliation between the stars of Ursa Major and the Pleiades as revealed in world mythology and as it exists in astronomy. While honey may be a naturally desired food of bear, its phenomenological link with this star cluster as a major theme cannot be overlooked. Within this context the biblical reference to the 'sweet influence' of the Pleiades provides a key to unlocking the mysteries. In Native American cultural traditions and specifically through bear medicine, the search for honey becomes a compelling metaphor for the personal and spiritual quest for the Holy Grail containing the elixir of life. The magical drink of immortality endlessly sought by seekers and sinners alike is not a beverage as such but rather the 'sweetness of truth,' which is none other than human understanding and spiritual enlightenment. In Christian terms this realisation is ritually enacted in the Catholic sacrament of Holy Communion where the Blood of Christ is shared with the faithful through the partaking of wine with the wafer.

Mystery caves, lost gold and dream lodges

The association of bears with caves is another natural phenomenon for obvious reasons: it is large enough to house the animal and strong enough to protect it during its long winter of hibernation, rather than an artificial structure that it might have to build. Once again, in caves we see another familiar Pleiadian theme directly and indirectly connected with Devils Tower. As with the Australian Aboriginal stories of the Seven Sisters, some of the Native American legends refer to the alleged existence of a secret cave at Devils Tower.[107] Like tales of the fabled lost gold mines of El Dorado[108] this cave is supposed to contain 'an untold wealth' of the precious metal, together with a crystal clear lake 'of the purest water.'[109] Legend tells that long ago, three warriors were hunting near the area when 'they discovered a passage-way, which led directly under Devils Tower'.[110] Exploring the cave with pine torches, the warriors came across numerous bones that they believed to be human. Venturing on, they came across the lake, which they estimated to be about 'seventy five feet long by fifty feet wide.'[111] Upon its shores they saw vast quantities of gold. Afraid to take any of it with them, the warriors resurfaced and covered the entrance to prevent others from discovering the cave and its valuable contents. Legend says they had hoped to return at some later date to retrieve some of the treasure but this never eventuated. The surviving member of this trio reportedly told of this discovery on his deathbed.

Like many a good story, it entered the world of folklore to be handed down over many generations where it still persists even today, despite the fact that such a cave has never been found. Whether the cave exists or not is not important; its real significance lies in the various meanings ascribed to the cave by Native Americans in their traditions. For instance, some claim that the Sacred Pipe of White Buffalo Calf Woman and the Four Arrows of Sweet Medicine are still preserved within its subterranean chambers. In any event, the suggestion of a hidden cavern near Devils Tower is not such a strange idea, given that there are many caves in the Black Hills of South Dakota and Wyoming, including the better known Wind Cave whose original entrance was nothing but a small hole in the ground.[112]

Caves have always been associated with the female principle, for obvious reasons. In many cultures it symbolises the womb of the Mother Goddess and so served as an appropriate venue for rebirth and other rituals.[113] That caves feature in world mythology of the Seven Sisters of the Pleiades is not surprising, given their connection with creation in many of these stories and their role in many of the world's initiation ceremonies. In Pleiadian mythology the Seven Sisters use caves as a passageway, not only for their entry into this world but as a point of departure into the heavens. In this respect the cave acts as a virtual conduit between the worlds that goes beyond its physical limitations.

What might the significance of this be? Is this just another metaphor for spiritual transcendence or could it be another example of science encoded within myths? Could the two-way traffic indicate the existence of parallel universes or other dimensions? For if such entities exist, how do we pass from one sphere to the other? Does it suggest the possibility of space travel between the worlds through the existence of wormholes in space as mini-travelling wombs that enable such encounters? We need not restrict our imaginations as we ponder and reflect upon the hidden meaning of world mythologies, for imagination inspires creativity, which can open up many different levels of knowing and understanding that contributes to the good of all.

Finally, the question remains, as Chief Max-Big-Man claimed in the opening quote of this chapter, did actual dream lodges made of stone ever exist at Devils Tower? Or are they only metaphors for the kind of inner temple of the soul as described by Sams and Carson, a sacred mental space that we enter only through the power of meditation? Whether they actually existed or not is not known and in the absence of archaeological evidence to the contrary it remains open to conjecture. In any event, it is not difficult to imagine, given the profuse amount of stone slabs and boulders that surround the base. Geologists maintain that the slabs have merely fallen off the tower itself. But could these very slabs and boulders be the remains of the ruins of these alleged stone lodges? Without archaeological evidence it is difficult to say with any degree of certainty. Furthermore, although Sams and Carson refer to our inner mental dream lodges, this does not discount the very real existence of actual dream lodges or sweat lodges, as they are also known. As well as meeting with supernatural members of the Council of Elders in meditation, one can do the same in the flesh with very real, live elders! Without any definitive scientific evidence to the contrary, I much prefer to defer to our Elders until proven otherwise.

The Great Bear in the sky

A further significance of the association of bears with Devils Tower is the celestial relationship between the stars of the Pleiades and Ursa Major, the Great Bear. During the winter months the cluster can be seen directly above the tower and on top of them are the stars of the Great Bear, including the distinctive pattern of seven stars popularly known as the 'Big Dipper' in the Americas or 'the Plough' in several European countries. Several other cultures besides Native Americans have interpreted the pattern of stars in Ursa Major as a bear, including the ancient Greeks and the Indigenous people of Japan, the Ainu, who are renowned for their bear dances. These cultures originate predominantly in the Northern Hemisphere where the animal has 'always been a symbol of frozen wastes.'[114] The Great Bear is 'one of the oldest' and 'perhaps the best known' of all constellations, says David Levy in *Skywatching*,[115] although how far back in history is 'lost in antiquity' say Gertrude and James Jobes.[116] They note that 'almost everywhere' the constellation has been depicted as a she-bear;[117] hence in the Greek legends Ursa Major represents Callisto, one of Zeus' many paramours.[118]

But as we have already seen, in Native American traditions the bear is largely masculine. However its gender is perceived among the world's diverse cultures, what is more important perhaps is the underlying symbolism peculiar to bears, and Native American mythology surrounding these animals and its relationship with the stars of Ursa Major is especially revealing. The Blackfoot, for example, regard the seven bright stars of the Big Dipper as seven brothers who were chased into the sky by a bear in much the same way as the seven star girls of the Kiowa legend.[119] Levy recalls a similar tale by the Micmacs of Nova Scotia and the Iroquois people of the St Lawrence River Valley in Canada. In both legends the bowl of the Big Dipper represents a bear that is hunted by seven warriors and Corona Borealis (the Northern Crown constellation) signifies his den. The bear's wanderings and its pursuit by the warriors takes place over an entire year, so the story effectively acts as a calendar recording the seasons and yearly cycles, for the event is endlessly repeated. Says Levy:

> Each spring the hunt begins when the bear leaves Corona Borealis, his den. The Bear isn't killed until fall, and the skeleton remains in the sky until the following spring. Then a new bear emerges from Corona Borealis and the hunt begins again.[120]

Both the Jobes and Edwin Krupp provide a full version of the legend in their respective publications where the seven warriors appear in zoological form as seven individual species of birds. They include the robin, chickadee, moose or cowbird, blue jay, pigeon, owl and saw-whet (a small owl whose head is covered in red feathers).[121] In *The Glorious Constellations*, Giuseppe Sesti provides their astronomical identities as: Robin (Epsilon Ursae Majoris), Chickadee (Zeta Ursae Majoris), Moose or Cowbird (Eta Ursae Majoris), Blue Jay (Epsilon Bootis), Pigeon (Gamma Bootis), Owl (Alpha Bootis or Arcturus) and Saw-whet (Eta Bootis).[122] This is intriguing because of the alleged connection between the stars of Ursa Major and the Pleiades in world mythology, and in particular the common characterisation of the Pleiades as birds. In this respect it bears a striking similarity with the Mayan legend of Seven Macaw in the *Popul Vuh*. In Mayan mythology the Big Dipper is seen as a parrot but, as Krupp points out, it is called Seven Macaw because it refers to the seven stars of Ursa Major.[123] Here in both stories we have a living and heavenly manifestation of the emanation of deity as expressed by the spiritual axiom of the 'One in the Many and Many in the One'.

Several writers have attempted to explain why so many cultures have depicted the stars of Ursa Major as a bear, including Max Muller, Sir James Frazier and Robert Graves. None are as satisfactory as the astronomical explanation provided by Edwin Krupp in *Beyond the Blue Horizon*. Beyond its mythic and spiritual relationship with rebirth and cyclic change, the bear's symbolic value has just as much to do with its location in the night skies, where its placement above the celestial North Pole is quite deliberate. 'We don't have a bear in the northern sky because the stars there look like a bear,' says Krupp.[124] We have a bear there because it 'prowls around the pole, performs a seasonal death, and returns to life when the world does.'[125] The bear's action, says Krupp, may be compared to a 'celestial powerhouse' whose energetic churning or turnings of the heavenly wheel 'seems to activate the entire world.'[126] Its 'death' is perceived as a necessary sacrifice to ensure the continuity of life, so that by dying it lives forever and therefore acts as a symbol of eternity.[127] Krupp's interpretation of the bear's symbolism — in particular the stars of Ursa Major — are not that different from certain views proposed by Madame Blavatsky, particularly those in regard to that constellation's connection with time. Her main assertion is that the stars of Ursa Major (the Great Bear), together with the Pleiades, control the various cycles of time. Within this context the bear in the story of the Seven Little Star Girls may signify the dark Kali Yuga that devours and destroys the end of one world age in order to bring about a new one.

This argument makes complete sense when we consider the obliquity of the ecliptic. The fact that the Pleiades star cluster lies within four degrees of the ecliptic[128] and that the stars of Ursa Major lie close to the northern celestial pole make them the perfect targets to symbolically represent this division of time, for — as the authors of *Hamlet's Mill* point out — before this separation 'time did not exist.'[129] As a scientist, however, Krupp would most likely want to distance himself from any 'non-scientific' explanations such as those offered by theosophy. This is unfortunate because the two perspectives are not mutually exclusive and it is possible that both can inform and complement the other in many instances. An exemplary illustration of the combination of spirituality and science furthering human knowledge and understanding of the workings of our universe are evident in the writings of Fritjof Capra, including the popular *Tao of Physics*. This is just one example of a new breed of scientists seeking to find common ground between diverse knowledge bases such as mythology, spirituality and science that serve to validate different ways of knowing.

Lessons from our Pleiadian teachers

Turtle Island (USA) and Bandaiyan (Australia) share many similarities. The two countries are roughly the same size in area and both have a shared history of invasion and colonisation by non-Indigenous peoples. Further comparisons may be made between Native Americans and Aboriginal Australians that reveal commonalities of spirituality and relationships to the land and all within. As previously noted, our dreamings may be likened to medicines contained within various animals, plants and other totems that serve to empower people and create kin relationships. In times of disharmony and disruption our shamans, men and women of high degree, intercede on our behalf to placate and commune with the Spirit world and to provide guidance during our individual vision quests. In times past Councils of Elders guided and governed our people while tales of our ancestors taught us human virtues and provided the foundation of our laws. Within this cultural milieu there was room for diversity among the many tribes or nations who shared the land with one another that enabled different languages, cultural modes of expression and spiritualities to flourish.

Above all there was deep respect for the precious gift of life in all its forms. To an Indigenous person there is no such thing as the 'inanimate' world for the entire world is *alive* with spirit, including rocks, stones, crystals and stars. Native Americans have a beautiful way of expressing this belief by referring to these so-called inanimate objects as 'peoples'; hence the popular phrase 'stone people', 'crystal people' and 'star people'. The use of the term people in this context is not so much about humanising these objects but rather the expression of an innate sense of connection or kindredness with all of nature. For Indigenous peoples there is no sense of separation from our environment or from one another. There is no talk of 'cutting the ties that bind' for it is not our role to undo what Grandmother Spider has created. All relationships are sacred and not to be severed, no matter how difficult they may become. To assist us on this journey of relating we are taught more honouring ways of resolving and dealing with conflict, which arises from our imperfections. The spinning and unravelling of webs is the sole privilege of She who creates and destroys. To even dare to presume to do so is the height of human arrogance and a contemptuous act of sacrilege.

With this egalitarian philosophical approach to the universe and the world around us, Indigenous peoples view all things, animals, plants and peoples, as mutual teachers. Just as the buffalo and bear may impart medicines and teachings of wisdom in Native American traditions, so do the stars, especially those of the Pleiades. Many of their stories of this particular star cluster feature a group of young children, usually orphaned, either girls or boys. The Blackfoot legend of the Pleiades, for instance, tells of six young orphan boys who were neglected and abused by their tribe here on Earth, and so decided to become stars to escape their lot. There they remain in the night skies, huddled close together and shining down on the people as six beautiful stars. Today they are commemorated as designs painted on the southern smoke flaps of Blackfoot tipis to serve as 'a visible reminder of this story and of the need to look after all the little children.'[130] The seven stars of Ursa Major that represent the Seven Brothers are on the northern smoke flap and on the very top of the tipi on the back is the Maltese cross that represents the morning star.[131]

That the stars of Ursa Major and the Pleiades feature on the two smoke flaps of Blackfoot tipis is particularly significant, as is the placement of Venus. The Blackfoot explanation for this particular combination of stars is not known, but clearly there is an emphasis placed on an existing complementary joint relationship, which may have some bearing on the obliquity of the ecliptic. Venus, as we will see in the last chapter on the Pleiades calendar, was of enormous importance to Mesoamerican cultures such as the ancient Mayans, Aztecs and Incas. What is especially revealing is that Venus symbolised the feathered serpent god Quetzalcoatl, whose return will herald the end of the Fifth Sun and the dawning of a New Age. The stars of the Pleiades and of Ursa Major are implicated in this celestial handover in these cosmogonies, and in the Blackfoot tipi whose structure resembles the world axis that enables the viewing of the Pole Star through its smoke hole.[132] Furthermore, whether intentional or coincidental, the two smoke flaps have seven and six stars on either side, thereby emphasising the interplay between six and seven — a constant theme of the Pleiades identified in the first chapter.

Astronomer Von Del Chamberlain, who has written extensively on Native American astronomy, says we have much to learn from Native cultures and from the stars themselves. In particular, we can seek to 'emulate the harmony and balance' of the Pleiades like many Native peoples in their various rituals, prayers and ceremonies and, like the Pawnee, could do well to look to these stars 'for a symbol of unity.'[133] The Pawnee pray to the Pleiades to teach them how to become united like the bunched stars of the night skies. This is one such prayer:

Look as they rise,
Rise up over the line
Where sky meets the Earth.
Seven Stars

Lo!
They are ascending,
Come to guide us,
Leading us safely,

Keeping us one.
Oh Seven Stars,
Teach us to be like you,
United.[134]

Native Americans, like other Indigenous peoples all over the world, have always looked to the Star People for guidance. We see among them our relatives, ancestors and supreme beings, not strangers, unknowns or potential enemies. And while they may be hundreds of light-years away from us our hearts know no distance. We share with them an enormous sense of kindredness and common destiny, these great celestial teachers.

Notes

1. Rathbun, *First Encounters: Indian Legends of Devils Tower*, Sand Creek Printing, USA, 1982. A small booklet of Indian legends told to Dick Stone. Available from the Visitors' Center at Devils Tower National Monument, or write to PO Box 10, Devils Tower, Wyoming 827140010.
2. Ibid.
3. Ibid, p. 4.
4. Buchanan, *Discovering the Wonders of Our World*, p. 319. See also the website 'Devils Tower History' at <www.nps.gov/deto/first50.htm>.
5. Rathbun, *First Encounters: Indian Legends of Devils Tower*, p. 26.
6. Eliade, *Images and Symbols*, p. 39.

7. Ibid.
8. Ibid.
9. Ibid, p. 42.
10. Campbell, *The Power of Myth*, p. 89.
11. Ibid.
12. Ibid.
13. Krupp, *Skywatchers, Shamans and Kings*, p. 16.
14. Eliade, *Images and Symbols*, p. 43.
15. Hancock, *Heaven's Mirror*, p. 252.
16. Krupp, *Skywatchers, Shamans and Kings*, p. 16.
17. Lippincott et al., *The Story of Time*, p. 234.
18. Hand Clow, *The Pleiadian Agenda*, p. 4.
19. Hand Clow, *Catastrophobia*, p. 109.
20. Rathbun, *First Encounters: Indian Legends of Devils Tower*, p. 23.
21. Lowenstein in *Mother Earth, Father Sky*, ed. S. Adamson, p. 88.
22. Ibid.
23. Ibid, p. 25.
24. Ibid.
25. Ibid, p. 21.
26. Ibid.
27. Ibid.
28. The case in question is known as *Babbitt's* case. While some legal commentators have labelled it a *major* victory for Indian rights I believe it is a minor victory in that it only supports a *voluntary* ban on climbing Devils Tower during June, not a *mandatory* one. The inequality of the protection of Native American religious rights is patently borne out by the irony that it is a federal crime to climb Mt Rushmore, known as 'the shrine of democracy.' For further information see the initial application for a preliminary injunction in *Bear Lodge Multiple Use Association v Babbitt*, _ F. Supp_(D.Wyo., 1998), No. 96-CV-063-D.
    Christopher McLeod and Malinda Maynor explore the conflicting views of climbing Devils Tower in the film documentary *In the Light of Reverence*. The film explores the issue of competing interests between Indians and non-Indians in regard to three major Native American sacred sites including Devils Tower, Mount Shasta and the Four Corners. For further information on the documentary see the American PBS (Public Broadcasting Service) website at
    <www.pbs.org/pov/pov2001/inthelightofreverence/thefilm.html>.
29. Hand Clow, *Catastrophobia*, p. 109.
30. Gunn Allen, *The Sacred Hoop*, p. 107.
31. Lowenstein in *Mother Earth, Father Sky*, ed. S. Adamson, p. 84.
32. Ibid.
33. Ibid, p. 87.

34. Ibid.
35. Gunn Allen, *The Sacred Hoop*, p. 107.
36. Lame Deer and Erdoes, *Gift of Power*, p. 201.
37. Adamson (ed.), *Mother Earth, Father Sky*, p. 129. See also chapter 1, pp. 3–9 of *The Sacred Pipe* (ed. Joseph Epes Brown) which tells of the coming of White Buffalo Calf Woman and her gift of the pipe.
38. Brotherston, *Book of the Fourth World*, p. 182.
39. Lame Deer and Erdoes, *Gift of Power*, p. 132.
40. See *The Sacred Pipe* by Joseph Epes Brown for more information on the seven sacred ceremonies as recounted by Black Elk. In *Animals of the Soul*, Brown maintains that White Buffalo Calf Woman only taught one of these — namely that of the Keeping of the Soul ceremony; however she did predict that six others would come 'through the individual vision experience.' See p. 74.
41. Adamson (ed.), *Mother Earth, Father Sky*, p. 87.
42. Brown, *Animals of the Soul*, p. 74.
43. Ibid, p. 75.
44. Ibid, p. 76.
45. Ibid, pp. 75–76.
46. Sams and Carson, *Medicine Cards*, p. 113.
47. Lowenstein in *Mother Earth, Father Sky*, ed. S. Adamson, p. 87.
48. Ibid, p. 129.
49. Sams and Carson, *Medicine Cards*, p. 113.
50. Ibid.
51. Gunn Allen, *The Sacred Hoop*, p. 17.
52. Ibid.
53. Ibid.
54. Ibid, p. 32.
55. Lame Deer and Erdoes, *Gift of Power*, p. 201.
56. Ibid.
57. Krupp, *Skywatchers, Shamans and Kings*, p. 39.
58. Ibid, p. 38.
59. Ibid.
60. See the following website on the history of the Sun Dance at <www.star-knowledge.net/sundance_history.htm>.
61. Krupp, *Skywatchers, Shamans and Kings*, p. 39.
62. Ibid.
63. Brotherston, *Book of the Fourth World*, p. 301.
64. Ibid.
65. See Eliade's discussion of 'The Doctrine of the Yugas' at pp. 62–67 in *Images and Symbols*.
66. Brotherston, *Book of the Fourth World*, p. 183.
67. Rathbun, *First Encounters: Indian Legends of Devils Tower*, p. 5.

68. Ibid, p. 5.
69. Adamson (ed.), *Mother Earth, Father Sky*, p. 15.
70. Rathbun, *First Encounters: Indian Legends of Devils Tower*, p. 10.
71. Lowenstein in *Mother Earth, Father Sky*, ed. S. Adamson, p. 15.
72. Ibid.
73. Vine Deloria Jr, *Red Earth, White Lies*, p. 205.
74. Ibid.
75. Hand Clow, *Catastrophobia*. See, in particular, her discussion of the devaluing of Egyptian civilisation by some writers that suggest it was founded by extraterrestrials at pp. 187–89.
76. Rathbun, *First Encounters: Indian Legends of Devils Tower*, p. 7.
77. Graves, *The White Goddess*, p. 182.
78. Hand Clow, *The Pleiadian Agenda*, p. 4.
79. See Chris Dolan's astronomy website at <www.astro.wisc.edu/~dolan/constellations/hr/1165.html>.
80. Rathbun, *First Encounters: Indian Legends of Devils Tower*, p. 8.
81. Ibid.
82. Ibid.
83. Ibid.
84. Adamson (ed.), *Mother Earth, Father Sky*, p. 94.
85. Sams and Carson, *Medicine Cards*, p. 13.
86. Ibid.
87. Ibid.
88. Lowenstein in *Mother Earth, Father Sky*, ed. Adamson, p. 68.
89. Ibid, p. 99.
90. Ibid.
91. Ibid.
92. Ibid.
93. Rathbun, *First Encounters: Indian Legends of Devils Tower*, p.6.
94. Fire Lame Deer and Erdoes, *Gift of Power*, p. 14.
95. Ibid.
96. Brown, *Animals of the Soul*, p. 31.
97. Ibid, p. 32.
98. Ibid.
99. Sams and Carson, *Medicine Cards*, pp. 57–59.
100. Ibid, p. 57.
101. Ibid.
102. Ibid.
103. Ibid, p. 57.
104. Ibid, p. 21.
105. Ibid.
106. Ibid, pp. 58–59.
107. See, for instance, the 'Legend of Untold Wealth in Gold and Crystal Water

Lake beneath the Devils Tower' in Rathbun, *First Encounters: Indian Legends of Devils Tower*, at pp. 31–32.

108. El Dorado was 'a mythical land filled with gold, jewels, and other great riches,' says Robert Hendrickson at p. 231 in *Word and Phrase Origins*.

109. Rathbun, *First Encounters: Indian Legends of Devils Tower*, p. 31.

110. Ibid, p. 32.

111. Ibid.

112. Versluis, *Native American Traditions*, p. 36.

113. Walker, *The Woman's Dictionary of Symbols and Sacred Objects*, p. 335.

114. Jobes and Jobes, *Outer Space*, p. 265.

115. Levy, *Skywatching*, p. 220.

116. Jobes and Jobes, *Outer Space*, p. 258.

117. Ibid, p. 265.

118. Graves, *The Greek Myths*, p. 84.

119. Krupp, *Beyond the Blue Horizon*, p. 227.

120. Levy, *Skywatching*, p. 220.

121. See the Jobes' version in *Outer Space* at pp. 263–64 and in Krupp's *Beyond the Blue Horizon* at pp. 239–40.

122. Sesti, *The Glorious Constellations*, p. 467.

123. Krupp, *Beyond the Blue Horizon*, p. 237.

124. Ibid, p. 240.

125. Ibid.

126. Ibid.

127. Ibid.

128. Burnham Jr, *Burnham's Celestial Handbook*, p. 1884.

129. De Santillana and Von Dechend, *Hamlet's Mill*, p. 153.

130. There are no page references but see 'Author's Note' in *The Lost Children* by Paul Goble.

131. Ibid, see his 'Note about the Tipis' at the very end of the book.

132. Ibid.

133. See Von Del Chamberlain's website Project ASTRO UTAH at <www.clarkfoundation.org/astro-utah/vondel/dilyehe.html>.

134. Ibid.

CHAPTER FIVE

Krittika

Seven Wives of the Seven Rishis

As he was returning, he saw the wives of those noble sages bathing happily in their own hermitages. They shone like golden altars, like spotless slivers of the moon, like rays of the oblation-devouring fire, like marvellous stars. As Agni saw the wives of the supreme Brahmins, his heart went out to them, his senses agitated, and he was in the power of desire. But he thought, 'it is not proper for me to be agitated, I desire the virtuous wives of the supreme Brahmins, and they are without desire. I cannot look upon them or touch them without cause, but by entering into the household fire I will look at them constantly.' He entered into the household fire and rejoiced to look upon them and to touch, as it were, all those golden women with his flames. Agni lived there for a long time in their power, setting his heart on them, desiring those supreme women. But then, as his heart was still enflamed with desire, Agni went into the forest and resolved to abandon his body, for he could not obtain the wives of the Brahmins.

— *Hindu Myths* translated by Wendy Doniger O'Flaherty[1]

In Hindu mythology, the Seven Sisters of the Pleiades are collectively known as the Krittika who were once married to the Seven Rishis or

Seven Sages.[2] They were the seven wise men of India who sailed in the ark with Manu to escape the Deluge, and who now form the constellation of Ursa Major, the Great Bear.[3] The Krittika were the seven daughters of Brahma and Savitiri whose individual names were Amba, Dula, Nitatui, Abrayanti, Maghayanti, Varshayanti, and Chupunika,[4] although there are other names for them as well. For instance, the eldest daughter Amba was known as Arundhati, the spouse of Vasistha, Chief of the Seven Sages.[5] Some say he is another form of Brahma who fled the rising waters with his six sons, thereby making seven wise men. Like the Greek goddess Athena who sprung from her father Zeus' head, the Rishis are often described as the 'mind-born sons' of their supreme god.[6] As befitting holy men, the sacred texts of the Vedas were imparted to them to share with humanity, and whose hymns are said to cause 'dawn to rise and the sun to shine.'[7] The association of the Krittika as wives of the Seven Sages might explain the custom of many newly married couples in India worshipping and praying to the Pleiades before entering their matrimonial homes.[8] According to theosophist Madame Blavatsky, the stars of Ursa Major, together with the stars of the Pleiades 'constitute the greatest occult mystery.'[9] This mystery, as we will see, is linked to the great cycles of time and the unfolding of human destiny.

Seven Bears in Ursa Major

The Hindu name for the constellation of Ursa Major the Great Bear, is Saptarshi, which refers to Seven Rishis as Seven Bears.[10] Some writers have argued that the connection between bears and the holy men is a linguistic mistake because the bear is not a familiar animal to India. Gertrude and James Jobes, for instance, maintain that the Sanskrit root word riksha means 'star' or to 'shine' but when used 'in a different gender' it means 'bear.'[11] This word, they say, was confused with rishi, which means 'sage' or 'poet', and so India 'had seven wise men to form their bear.'[12] This line of reasoning is inherently racist because it implies and assumes that Hindus know nothing of their language or culture and are mere idle cultural participants and non-savants. Although Edwin Krupp argues the same in *Beyond the Blue Horizon*, he does at least concede that we have no way of knowing if this etymological explanation is necessarily true.[13] Madame Blavatsky, on the other hand, argues there is no mistake or confusion because Hinduism is essentially an Aryan religion, whose people originally came from the northern regions of Europe.[14] Therefore the bear would have been a familiar icon in their cultural traditions, even if in more remote times the constellation was much closer to the North Pole due to precession. It is interesting that the modern term 'arctic', which refers to the colder, northern regions is derived from the Greek word *arkos* meaning 'bear.'[15]

In the previous chapter on Mateo Tipi, we saw that many Northern Hemisphere cultures assigned the image of a bear to the stars of Ursa Major because of the way the stars imitated their prowling actions around the North Pole.[16] The Aryan peoples who migrated to India from northern Europe and Asia would have brought similar traditions and beliefs with them to the subcontinent, and this may account for the inclusion of bears in their mythologies. Generally speaking, this bear was more often portrayed as female, such as in ancient Greece where she represented Callisto, one of the many love interests of Zeus.[17] This notion of the she-bear is retained in the girl's name Ursula, the Latin word for 'bear.' That this animal becomes multiple in India to represent the Seven Sages is not surprising given the existence of seven prominent stars in Ursa Major, but at what exact point in history they became 'male' is not known.

It may be that the Aryan Hindus were culturally influenced by the ancient Greek civilisation where the Lesser Bear, Ursa Minor, represented Arcas, son of Zeus and Callisto.[18] Or it may have been the other way around. Whatever the explanation, there is no doubt that the seven stars of Ursa Major, and those of the Pleiades, play a prominent role in Hindu beliefs as they do in other cultures. The mythic overlapping of gods and goddesses in these stories (such as Brahma and Vasistha, then Brahma/Vasistha with the Krittika) suggests a deeply entwined relationship between these groups in history and the cosmos. Apart from their alleged occult nature, the change in characters from male to female could possibly reflect the alternating patriarchal and matriarchal cycles of human societies that Robert Lawlor has identified in *Voices of the First Day*.[19] Just what these cycles are and how they impact on human cultures and inter-personal relationships is explored later on in this chapter.

Nakshatras (housing the Moon)

As in many other cultures around the world, the Pleiades once began the New Year in ancient India.[20] In addition to this momentous role, they signalled the commencement of the lunar cycle known as the nakshatras — the procession of the lunar mansions that mark the positions or stages in the night sky where the Moon is housed during its monthly sojourn around the Earth.[21] As it circles the Earth in 27.3 days, this means there are twenty-seven nakshatras in all.[22] These lunar asterisms are named for the twenty-seven daughters of Daksa (one of the Seven Sages), who 'were given in marriage to Soma, the Moon.'[23] Of the twenty-seven, Rohini, the fourth daughter, was the favourite wife of the Moon.[24] Richard Allen tells us in *Star Names* that Rohini is the Hindi name for the red giant star, Alpha Tauri, or Aldebaran, that marks the eye of the bull in the constellation of Taurus.[25] Her name literally denotes 'a female gazelle or red cow,' says Wendy Doniger O'Flaherty, and was applied to the entire constellation of Taurus, not just to its leading star.[26]

The story relates that Daksa, their father, became concerned about Soma's affections for Rohini in particular. It seems the Moon favoured her to the exclusion of the other daughters. The sage took immediate action by cursing Soma 'to be childless and to suffer from consumption (the cause of the Moon's waning).'[27] Soma's other wives became concerned at the severity of this curse and jointly agreed to intercede on his behalf, whereby Daksa modified it so that the waning would only be periodical and not a fixed phenomenon.[28] Egyptian mythology has a comparable tale of a father's curse surrounding the procreative powers of his progeny. In ancient Egypt, as in India and Bandaiyan (Australia), the Moon was depicted as a man, although the curse did not involve the Moon *per se* but the daughter instead.

The legend states that Ra, the Sun god, feared that Nut the sky goddess and her future children would one day usurp his power, and so he placed a similar taboo on her. Thoth, the Moon god came to the goddess' rescue by challenging her father to a board game using dice, whereby he won an extra five days in which she could conceive, outside the remaining 360 days of the original Egyptian year.[29] This tale not only provides a mythical explanation for the emergence of the 365-day Egyptian calendar we now follow, but it also connects one of their primary goddesses with dice, an ancient Indian tradition relating to the Krittika and their involvement with time.[30]

Although the legends do not name the mother of the twenty-seven daughters, she may be one of the Krittika, given that they were once married to the Seven Sages. It is interesting in this regard to note that the equinox month was given the name, Kartika, in which 'the vernal equinox moon was itself known as the Child of the Pleiades.'[31] Edwin Krupp says it received this name 'by observing the full moon among the Pleiades' during the vernal equinox in the Northern Hemisphere when the Krittika marked the New Year.[32] The fact that the brightest star of the Pleiades, Alcyone or Amba, is referred to as the 'junction star of the Nakshatras Krittika and Rohini,' may explain Robert Graves' reference to her as the group's 'mystic' leader, as we saw in the last chapter.[33] In some Hindu myths Amba or Arundhati is often referred to as the Great Mother, whose name becomes Ambika among the Jains.[34] Some writers have suggested that she is the Indian equivalent of Alcyone, the second eldest daughter of the Pleiades in the Greek legends, although linguistically this would have to be Maia whose name means 'mother.'[35] Although there is no explanation for this suggestion, it may be inferred from her prominence in the lunar cycle and therefore her comparison with Alcyone, who has been described and considered by many as the 'central or leading star of the Pleiades.'[36]

The nurse maidens of the God of War

Besides their role as the wives of the Rishis, the Krittika are renowned as the six nurses of Karttikeya, the Hindu equivalent of Mars who was named after them.[37] As befitting a god of war, Karttikeya bore the title of Commander of the Celestial Armies, or Commander of the Siddhas — the 'yogis in heaven and holy sages on Earth.'[38] Thus another name for him was Siddhasena meaning 'the leader of the Siddhas.'[39] This, says Blavatsky, effectively makes Karttikeya the equivalent of Saint Michael as the 'leader of the seven celestial hosts.'[40] In Hindu iconography Karttikeya is often depicted with six heads, or as a boy god walking alongside or astride the peacock, India's national bird, holding a spear in his hand to show his warrior status. Legend says he was created 'to defeat the powers of evil, represented by the demon Takara.'[41] As the Guha, 'the mysterious one' he, like his nurses the Krittika and their husbands the Seven Sages, hold the key to 'the greatest mysteries of occult nature' associated with time and the creation of the universe.[42] Like many Hindu gods Karttikeya had several names 'such as Skanda, Kumara and Subrahmanya.'[43] This suggests 'he is probably an amalgam of different cults,' says John Bowker in *World Religions*.[44] And as one would expect, there are many diverse stories and conflicting accounts as to his birth and parentage. Both Shiva, the destroyer and lord of time, and Agni the god of fire, are said to be his father although it is more likely that this was really Shiva in his aspect as Agni. Likewise, his mother was thought to be the goddess Ganga or else Parvati, one of Shiva's wives who (like Amba or Arundhati, the eldest of the Krittika) were considered proper and virtuous, unlike his other female partners, the powerful and ferocious goddesses Durga and Kali.

Others say the Krittika were the mothers of Karttikeya, or else Svaha who seduced Agni in their form; all but one, Amba or Arundhati. As a consequence, soon after this deception, the (not so very wise) Seven Sages wrongfully accused their wives of adultery and despite their pleas of innocence divorced them, except for the eldest sister, Amba. In the post-Vedic *Mahabaharata*, composed about 500 BCE, she remained married to her husband and went to live with him in the constellation of Ursa Major.[45] There they can still be seen together, side by side, happily married as the double binary star Alcor and

Mizar in the middle of the Big Dipper's handle. Several other cultures, as previously noted, have placed the lost or missing sister of the Pleiades within this constellation. The Greek poet Aratus, for example, named Electra specifically as that star. This tradition is quite deliberate, say the authors of *Hamlet's Mill*, and extends as far back as Mesopotamia where Arundhati, or Amba, was better known as Elamatic Narundi, sister of the Sibitti, the Babylonian version of the Seven Sages.[46] Its significance rests in Alcor's appellation as that of the 'fox star,' which plays a key role in certain mythologies in the turning of the celestial mill that dictates the tide of ages that make up time, as we know it. Proclus himself referred to this star as that which 'nibbles continuously at the thong of the yoke which holds together heaven and earth.'[47] What is more, many prophecies about the end of the world refer to the stars of Ursa Major as playing an active role. German folklore, in particular, makes specific reference to Alcor by stating, 'that when the fox succeeds, the world will come to an end.'[48]

In any event, whether the Krittika are his mothers or simply his nurse-maidens, the legends suggest they have an intimate relationship with their namesake Karttikeya, as evidenced by his name and the fact that he has six heads. Yet despite this reference to *six* sisters in the *Mahabaharata*, the older Sanskrit texts refer to the Krittika as the 'Seven Mothers of the World'. Once again, this has implications for Blavatsky's theory of humanity's seven root races, or in less inflammatory language, the seven rounds or strands of human beings through the ages. The mythological tradition of seven mothers or seven creator beings has been revisited by science through the findings of geneticist Professor Bryan Sykes in the *Seven Daughters of Eve* (as discussed in earlier chapters). The constant switching between seven wives then six nurses in the Hindu stories of the Pleiades once again reflects aspects of the Lost Pleiad. These numeric references, according to the renowned theosophist, are secret codes that are part of occult mysteries related to our creation and are tied up with the various cycles of time that are explored further on. Finally, some claim Karttikeya 'was born without a mother'[49] and therefore referred to as Kumara, meaning 'virginal youth.'[50] That there were seven kumaras re-emphasises the septenary or sevenfold nature of creation, and in this regard relates back to the Krittika.

Kumbh Mela the Pitcher Festival and Diwali the Feast of Lanterns

The various accounts of Karttikeya's conception reveal interesting information about the Krittika and their connection with the creative elements of fire and water. In fact, many overlays between the Krittika and these other goddesses seem to suggest that they either represent different aspects of the individual Sisters, or they are earlier or later images of them. Parvati, for one, is easily linked with Amba because they share similar personal traits and virtues such as fidelity and devotion to their respective spouses. Kali's relationship to the Sisters is twofold, firstly through their common affiliation with fire and secondly through time, including notions of fate and destiny. For instance, John Bowker tells us Agni the god of fire has seven tongues of fire that are known as the 'seven red sisters,' one of whom is called Kali.[51] In *The Secret Doctrine*, Madame Blavatsky says the goddess' own tongue 'dripping with blood represents Rajas, the material force in the universe that gives rise to effort, as well as passion and suffering.'[52]

Although Kali is often referred to as the consort of the lord of time, her name is actually derived from the Sanskrit kala, which means 'time.'[53] This means she is a goddess of time in her own right and, as befits a monarch, she reigns over all of it, especially the destructive period known as the Kali Yuga that is named in her honour.[54] Kali has many characteristics comparable with the Lion Goddess Durga, and both are often interchangeable in Hindu traditions and therefore embody the same destructive and creative forces of fire. Ganga's connection to the Krittika, on the other hand, is through her role as goddess of the Ganges, the most sacred and holiest of all of India's rivers.

Legend tells that the river originally flowed in heaven until Bhagiratha, one of the Seven Sages, asked the gods to allow the Ganges to flow on Earth.[55] His special request was granted because of his good karma and standing and as Shiva stood 'beneath the sacred waters of the Ganges' the river flowed down through his tresses where it divided into seven, forming the seven holy rivers.[56] As it poured down over the god's head, Ganga became caught in one of Shiva's dreadlocks that broke her fall to Earth and thereafter became his partner.[57] The Krittika are implicated in this legend because of their association with water, the number seven and their involvement in the ritual bathing festival of Kumbh Mela where they are celebrated once every twelve years at Allahabad, 'where the waters of the Ganges and Jumna combine.'[58] Whether this festival coincides with a particular conjunction of Mars (Karttikeya) and the Pleiades (Krittika), or the Pleiades and some other planet is not known, but it is highly likely there is some correlation between the two. Each day during this festival, hundreds of thousands of people flock to the banks of the Ganges to 'perform their ritual ablutions in its waters and sip from its life-giving stream,' says Raymond Hammer.[59] This can be from anywhere 'up to as many as 10 million people' who share in this communal water ritual.[60] The fact that the Pleiades are the primary stars that are celebrated at this festival suggests the Krittika are linked with the goddess Ganga, because of the significance of the Ganges as one of India's seven holy rivers. This, in turn, re-emphasises their status as water maidens, a common theme in much of the world's mythology on the Pleiades.

In addition to this water festival, the Pleiades are related to another well known Indian celebration, Diwali or the Feast of Lanterns.[61] Although it is held in honour of the goddess Lakshmi, there are more ancient ties to these stars, as will be seen. To begin with, this five-day festival that is a central feature of the Hindu calendar is held every year between late October and mid-November, which is known in India as 'the Pleiad month, Kartik,' after the son of the Pleiades.[62] This alone should indicate that the Krittika are intimately linked to this particular event, but their relationship goes even further. Diwali is an actual 'corruption of the Sanskrit word Dipawali,' says Brijendra Sharma in *Festivals of India*, 'which literally means a row of lights.'[63] Apart from its obvious meaning as symbolising stars glowing in the night skies, the Pleiades are specifically implicated because of their depiction in Hindu mythology and beliefs. For instance, the Indian poet of the fifth century CE, Kalidasa, described the Krittika as having 'the shape of a flame.'[64] Some writers have played down this portrayal as having any significance and say the Sisters were only depicted in this way because of their liaison with Agni, the god of fire, who desired them with a consuming intensity. This attempt to silence the voice of the Sisters by minimising their crucial role is an obvious example of the way in which patriarchy often erodes women's history and spirituality. The more ancient religious texts of the Vedas and the Purana, on the other hand, clearly state the Krittika are flames in their own right. They point out that the 'forty-nine original fires' of Hindu traditions include both the Krittika and the Rishis, along with Agni and several other Indian deities. Blavatsky suggests these forty-nine fires 'are an allusion to the forty-nine Manus, the seven rounds, and the seven times seven human cycles in each round on each globe.'[65]

Several writers have suggested the diverse feasts of lanterns in several Asian countries, including those of China and Japan, may also have originated from this earlier association of the Krittika.[66] The underlying belief common to all these traditions lies in the notion of a divine flame that 'represents God, the source of light and life' which must always be tended and never allowed to go out.[67] Hence there were seven vestal virgins to guard the eternal flame in Hestia's temple in ancient Rome.[68] In contemporary times, Zoroastrians do the same for their god, Ahura Mazda, according to the teachings of their prophet, Zarathustra.[69] Barbara Walker tells us in *The Woman's Dictionary of Symbols and Sacred Objects*, that this early connection of the Pleiades with lantern festivals may even be commemorated in the Jewish seven-branched candlestick, the menorah.[70] This sacred object is central to teachings of the mystic Tree of Life in the Kabbalah. Although the seven branches were in later times named after seven male archangels, Walker says they were originally given female names to represent the seven celestial spheres of which the totality 'stood for the Shekina, the female presence of

God.'[71] In addition to this divine flame having a 'distinct function and meaning in the words of the physical and the spiritual,' Madame Blavatsky reminds us that it also 'symbolises the *psychic* faculties.'[72]

Seven mother judges

As well as being portrayed as a flame, the Krittika are sometimes depicted as a razor or knife 'with a short handle.'[73] Once again the Jobes provide a linguistic explanation for this illustration. They suggest the name for the Sisters derives from the Sanskrit *kart*, which means 'to cut', although they remain not entirely convinced and so fail to see the hidden symbolism of the Pleiades contained in this very powerful image.[74] In fact, this linguistic and pictorial association goes to the very heart of many of our notions and perceptions of justice that ultimately derive from the idea of the Mother Goddess as the quintessential embodiment of fairness and equality. In her dual aspect as the goddess of fate, she must cut the threads of destiny equally and fairly as she cuts the umbilical cord in her role as the crone midwife, Hekat or Hecate.[75] It is therefore fitting that the fierce warrior goddesses Kali and Durga, who reflect different aspects of the Krittika, should brandish swords to sever heads and right wrongs to level and stabilise the world. Apart from this physical use of the sword, what might they represent emotionally, intellectually and spiritually? Barbara Walker suggests that razors, like knives and swords, epitomise 'the keenness of female judgement.'[76] The Krittika, she says, were the Seven Mothers who 'judged and approved candidates for the office of a sacrificial god, a solar hero destined to be slain, wounded in the side with a spear, and resurrected.'[77] Their cutting tools 'may have been the moon sickles signifying castration of the sacred king, in order that his life might be reborn.'[78]

This perception of the Pleiades as mother judges or female justices is similar to ancient Greek and Egyptian mythologies where the Goddess meted out punishment and decrees to wrongdoers. The archaic matri- archal image of Justice portrayed as a blindfolded woman balancing a set of scales still survives in the modern day Anglo-legal system today. This notion is explored in further detail in the Egyptian chapter on the

Pleiades, who were known as the Seven Judges of the goddess Hathor. It is from the Sanskrit Krittika that we get the Greek word *krittikos*, to 'judge', and also the English terms critic, critical, criticism and critique, which variously mean to analyse, to find fault with and to examine as any good judge is expected to do. Joseph Campbell offers further insight into this depiction of the Krittika as razors. In *The Power of Myth*, he refers to an image of the Buddha holding a sword above his head. Viewed in a similar but different vein as a sword of discernment, Campbell suggests it separates 'the merely temporal from the eternal'[79] and in this sense acts as another 'mask of eternity.'[80]

This has enormous implications for the Krittika and their connection with time, especially that of the goddess Kali, who represents an aspect of the Krittika as one of Agni's seven tongues of fire. Kali slices the monad in equal halves with her sword of time, as did the goddess Tiamat, who gave us the word 'diameter' to express this mathematical equation.[81] In doing so she separated the unit 1 from the cipher — the arithmetical representation for zero (the Cosmic Egg) to give us the number 10 or the Decad, whose expression in binary numbers forms the basic language of computers. This sacred number of the universe was so secret it found its expression and revelation in many arcane teachings throughout history. Even today its complete mystery is not fully comprehended.[82]

The begetting of Karttikeya by Agni and the Krittika

There are many different accounts that relate to Karttikeya, or Skanda's conception, as previously mentioned. In the *Mahabaharata*, for instance, Agni the god of fire 'agrees to beget' Skanda in the story of Svaha's dishonest impersonation of the sisters.[83] However, there is no mention of this trickery in the *Siva Purana* version where six of the sisters (except the eldest) are said to willingly engage in an illicit affair with Agni. At first glance this appears to be another version of the story, but it is in fact a variant of the cheating theme, as the sisters must appear willing to commit marital infidelity and thereby lend to the illusion if it is to work. Both tales are beautifully translated and recorded by Doniger O'Flaherty in *Hindu Myths*. In the story where Agni consents to participate in Karttikeya's conception, the *Mahabaharata* tells how Indra, the king of the gods, played a pivotal role in the circumstances leading up to the begetting of Karttikeya.[84] In this version, Indra desires a great general to lead his army, and by chance his search for a great military commander coincides with the related desire of his maternal cousin, Army of the Gods, to find a husband worthy of her status as 'the daughter of Prajaparti' (Brahma) and a goddess in her own right.[85] However, her desired qualities of a suitable husband far exceeded those of any ordinary man; apart from the usual virtues sought in a potential lifetime partner, he had to have great strength and heroic powers to conquer wicked gods and demons. In addition, he had to be 'pious and widely famed.'[86]

Indra was at a loss as to who could fit such enormous shoes of expectation until he witnessed a great event in the skies. On this auspicious occasion, Indra saw Soma (the Moon) 'entering the sun, the maker of the Day.'[87] On this 'day of the new moon', Agni also entered the Sun, along with Soma, carrying an offering to the gods.[88] Upon his re-emergence he materialised fire-born, ready to sire the woman's husband who would later become the leader of Indra's army. After completing a number of rituals, Agni then set out on his journey where he stumbled across the Krittika bathing in the waters (as in the opening quote of this chapter). It is at this juncture that Svaha, the daughter of Daksa (one of the Seven Sages), embarks on her deadly deception to seduce Agni by taking the form of six of the seven wives of the Seven Sages. In this version, Svaha first took the form of Siva, the wife of Angiras, a woman of virtue, beauty and merit.'[89] After coupling with Agni, and with 'his seed in her hand,'[90] she turned into a Garudi bird and flew up into the mountain where 'she threw the seed hastily into a golden pot.'[91] Upon her return, she then assumed the forms of the remaining Krittika to satisfy 'the desire of the Purifier.'[92] But no matter how hard she tried, Svaha 'could not take the divine form of Arundhati' because of her 'powerful asceticism and devotion to her husband.'[93] The legend states that sex took place six times and on each occasion Svaha threw Agni's seed into the same pot 'on the first day of the lunar fortnight'[94] where:

The seed shed there, full of energy, engendered a son who was honoured by the sages as Skanda because he had been shed (skannam). The Youth had six heads, twice that many ears, and twelve eyes, hands and feet; he had one neck and one body. On the second day he assumed a distinct form, on the third he became a child, and on the fourth he became Guha, with all his limbs developed. As he was enveloped in a great red cloud ablaze with lightning, he blowed like the rising sun in a great red cloud.[95]

In the version where the Sisters engage in consensual sex with Agni, there is a striking resemblance with the Bundjalung myth of Aboriginal Australia in the story of the Maimai who are equally desired and pursued by the fiery ancestor, Wurrunah. And like the Berai Berai who died pining for them, Agni succumbed to a similar fate, that of abandoning his body because of the unattainability of the Brahmin's wives. Here in the *Siva Purana* version the Krittika, like the Maimai, are said to be 'pained by the intense cold' of bathing and are therefore amiable to embrace the flames of Agni's fires 'in order to dispel their chill.'[96] Doniger O'Flaherty describes their sexual union that begat Karttikeya and the events that ensued:

All the tiny particles of his seed immediately entered their bodies through their hair follicles, and Agni was free of his feverish burning. Then Fire quickly vanished in the form of a flame and went happily back to his own world, thinking in his heart of Sankara and Brahma. All of the women but Arundhati became pregnant and they went home, tortured by the feverish burning and suffering because of Agni. When the Husbands saw the condition of their wives, they were quickly overcome by anger, and they took counsel together and abandoned their wives. When all the six women realized their transgression they were afflicted with great misery and their hearts were deeply troubled. The sages' wives left the seed of Siva in the form of an embryo on a peak of Himalaya, and then they were free of the feverish burning.[97]

But even the great mountain could not sustain Karttikeya, for the seed proved too hot and so she threw it into the river Ganges,

This Roman marble statue of Atlas holding the globe on his shoulders is known as the *Atlas Farnese*. It dates from c. 150 CE and is a late copy of a Greek original made c. 200 BCE (artist unknown). It is housed in the Museo Archeologico Nazionale, Naples, Italy.
Credit: akg-images, London

El Castillo pyramid at Chichén Itzá in the Yucatan, Mexico. Each year thousands flock to this site on 20 May at the Spring equinox to view the famous descent of the serpent caused by the Sun's shadow on the steps. On 20 May 2012 there will be a conjunction of the Sun, Moon and the Pleiades accompanied by a solar eclipse that will sweep across central and western North America.
Credit: Susan Hawthorne

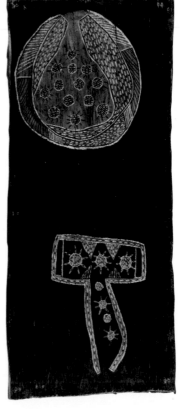

Minimini Mararika, Australia, 1904-1972, *Orion and the Pleiades*, 1948, Umbakumba, Groote Eylandt, Northern Territory, natural pigments on eucalyptus bark, 77.0 x 32.5 cm (irreg); South Australian Grant 1957. This traditional Aboriginal bark painting depicts the Burumburumrunja fishermen (Orion) and their wives, the Wutarinja (Pleiades).
Credit: Art Gallery of South Australia

Karttikeya/Skanda/Murugan, son of the Pleiades in Hindu mythology and the equivalent of Mars/Ares as the Boy God of War.
Credit: Murugan Bhakti Association, Sri Lanka

Hylas and the Nymphs by John William Waterhouse. Although the artist refers to them as 'nymphs', the young women are very likely to be the Hyades, half-sisters to the Pleiades through their father Atlas, named in honour of their brother Hylas. It reveals their mythic association with water.
Credit: Manchester City Art Gallery

Ice Maidens by Ainslie Roberts. A beautifully illustrated (non-Indigenous Australian) interpretation of the Pleiades in Aboriginal mythology. Again, depicting the Pleiades in water form.
Credit: Viscopy, Sydney

The Legend of Devils Tower by H. Collins. This painting is an interpretation of the Sioux legend of the Seven Star Girls that are chased by a bear. Credit: National Parks Service, USA

Dance of the Pleiades by Elihu Vedder
Credit: Metropolitan Museum of Art, New York

Pleiades star Merope enshrouded in nebula.
Credit: Ted Simon of the University of Hawaii
and the Hubble Heritage Team.

Kahuna Kilo Hoku (Polynesian navigator) by Herb Kane. The stars of
the Pleiades played a key role in the peopling of the Pacific.
Credit: Hawaiian Paradise Trading Company (Hawaiian Eyes Images)

Carved ivory statue of Schichi Fujukin in their treasure ship.
Although they are referred to as the 'Seven Gods of Happiness'
in Japan, one of them is the Sea Goddess, Benten.
Credit: Victoria and Albert Museum, London

although some say it was the ocean where Karttikeya was born.[98] The *Puranas* state this occurred 'on the sixth lunar day in the bright half of the month when the moon is in the constellation Deer's Head' (Capricorn).[99] And so it came to pass that news of Skanda's, or Karttikeya's phenomenal birth, were 'accompanied by rumours that six of the Rishi's wives were his mother.'[100] Despite their protestations and Svaha's own confession, six of the Seven Sages promptly 'divorced their wives and sent them away.'[101] Once spurned, the Krittika took to their familiar position in the skies close to the ecliptic; all but one, the eldest sister Arundhati who, as we know, went to live with her husband in the constellation of the Great Bear.[102]

Astronomical symbolism of Karttikeya

In the tradition of *Hamlet's Mill*, whose basic premise is that myths are astronomically coded, Doniger O'Flaherty acknowledges these factors are central to these stories. For instance, she notes that Karttikeya's 'unusual features may represent the birth of the year (Skanda) with its six seasons (six heads) during the new moon at the spring equinox, when the sun is in the Pleiades.'[103] The mythical parlance for this celestial conjunction is referred to as 'when Agni is *in* the Krittika,' which is another reference to sexual intercourse.[104] This is further accentuated by the reference to *six* mothers instead of seven and the fact that sex takes place six times. In theosophical terms, therefore, Karttikeya's conception embodies the theosophical principle of the sevening of creation, including the shedding of six rounds of the creator's seven skins as implied by his name Skanda, which derives from the Sanskrit *skannam* to shed.[105] Another interpretation of Skanda's name means, quite literally, the 'jumping one' or the 'hopping one', a typical action of someone jumping out of their skin.[106] The symbolism is made complete with his six heads that represent six ages, six periods, rounds or counts of humanity. The relationship between Agni, Indra (as an aspect of Brahma) and Svaha relate to the notion of the three fires of Hindu beliefs — solar, electrical, and fire produced by friction —that together make up the 'creative spark, or germ' that creates life.[107]

Although some writers have commented on the 'multiple wombs, multiple mothers' aspect of Karttikeya or Skanda's birth and conception, no further explanation or analysis is proffered.[108] Giorgio De Santillana and Hertha Von Dechend merely comment on the rarity of the mythological distinction of children being born to several mothers,[109] whereas Doniger O'Flaherty observes a similar theme in the birth of Krishna. On the other hand, she does acknowledge linguistic complexities may hinder any explication. For instance, she points out that the single Sanskrit word *garbha* in Hindu texts 'may refer to womb, mother, embryo, or offspring, depending upon the context.'[110] Given that recent genetic findings were not available to these authors at the time of writing — such as that relating to the 'Seven Clan Mothers' of the world — it is not surprising that these observations were not fully explored. In any event, interesting parallels may be drawn between these mythological statements, recent genetic discoveries, and the theosophical theories of Madame Blavatsky.

Doniger O'Flaherty makes one other interesting but major observation of the Krittika; notably their 'two contrasting aspects' as both 'loving nurses' and 'hideous mothers' who attempt to kill their son Skanda.[111] In the latter aspect they represent Kali, the goddess of death and equally 'the dark side of Gauri who is Skanda's other mother.'[112] Although Doniger O'Flaherty notes this contrast, she does not explore this further. Perhaps she did not realise that this relationship between the Krittika and the black goddess of death and destruction go to the very heart of the occult mysteries connected to the great cycles of time, including our present age, the Kali Yuga. For it is the Seven Rishis and their wives, the Krittika, 'who mark the time and periods of Kali-yuga, the age of sin and sorrow.'[113] The reference to the bird goddess Gauri as Karttikeya's 'other mother' emphasises in turn the avian nature of the Krittika and reveals their status as bird goddesses of the Pleiades. Agni's relationship with the male Indian firebird, Garudi that carries the seed or ambrosia to the gods provides another link, as does Karttikeya, who is often depicted riding on a peacock. This mythical bird is commemorated in Indonesia's state airline Garuda, which has made it their official logo.

Other astronomy references are also present in these legends. The representation of binary stars as metaphors for our intimate personal relationships are suggested by the marital status of Arundhati and Vasistha who reside together among the stars of Ursa Major, and vice versa. It certainly gives added meaning to the Greek concept of our significant other being described as our 'other half.' The fact that the Pleiades are situated near the ecliptic along with the band of zodiac is not lost to several writers, who suggest it is one reason why they are regarded as important to humanity. Once again, Blavatsky and some other writers see beyond the superficiality of this observation and suggest much deeper, esoteric reasons for their location. Apart from assuming the form of the Sisters, Svaha's true stellar identity is especially revealing for she is none other than 'the star Zeta Tauri, which marks the tip of one of the horns of Taurus the bull.'[114] In this regard she resembles Hathor, the Egyptian cow goddess whose son Horus (Karttikeya) lives with his mother in her celestial mansion. As the boy god, Karttikeya — like the boy god Horus in ancient Egyptian folklore — is a boy child, the eternal Peter Pan of Hindu mythology, who comes to visit his mother when Mars is conjunct the Pleiades, the warrior knight Lancelot would often visit his mother and her priestesses on the sacred Isle of Avalon.[115]

Karttikeya's warrior status is affirmed by the description of him being enveloped in a 'great red cloud.'[116] This not only links him with the blood of life coursing through our veins but with blood which is shed in battle; hence he is identified with the red planet, Mars the Roman god of war and that of his Greek counterpart, Ares.[117] The fiery energies and equally fiery temperaments of both gods are legendary. The timing of Karttikeya's birth and investiture is also suggestive of astronomical events, say the authors of *Hamlet's Mill.* Hence the boy god's installation as commander and general before the assembled gods in the *Mahabaharata* represents a conjunction of Mars and several other planets.[118] This is evidenced by a key phrase in the Hindu text, which stresses that the gods 'poured water upon Skanda, even as the gods had poured water on the head of Varuna, the lord of waters, for investing him with dominion.'[119] This investiture, says De

Santillano and Von Dechend, 'took place at the beginning of the Krita Yuga, the Golden Age.'[120] Here we see the interplay of opposite elements of fire (male) and water (female) that are central to creation. The *Rig Veda* states that Agni the god of fire was born from fire and water, as was his son Karttikeya.[121] Fire and water feature quite prominently in many creation stories as the very basis of life. This is hardly surprising as we carry these very elements — fire and water — inside our bodies, along with the other two — air and earth. Specifically, these are identified as electricity (fire) in the nerve impulses, water (in our blood and cells), air (breath) and earth (minerals) in our bodies. Our myths reflect the very chemistry of our creation.

Kalacakra, wheel of life

On a macrocosmic level the Krittika are implicated once again in the keeping of time and in the destiny of human beings through their connection with the Hindu wheel of life, the Kalacakra,[122] and in particular with our current age the Kali Yuga.[123] Essentially, Hinduism views time (and no-time) as a series of unending cycles of life, death and rebirth. The turning wheel of the Kalacakra, the wheel of life, represents these phases. Within these cycles, the lives of human beings, gods and goddesses and multiple universes are measured and amplified into what Mircea Eliade refers to in *Images and Symbols* as 'terrifying proportions.'[124] The reference to 'One Day in the life of Brahma', for instance, would require some fifteen digits or more to express mathematically. Hindus therefore speak of time in symbolic terms and mythic imagery to dispense with such complex computations in everyday general speech. Even the game of dice embodies Hindu concepts of time and fate, for the smallest of cycles and the basic unit of measurement on the Kalacakra — the yuga or world ages, are named after its throws.[125] These are not ages in the sense of historical centuries, as we may understand that term, but are much longer periods of time, marked by certain characteristics that impact on our collective unconscious and individual lives, including our environment. Each age, says Eliade, 'is preceded by a *dawn* and followed by a *dusk*, which fill the intervals between the successive ages.'[126] Unlike our earthly dawn and dusk, these intervals are much longer and occur over a period of some hundreds of years. Curiously, although there are six sides to a die, there are only four ages that make up a complete cycle known as a mayayuga or mahayuga.

One explanation for this fourfold division is that it implies the Hindu caste system. Surprisingly, these ages are not of equal duration as one might expect. They do, however, follow each other sequentially in accordance with their length. Hence the cycle begins with the longest age *Krita Yuga*, followed by the second longest *Treta Yuga*, then the third longest *Dvapara Yuga* and finally by the shortest but no less momentous age of *Kali Yuga*. To give an idea of this time frame, the respective ages of each yuga in order of sequence are 1,728,000 years, 1,296,000 years, 864,000 years and 432,000 years. However, Hindu calculations of time are based on whether divine or human years are counted; thus in divine years each yuga translates to 4,000, 3,000, 2,000 and 1,000 years respectively, which are named after the throws of dice. Numerologically these numbers add up to ten, the sign of completion. The cycle, however, does not end there for after a mayayuga is completed it dissolves into a mini-dissolution known as a pralaya only to continue counting. In addition to differentiating between divine and human years one must take into account the number of years of each dawn and dusk period in each age. Thus one mayayuga consists of 12,000 divine years or 4,320,000 human years.

Yet even this cycle is diminutive in comparison to others on the Kalacakra as the wheel continues to turn and begins to clock up larger and vaster cycles. A thousand mayayuga for instance, make up a kalpa, which is equal to 'one day in the life of Brahma, another kalpa to one night.'[127] Fourteen kalpa make up one manvantara, that ends in a great dissolution known as the mahapralaya, only to recommence the cycle all over again! From our limited human perspective these vast cycles of time may appear to be incomprehensible, but as Eliade points out, all we need to remember from this vast array of numbers is 'the cyclic character of cosmic time.'[128] Ultimately we are left with infinite cycles of life, death and rebirth. From the perspective of the universe, which knows no time, whole galaxies and worlds are created and destroyed in the mere blink of the eye or stir of the Creator. What these phases tell us is that the universe and all within are infinite, eternal beings. Scientists are coming more and more to this realisation through their cosmological theories that acknowledge how the universe is continuing to expand and that stars are created in cycles of life, death and rebirth. Those who are more spiritually inclined might say that it is all part of the cosmic plan of the Divine Mind, the seemingly Incomprehensible Mind of the Universe that is One and Many at the same time.

The reason for the disparity of time among the four world ages or yuga is a direct result of apportionment of dharma, which in turn affects the amount of light and therefore knowledge accompanying the ages. The basic idea is that the cycle begins with complete dharma, light and wisdom that become increasingly diminished through the ages. Thus the Krita Yuga, the 'Golden Age' or 'Age of Perfection' named for the number four die (the winning throw because it represents totality), is the first age of the cycle because of the belief that dharma 'is observed in its entirety.'[129] The Treta Yuga or 'Age of Ritual' is named for the number three die, for human beings are considered no longer observing 'more than three-quarters of the dharma.'[130] With the succeeding age, the Dvapara Yuga, represented by the number two die, 'only half of the dharma' is operating.[131] This translates to the beginning of human corruption on Earth, which becomes complete during the Kali Yuga, the 'Black Age' or 'Age of Sin and Sorrow', signified by the number one die where 'only a quarter of the dharma remains.'[132] This lessening observation of dharma by human beings and diminished light and wisdom may account for the following Purana description of the fourth and final age of the Kali Yuga:

> According to the Vishnu Purana (IV, 24) the syndrome of kali yuga is marked by the fact that it is the only age in which property alone confers social rank: wealth becomes the only motive of the virtues, passion and lust the only bonds between the married, falsehood and deception the first condition of success in life, sexuality the sole

means of enjoyment, while external, merely ritualistic religion is confused with spirituality.[133]

This description of the Kali Yuga is perfectly apt to describe our present situation and worldly disposition. The good news, however, is that according to Hindu calculations, we are presently living during its dusk period. In other words, we are nearing its completion and are soon to re-enter the dawn of the Golden Age of the Krita Yuga. What this means spiritually is to have a deeper appreciation and understanding of the Kali Yuga; where it fits within the revolving wheel of life and its relationship to other time series which make up the Kalacakra, including the zodiac ages and precession. One way to comprehend the complexity of this multilayered, multifaceted concept of time is to imagine a series of wheels within wheels, or cycles within cycles, all turning simultaneously. As you start to visualise the concept it becomes easy to appreciate the biblical tale of the prophet Ezekiel who had an alarming vision in 592 BCE of a huge moving, burning monstrosity that he described as chariots of fire. This 'chariot' consisted of a huge, turning wheel comprising a series of smaller wheels turning within it.

From this description we can visualise a huge cosmic orrery, resembling that depicted in the 2001 sci-fi movie *Lara Croft Tomb Raider 1*, starring Angelina Jolie as the intrepid British aristocratic heroine. Ezekiel's chariot, however, appeared to him as a monster, larger than life, spewing fire, chortling with smoke and shaking the ground like thunder, which made it all the more frightening to the prophet. Nonetheless it is an interesting analogy, for the driver at the reigns of this chariot of time is none other than the Black Goddess Kali, whose fearsome face engenders equal terror and revulsion by those who see her for the first time or fail to acknowledge her divine presence in fires, in death and destruction or in dreams and visions like Ezekiel.[134]

The end of the Kali Yuga and the dawn of a New Age

Hindu calculations reveal the Kali Yuga commenced in 3100 BCE and will end on 23 December 2012.[135] Although it is the shortest of the yugas, it is a time of enormous change, uncertainty, and confusion primarily because it is the Age of Darkness. Named after the number one pip in the game of dice, the Kali is considered the 'losing throw' because it personifies 'an evil spirit.'[136] The association of the Krittika with the game of dice not only confirms their connection with fate and destiny, but also suggests that the goddess Kali is another aspect of the sevenfold deity. In addition to the personification of evil, the term Kali signifies dispute and discord, says Eliade.[137] A more positive view suggests this is no evil spirit as such but merely the workings of the goddess as the Mistress of Time. She is the hard, efficient taskmaster who cracks her whip with precision and exactitude, for the divisions of time, the ages of the world and the turning of the Great Wheel depend on her self-discipline and reliability as the Charioteer of the Kalacakra.[138]

Kali's relationship with darkness is twofold, says Mircea Eliade. Firstly, the transition from one age to another 'takes place during a twilight interval which marks a decline even within each yuga, every one of them terminating in a phase of darkness.'[139] As we approach 'the end of the cycle, the fourth and last yuga, the darkness deepens.'[140] Thus our current age, the Kali Yuga, is considered the darkest of all ages. Secondly, the fourth age has become linked with Kali the Black through 'a play on words,' says Eliade.[141] 'Time is black,' he says, because 'it is irrational, hard and pitiless.'[142] Thus Kali 'not only becomes the black but also the personification of time.'[143] This may explain the symbolism of the black bear that chases the seven star girls in the Native American Kiowa myth in the previous chapter on Devils Tower. Perhaps the she-bear in that story is another incarnation of the black goddess Kali as harbinger of the dawning of a New Age.

This interplay between the opposing forces of light and darkness, good and evil, provide the perfect imagery for the dawning of a New Age, whether it is the Age of the Fifth Sun, the Krita Yuga or the much anticipated Age of Aquarius. Several writers, including Graham Hancock, have noted that the dates of the most recent Kali Yuga happen to coincide with the beginning and end of the Mayan calendar, which records the birth and death of the Fifth Sun, our current epoch.[144] Others like John Major Jenkins disagree with these dates and are keen to separate the Age of Aquarius from the Mayan end-date, on the basis that most astrologers favour the twenty-third century as its commencement date, which is expected to occur 'over 200 years *after* the Maya end-date.'[145] Although the Hindu and Mayan calendars have identified specific dates when we enter the New Ages, this need not preclude the Age of Aquarius from having an associative effect, for all of these cycles are identified as a time of enormous spiritual growth and enlightenment. Furthermore, the notion of periods of dawn and dusk that accompany the various yugas also applies to the zodiac ages, so even though our actual entry into the Age of Aquarius is not expected to occur for another two hundred years or so, we can correctly speak of living in its dawn. What this means is that the experience can be just as spiritually exhilarating, influential and transformational as when entering it officially for the first time or when completely immersed in it. Ultimately, what matters more than precise coordinating dates is the wisdom we will gain during these times.

A more easy way of comprehending how these cycles interact with one another is to visualise a cosmic orrery and the mechanics of wheels within wheels and cycles within cycles. As these wheels start to turn simultaneously we can begin to see and appreciate that there are certain points along the wheels where all the cycles intersect with one another in time and space. From a spiritual perspective these shared intersection points mark important periods in human evolution and consciousness. The language of 'dawn' and 'dusk' of the various ages is particularly useful as highlighting these junctions. Even more so they are symbolic of this evolving consciousness, whereby we are spiritually awake at dawn and spiritually asleep at dusk. In *The Biggest Secret*, David Icke tells us that the Ancients divided the mayayuga into two halves to highlight two separate cycles of spiritual awareness and spiritual amnesia. So as our planet faces the centre of our galaxy, it 'is bathed in positive light for 13,000 years' but when tilted away from the centre, it 'moves into darkness for the next 13,000 years.'[146] This phenomenon has an enormous effect on our lives, says Icke, and as we approach the dawning of a New Age, 'there is a rapid global spiritual awakening and incredible events await us in the next few years' as we re-enter the light once again.[147] Bob Frissell narrows these periods to just 'two points located 900 years on either side where we fall asleep and where we wake up.'[148] It is unclear whether both writers are counting in divine or human years, but presumably it is the latter.

Barbara Hand Clow, on the other hand, relates these periods of light and darkness specifically to the zodiac cycle. She says we are more spiritually enlightened as we travel in the light of the 'photon band', which she claims only occurs during the two ages of Leo and Aquarius. At all other times we travel through the long 'Galactic Night' of diminished awareness 'during all the other zodiacal polarities of the Great Ages.'[149] Once again, the exact points in time when we supposedly enter these periods of light and darkness are not precise; rather we should become aware of their significance and impact upon our spiritual consciousness and growth. Whether it is the promised Age of Aquarius, the Krita Yuga or the Age of the Fifth Sun, the common message seems to be that we are now at a turning point in our history where we are beginning to move back toward the light to become spiritually reawakened.

Esoteric symbolism of the world-supporting tortoise

One of the oldest symbols for the various cycles of time is the cross and circle. This includes its many variants such as the Egyptian ankh, the ansated cross of medieval Christianity and the Hindu swastika. The circle represents the wheel of time, and the cross with its four quarters represents the four cardinal points — north, south, east and west, the solstices and equinoxes as well as the four yugas. In many Eastern traditions these cycles are embodied in the turtle or tortoise. Thus in Hinduism, the turtle represents the second incarnation of their god Vishnu the Pervader, the third member of the divine trinity. Vishnu's earthly manifestations are known as avatars that arrive 'to prevent a great evil, or to effect good upon the Earth.'[150] In his incarnation as Kurma or Sisumara the turtle, Vishnu 'helped create the world by supporting it on his back.'[151] Because of this selfless act, Kurma was affectionately called Old Man Turtle and he is regarded as 'the father of all creatures and the mate of the primal goddess of the watery deep.'[152] The image of the world-supporting tortoise is one of the oldest eastern motifs, says Barbara Walker. In *The Woman's Dictionary of Symbols and Sacred Objects* she dubs the icon an 'animal Atlas' that underlies 'all the abysses of the sea and the roots of the mountain.'[153]

This notion of the Earth being supported on the back of a giant turtle is not only reminiscent of the Atlas myth, but is integral to it. Atlas is said to have descended down into 'the depths of Tartarus' to join his brother, or else to remain standing upon it while shouldering the heavens.[154] It is interesting that the Greek god of the underground was identified with the turtle whose realm, Tartarus, is derived from *tartaruga*, which means turtle or tortoise.[155] The memory of Sisumara supporting the world on his back is forever imprinted on the back of all turtles, where the cross, stars, planets and constellations can still be seen to this day.[156] The *Bhagavata Purana* refers to this celestial turtle map, where it describes the physical attributes of the turtle and its relationship to the cosmos. Its circular body represents the night skies, where the head faces south and the tail faces north. At the extremity of its tail lies the ex-Pole star, Dhruva (Thuban in Draco); along its tail are the Prajarparti, Agni, Indra and Dharma, and across its side are the Seven Rishis.[157] This image of the turtle, says Blavatsky, is 'the first and earliest cross and circle.'[158] It denotes 'the external circle of boundless time, *Kala*, on whose plane lies crossways all the gods, creatures, and creations born in space and time' and who 'all die at the Mahapralaya.'[159]

Among many Indigenous peoples the turtle or tortoise is much revered. In some areas of the Pacific, such as the Tuamotus in French Polynesia and Pukapuka in the northern Cook Islands 'there was a symbolic association between the Pleiades and turtles,' say Patrick Kirch and Roger Green in *Hawaiki, Ancestral Polynesia*.[160] To some extent this was connected to their Pleiades year and the reproductive cycle of the green sea turtles, 'which lay . . . eggs in the sandy beaches of Polynesian islands between June to September, a time when the star-cluster Pleiades becomes visible in the early morning sky before sunrise.'[161] Curiously, Emory reports that at Vahitahi, 'Honu (Turtle) was the child of Takero (Belt of Orion) and Matariki (Pleiades).'[162] Perhaps more revealing is the fact that 'Matariki, which stands for a female turtle, is the name of the Pleiades.'[163] In Native American teachings the turtle not only represents the Americas (both North and South) that are referred to as Turtle Island, but also the entire planet. As such, it is 'the oldest symbol for Planet Earth,' says Jamie Sams and David Carson, the creators of *Medicine Cards*.[164] Turtle signifies the Great Mother that feeds our spirit, and clothes our hearts.[165]

Like Hindus, Native Americans see time inscribed on the turtle's back, where the divisional markings represent their thirteen moons of the year.[166] This notion of the world-supporting turtle (like many other Native American cultural and spiritual beliefs) accompanied Native peoples as they migrated to the Americas from Siberia or north-central Asia, through the Bering Strait between Alaska and Siberia some twelve to thirteen thousand years ago.[167] What was once an anthropological hypothesis has since been affirmed by the recent genetic research findings of geneticist Bryan Sykes in *The Seven Daughters of Eve*.[168]

In Bandaiyan (Australia) in the Kimberley region, the great Wandjina creator beings left the imprint of their images in the skeletons of freshwater turtles to remind Aboriginal people that they were created from water.[169] Let us not forget either that the turtle was especially sacred to the Greek god Hermes, who fashioned the first lyre from a turtle shell so it could magically echo 'the harmonies of the spheres.'[170] As Hermes Trismegistus of the alchemists, he was known as an underground god who 'was called *spiritus tartari*, spirit of the underworld, whose essence was tartaric acid,' says Walker.[171] In China and Japan the world-supporting turtle is seen in many sculptures adorning renowned buildings such as in the Forbidden City in Beijing, where it symbolises eternal life and therefore acts as another mask of eternity.

Sixes and sevens, the mystery of the Pleiades

The interplay between six and seven stars of the Pleiades is played out in Hindu mythology where the Sisters are sometimes spoken of as

Karttikeya's six wet nurses or six mothers, and at other times as the seven wives of the Seven Rishis, or as the seven mothers of the world. Its symbolism is contained in the game of dice with its six sides, but where the opposite numbers amount to seven. This constant switching between six and seven in the legends might at first appear confusing or contradictory but a deeper examination reveals otherwise. What we are primarily dealing with are symbols and once the special language of symbolism is understood, these so-called 'contradictions' begin to make sense. Essentially, the interchange between six and seven, says Madame Blavatsky, is central to the 'occult significance of the Pleiades,' especially in relation to that of 'the six present, and the seventh *hidden*.'[172] As well as being cryptic references to the hidden face of divinity, this interplay speaks to the various cycles of time, including the prophetic ones. One of these, the 'Cycle of Naros' is particularly significant in Hindu traditions, to which Karttikeya the son of the Krittika holds the key. 'Composed of 600, 666, and 777 years,' its meaning differs depending on the context as to whether 'solar, lunar, divine or mortal years are counted.'[173] The six visible, or the seven actual Sisters are needed to complete 'this most secret and mysterious of all the astronomical and religious symbols,' says Blavatsky.[174]

Within this context, the symbolism surrounding Karttikeya and his mothers, the Krittika, become more comprehensible. It explains why Karttikeya is sometimes pictured as wearing 'a six-rayed star (double triangle), a swastika (the Wheel of Time), and a six and occasionally seven-pointed crown.'[175] It also explains why at other times he appears 'old as a Kumara, an ascetic' and depicted 'with six heads — one for each century of the Naros', to commemorate a particular event.[176] When a new sign was needed to denote another event such as the conjunction of Mars (Karttikeya) and the Pleiades (Krittika) he appears 'riding on a peacock', which symbolises 'his female aspect.'[177] This was the bird sacred to the goddess Sarasvati and another symbol, perhaps, of the Krittika as bird goddesses. Hindus regard the peacock as the 'bird of wisdom and occult knowledge' for it carries the All-Seeing Eye of the Mother Goddess and her starry skies on its feathers.[178]

The peacock's tail is considered especially sacred because 'it represents the sidereal heavens' upon which the twelve signs of the Zodiac are hidden on its body, and because of this it is called 'Dwadasa Kara ('twelve-headed') and Dwaadasaksha ('twelve-eyed').'[179]

Blavatsky refers to the peacock as the 'Hindu Phoenix' because she claims the legendary bird of Greek mythology was well connected 'with the 600 years of Naros.'[180] This cycle plays a vital role in St John's Revelation, and Blavatsky believes that one has to be initiated into the mysteries to comprehend its true meaning. 'It is the truly Apocalyptic Cycle,' she says, 'yet in none of the numerous speculations about it have we found anything but a few approximate truths, because of its being of various lengths and relating to various pre-historic events.'[181] Ultimately for Blavatsky, the Lost Pleiad aspect is not a figment of anyone's imagination. She believes that there were once in the skies above India — at the very least — seven visible stars of the Pleiades. Rather than offer a mythical explanation for the disappearance of one of the Sisters, as other writers have done, she offers an alternative view, arguing that the six and seven aspects are very much a part of the overall story of the Pleiades, of which the lost or missing sister is an integral component to its mystery. At the heart of this mystery lies the hidden face of divinity that controls the various cycles of time, the pulse and the heartbeat of the universe.

Another way of understanding these cryptic messages is through the mystical knowledge of the Kabbalah and the principle of the hidden name of God, the Tetragrammaton whose four-lettered sacred title is written as YHVH.[182] The Book of Zohar says, 'We six are lights shine forth from a seventh, thou art the seventh light.'[183] In other words the six are the *parts*, the seventh is the *whole*. The seventh light represents God and 'his seven companions,' says Rabbi Abba, which he calls 'the Eyes of Tetragrammaton.'[184] Tetragrammaton *is* Brahma Prajaparti, says Blavatsky. He who 'is re-born in the Seven Rishis, his Manasaputras, mind-born sons, who became 9, 21 and so on, who are all said to be born from various parts of Brahma.'[185] Given the patriarchal usurpation of female deities throughout history in various cultures around the world, there is no doubt that

Tetragrammaton is both the Krittika and the Saptarshis. For in all cosmologies the supreme creator being has always been bisexual and androgynous, from Bandaiyan to Purusha to Shekinah, the true face of Tetragrammaton and the synthesis of Hokmah (male) and Binah (female). This quote applies to the Krittika as the six and seventh aspect of the divine mysteries just as much as their former husbands, the Saptarshis. For all of creation is male and female, yin and yang, negative and positive, which represents the universal feminine and universal masculine energies of the All-Mother and the All-Father of the caduceus — there cannot be one without the other. Therefore the Seven Rishis, who are said to be responsible for marking time, including the Kali Yuga, can only do so with the assistance and cooperation of their seven wives, the Krittika — the Seven Sisters of the Pleiades.

Notes

1. Doniger O'Flaherty, *Hindu Myths*, pp. 108–109.
2. Jobes and Jobes, *Outer Space*, pp. 339–40. They list the crew of seven at p. 262 as Manu or Vashishtha, a form of Brahma (Zeta Ursae Majoris) and his six sons — Kratu (Alpha Ursae Majoris), Pulaha (Beta Ursae Majoris), Pulastya (Gamma Ursae Majoris), Atri (Delta Ursae Majoris), Angiras (Epsilon Ursae Majoris) and Marici (Eta Ursae Majoris).
3. Jobes and Jobes, *Outer Space*, p. 262, and Blavatsky, *The Secret Doctrine*, vol. 2, p. 139.
4. Blavatsky, *The Secret Doctrine*, vol. 2, p. 551.
5. Allen, *Star Names*, p. 404 and Jobes and Jobes, *Outer Space*, p. 294.
6. Blavatsky, *The Secret Doctrine*, vol. 2, p. 176.
7. Jobes and Jobes, *Outer Space*, p. 262.
8. Allen, *Star Names*, p. 404 and Jobes and Jobes, *Outer Space*, p. 294.
9. Blavatsky, *The Secret Doctrine*, vol. 2, p. 549.
10. Krupp, *Beyond the Blue Horizon*, p. 234.
11. Jobes and Jobes, *Outer Space*, p. 262.
12. Ibid.
13. Krupp, *Beyond the Blue Horizon*, p. 234.

14. Blavatsky, *The Secret Doctrine*, vol. 2, p. 768.
15. Krupp, *Beyond the Blue Horizon*, p. 232.
16. Ibid, p. 240.
17. Graves, *The Greek Myths*, p. 84.
18. Burnham et al, *Astronomy: The Definitive Guide*, p. 318.
19. Lawlor, *Voices of the First Day*, p. 92.
20. Krupp, *Beyond the Blue Horizon*, p. 250.
21. Ibid.
22. Doniger O'Flaherty, *Hindu Myths*, p. 211.
23. Ibid.
24. Ibid.
25. Allen, *Star Names*, p. 384.
26. Doniger O'Flaherty, *Hindu Myths*, p. 30. See the short tale on Rohini in Jobes and Jobes, *Outer Space*, pp. 294–95.
27. Ibid, p. 211.
28. Ibid.
29. Spence, *Egypt*, p. 65.
30. Eliade, *Images and Symbols*, p. 63.
31. Krupp, *Beyond the Blue Horizon*, p. 250.
32. Ibid.
33. Allen, *Star Names*, p. 404 and Jobes and Jobes, *Outer Space*, p. 294.
34. Bowker, *World Religions*, p. 52. See also Walker, *The Woman's Dictionary of Symbols and Sacred Objects*, p. 340.
35. Allen, *Star Names*, p. 404.
36. Ibid, p. 403.
37. Blavatsky, *The Secret Doctrine*, vol. 2, pp. 549 and 551. Also *Outer Space* by Jobes and Jobes, p. 339.
38. Blavatsky, *The Secret Doctrine*, vol. 2, p. 549.
39. Ibid.
40. Ibid.
41. Blavatsky, *The Secret Doctrine*, vol. 2, pp. 382–83.
42. Ibid, p. 549.
43. Bowker, *World Religions*, p. 28.
44. Ibid, p. 23.
45. Krupp, *Beyond the Blue Horizon*, p. 248.
46. De Santillana and Von Dechend, *Hamlet's Mill*, p. 157.
47. Ibid. 385.
48. Ibid.
49. Blavatsky, *The Secret Doctrine*, vol. 2, p. 550.
50. Ibid, p. 549.
51. Bowker, *World Religions*, p. 25.
52. Blavatsky, *The Secret Doctrine*, vol. 2, p. 443. Blavatsky describes this tongue as a 'black flickering flame.'

53. Eliade, *Images and Symbols*, pp. 64–65.
54. Ibid.
55. Bowker, *World Religions*, p. 34.
56. Ibid, p. 22.
57. Ibid, pp. 23 and 34.
58. Hammer, Raymond, 'The Eternal Teaching: Hinduism' in *The World's Religions*, Pierce Beaver (ed.), p. 171.
59. Ibid.
60. Ibid.
61. Bowker, *World Religions*, p. 38.
62. Jobes and Jobes, *Outer Space*, pp. 339–40.
63. Sharma, *Festivals of India*, p. 118.
64. Jobes and Jobes, *Outer Space*, p. 339 and Krupp, *Beyond the Blue Horizon*, p. 250.
65. Blavatsky, *The Secret Doctrine*, vol. 2, p. 617.
66. See Allen, *Star Names*, p. 39, Jobes and Jobes, *Outer Space*, pp. 339–40 and Tyler Olcott, *Starlore of All Ages*, p. 411.
67. Bowker, *World Religions*, p. 13.
68. Walker, *The Woman's Dictionary of Symbols and Sacred Objects*, p. 226.
69. Bowker, *World Religions*, p. 13.
70. Walker, *The Woman's Dictionary of Symbols and Sacred Objects*, pp. 97–98.
71. Ibid.
72. Blavatsky, *The Secret Doctrine*, vol. 1, p. 360.
73. Allen, *Star Names*, p. 393 and Jobes and Jobes, *Outer Space*, p. 340.
74. Jobes and Jobes, *Outer Space*, p. 340.
75. Walker, *The Woman's Dictionary of Symbols and Sacred Objects*, p. 77.
76. Ibid, p. 151.
77. Ibid.
78. Ibid.
79. Campbell, *The Power of Myth*, p. 226.
80. Ibid, p. xvii.
81. Walker, *The Woman's Dictionary of Symbols and Sacred Objects*, p. 351.
82. Blavatsky, *The Secret Doctrine*, vol. 1, p. 360. Blavatsky says this split also symbolises the separation of the sexes from the androgynous third root race.
83. Doniger O'Flaherty, *Hindu Myths*, p. 104.
84. Ibid, pp. 105–15. See Indra's role at pp. 105–108.
85. Ibid, p. 106.
86. Ibid, p. 107.
87. Ibid.
88. Ibid.
89. Ibid, p. 109.
90. Ibid, p. 110.

91. Ibid.
92. Ibid.
93. Ibid.
94. Ibid.
95. Ibid.
96. Ibid, p. 167.
97. Ibid.
98. Ibid.
99. Ibid, p. 168.
100. Krupp, *Beyond the Blue Horizon*, p. 248.
101. Ibid.
102. Ibid.
103. Doniger O'Flaherty, *Hindu Myths*, p. 104.
104. Ibid.
105. Ibid, p. 110.
106. De Santillana and Von Dechend, *Hamlet's Mill*, p. 157.
107. Blavatsky, *The Secret Doctrine*, vol. 2, p. 247.
108. Doniger O'Flaherty, *Hindu Myths*, p. 205.
109. De Santillana and Von Dechend, *Hamlet's Mill*, p. 157.
110. Doniger O'Flaherty, *Hindu Myths*, p. 205.
111. Ibid, p. 104.
112. Ibid.
113. Blavatsky, *The Secret Doctrine*, vol. 2, p. 550.
114. Krupp, *Beyond the Blue Horizon*, p. 248.
115. See Marion Bradley's interpretation of the Arthurian legend in her novel *The Mists of Avalon*.
116. Doniger O'Flaherty, *Hindu Myths*, p. 110.
117. Blavatsky, *The Secret Doctrine*, vol. 2, p. 382. See also *An Australian Geographic Guide to Space Watching* by Robert Burnham et al., p. 118.
118. De Santillana and Von Dechend, *Hamlet's Mill*, p. 157.
119. Ibid.
120. Ibid.
121. Doniger O'Flaherty, *The Rig Veda*, p. 75.
122. See Walker, *The Woman's Dictionary of Symbols and Sacred Objects*, pp. 16–17.
123. See Eliade's discussion of 'The Doctrine of the Yugas' at pp. 62–67 in *Images and Symbols*.
124. Ibid, p. 62.
125. Ibid, p. 63.
126. Ibid, pp. 62–63.
127. Ibid, p. 65.
128. Ibid.
129. Ibid, p. 63.

130. Ibid.
131. Ibid.
132. Ibid.
133. Ibid, p. 64.
134. For a scientific interpretation of Ezekiel's vision see the discussion by Von Daniken in his controversial book *In Search of the Gods* at pp. 38–46 where he contends that Ezekiel 'saw and described a space-ship.' He refers to American rocket engineer Josef F. Blumrich's technical analysis of the same biblical passage affirming his spaceship visitation theories in Blumrich's book, *And the Heavens Opened.*
135. Hancock, *Heaven's Mirror*, p. 150.
136. Eliade, *Images and Symbols*, p. 63. See the translation of *The Rig Veda* by Wendy Doniger O'Flaherty at pp. 239–42 on the use of dice to signify the four world ages.
137. Eliade, *Images and Symbols*, p. 63.
138. Walker, *The Woman's Dictionary of Symbols and Sacred Objects*, p. 16. It answers the question posed by the authors of *Hamlet's Mill* on p. 178 where they ask: 'But who are we to impose Mrs Grundy on the assembly of heaven?' Kali is that comic book school headmistress, 'Mrs Grundy' in charge of celestial affairs!
139. Eliade, *Images and Symbols*, p. 64.
140. Ibid.
141. Ibid.
142. Ibid, p. 65.
143. Ibid.
144. Hancock, *Heaven's Mirror*, p. 150.
145. Major Jenkins, *Maya Cosmogenesis*, p. xxxvi.
146. Icke, *The Biggest Secret*, p. 12.
147. Ibid.
148. Frissell, *Nothing In This Book*, p. 15.
149. Hand Clow, *The Pleiadian Agenda*, p. 27.
150. Bowker, *World Religions*, p. 26.
151. Ibid.
152. Walker, *The Woman's Dictionary of Symbols and Sacred Objects*, p. 392.
153. Ibid.
154. Ibid.
155. Ibid.
156. Blavatsky, *The Secret Doctrine*, vol. 2, p. 549.
157. Ibid.
158. Ibid.
159. Ibid.
160. Kirch and Green, *Hawaiki, Ancestral Polynesia*, p. 260.
161. Ibid.

162. Ibid. From Emory, K. 1947, 'Tuamotuan Religious Structures and Ceremonies', *Bishop Museum Bulletin*, 191, p. 61.
163. Ibid.
164. Sams and Carson, *Medicine Cards*, p. 77.
165. Ibid, p. 76.
166. See ch. 8, 'The Counting of the Moons', in *Early Man and the Cosmos* by Evan Hadingham at pp. 96–109. See also the Native American story in the children's book, *Thirteen Moons on Turtle's Back* by Bruhac and London.
167. Sykes, *Seven Daughters of Eve*, pp. 279–82.
168. Ibid, p. 90.
169. Mowaljarlai and Malnic, *Yorro Yorro*, pp. 80–82.
170. Walker, *The Woman's Dictionary of Symbols and Sacred Objects*, p. 392.
171. Ibid.
172. Ibid, pp. 618–19.
173. Ibid, p. 619.
174. Blavatsky, *The Secret Doctrine*, vol. 2, p. 619.
175. Ibid.
176. Ibid.
177. Ibid.
178. Ibid.
179. Ibid.
180. Ibid.
181. Ibid.
182. Walker, *The Woman's Dictionary of Symbols and Sacred Objects*, p. 196.
183. As per the teachings of Rabbi Abba in the *Zohar* as quoted in *The Secret Doctrine*, vol. 2 at p. 625.
184. Ibid.
185. Ibid.

CHAPTER SIX

Athurai

Seven Cows of Ancient Egypt

Seven propositions issued successively from her mouth
and these seven propositions became *seven divine beings.*
Thus began the genesis of the world which she, in the form
of a cow, meditated, from the moment of her own creation
to the creation of man: 'For all creatures came into existence
after she was born. It is she who touches the boundaries of
the entire universe under her bodily aspect of surface liquid
and under her real name of perpetual time.'

— Lucie Lamy and Serge Sauneron[1]

Ancient Egypt, land of the pharaohs, pyramids, temples and tombs of
another era presents as a modern enigma whose many mysteries remain
unsolved. Within this cultural and historical setting the story of the
Seven Sisters is shrouded by the passage of time. What we can learn of
them may only be gleaned from minuscule sources activated by the
human imagination and impressions of the soul. We do know the
ancient Egyptians divided their night skies into seven regions.
Therefore their heaven was sevenfold.[2] We know, too, they had their
own Seven Sages who fled the rising floodwaters of their original island
homeland to settle the Nile Valley where the Seven Builders established

an extraordinary civilisation.[3] Therefore the seven stars of the Pleiades would have made an impressionable mark upon their psyche and mythology. Their very name, Athurai or Atauria, desig-nated them the stars of Athy or Hathor the great Cow Goddess, so they were collectively referred to as the Seven Cows or the Seven Hathors.[4] Their link with the constellation of Taurus is twofold; through mythology in terms of their relationship with the Heavenly Bull and in the etymology of their name. Some writers such as William Tyler Olcott even suggest that the Latin term 'Taurus' for bull or cow is actually derived from the Arabic Al Thuraya, or the Egyptian names of the Pleiades.[5]

As the Seven Fates they had the power to foretell the destiny and fortune of every newborn child,[6] while at death they provided nourish-ment to the souls of the newly departed in their aspect as the Seven Cows.[7] They were the seven beings that the dead would meet on their journey 'through the seven spheres of the afterlife,'[8] whose names had to be uttered to provide the necessary password to enter into heaven. As the 'Seven Beings that Make Decrees'[9] they were therefore perceived as Seven Judges, for Hathor in one of her many other aspects was Maat, the Egyptian deity of truth and justice.[10] As goddess of the underworld she presided over the Inquest of the Dead, who would confess their sins to her before she placed her feather of truth against her heart in the balancing scales, during the weighing of the heart-soul.[11] The Seven Hathors are mentioned in numerous texts and papyri, including the celebrated Book of the Dead[12] and in the Litanies of Seker.[13] Such was their significance in ancient Egypt that they were adorned on Isis' temple walls on the island of Philae.[14] They can also be seen in zoological form on the south wall of Nefertari's tomb (in the Valley of the Queens at Luxor) as the Seven Cows with the Heavenly Bull.[15] Sometimes they were depicted 'as a group of young women carrying tambourines and wearing the Hathor head-dress of a disk and a pair of horns', as on the temple walls of Dendera.[16]

Apart from Hathor and Neith the names of the other goddesses that comprise the legendary seven are not known, but we can assume their names by association.[17] For example, the reference to Isis being Hathor's 'dark' twin makes her a likely candidate, as does Isis' other twin Nephthys, and so on. Together with the fact that many Egyptian goddesses shared similar attributes and bore identical epithets such as cow, serpent and vulture goddesses, meant that they were ultimately all aspects of the one Goddess. So it can be safely assumed that the Seven Sisters are embodied in seven primary Egyptian goddesses — Hathor, Isis, Maat, Nephthys, Neith, Nekhbet and Nut.[18] Like Native American and Polynesian legends, Nut represents the personified whole as the universal Sky Goddess, or else Hathor in her multiple incarnation as the heavenly cow. In *Myths and Legends of Ancient Egypt*, however, Lewis Spence maintains that the Seven Hathors were a selection of several forms of the one goddess and identifies them as: Hathor of Thebes, Hathor of Heliopolis, Hathor of Aphroditopolis, Hathor of the Sinaitic Peninsula, Hathor of Momemphis, Hathor of Herakleopolis, and Hathor of Keset.[19]

Similarly, George Hart gives their individual epithets as 'Lady of the Universe', 'Sky-storm', 'You from the Land of Silence', 'You from Khemmis', 'Red Hair', 'Bright Red' and 'Your name flourishes through skill'.[20] However that may be, if we apply the pagan principle that 'All the Goddesses are One Goddess', this means that all the other Egyptian goddesses are mere reflections and aspects of one another. Whatever the true situation within this mythological melange, we see in ancient Egypt as elsewhere, several familiar Pleiadian themes. These include the notion of the sevenfold deity or seven creator beings, water maidens (through their connection with the Nile River and the celestial waters of the heavens), bird goddesses and bulls (emblem of Taurus and the Pleiades), female justices, magical cords, cosmic mountains, and goddesses of time.

Hathor the Cow Goddess

Hathor, along with Isis, are two of the better known Egyptian goddesses, followed by Nut and Maat. Hathor, in particular, is often portrayed as the great Cow Goddess, either in zoological form as a cow, or in anthropomorphic form as a woman with cow ears or wearing a headdress of a pair of horns with the sun or moon disc between them.[21] In Egypt, as in ancient Greece, the cow represented the female characteristics of the Taurean bull. This is not because (as some writers have suggested) the Greek zodiac was imported into Egypt — for even that owes its development to Mesopotamia — but rather that the Cow Goddess is an archaic image indigenous to ancient Egypt. The reason both cultures placed the bull and the cow in Taurus is due to the particular symbolic qualities and characteristics ascribed to that animal, whose sheer power and fecundity embodied the highest of male and female principles as recognised in their most supreme deities such as the Greek Zeus and Egyptian Hathor. As the Cow Goddess, she represented many things, says Barbara Walker in *The Woman's Dictionary of Symbols and Sacred Objects.* Of her many attributes, she stood for life, food, sustenance, nourishment, milk, power and rebirth.[22] It is for these characteristics that the cow was 'one of the most common totemic images of the great Goddess in those regions of the world where that particular animal species was known'.[23] This was perhaps nowhere truer than in ancient Egypt, where many of the goddesses assumed her form, although Hathor easily remains the primary Cow Goddess of them all.

In her divine aspect Hathor personifies the night sky, whose stars appear along her body, where she is sometimes referred to as Nut or Neith the sky goddess.[24] As the celestial cow her body is often illustrated in Egyptian art as being covered with eyes that denote the stars, 'and wings to represent the air.'[25] This image of the flying Cow Goddess, says Walker, is commemorated in the modern day nursery rhyme 'The Cow Who Jumped Over the Moon.'[26] The Greeks knew this same heavenly bovine goddess as Europa, who created our galaxy by spraying her milk across the night skies.[27] In her role as the land-based great mother cow, Hathor nourishes and nurtures the world, hence her horns of plenty.[28] Cornucopia, as they were known, became household decoration items and were placed on temples and store-houses to invoke her blessings as the bountiful, 'All-giving Mother.'[29] Hathor not only nurtured the living but also the dead in the underworld, whom she refreshed 'with food and drink' from her sacred wild fig tree, the sycamore, says Rachel Storm in *Egyptian Mythology.*[30] This explains why dried figs were placed in the tombs of the dead 'to serve as womb symbols for the rebirth of the dead.'[31] These ties with reincarnation and rebirth led to royal coffins being made from sycamore trees, 'in the belief or hope that death was no more than a return to the womb.'[32] Thus in the *Books of Coming Forth by Day*, the deceased would invoke the Seven Hathors by saying:

> I know the names of the seven cows and their bull who give bread and beer, who are beneficial to souls and who provide daily portions; may you give bread and beer and make provisions for me, so that I may serve, and may I come into being under your hinder-parts.[33]

Hathor was also renowned as the goddess of love, joy and wisdom.[34] As the fun-loving patroness of song, music played a principal role in her life, as it did equally in all Egyptian ceremonies and rituals. An instrument made especially sacred to the goddess was the Egyptian rattle, the sistrum.[35] Made of metal, it was shaped to resemble the goddess, complete with horns 'bent round to form a loop.'[36] Its four sides signified the four primordial elements of creation, whose sounds were recreated by the rattling.[37] Barbara Watterson tells us in *Gods of Ancient Egypt* that Hathor's son Ihy, in particular, mastered the percussion instrument and was especially renowned for his playing. Like all young boys in ancient Egyptian society his head was shaven except for 'one long lock of hair,' indicating he had not yet attained puberty, 'at which time it was shaved off and offered to a favourite god.'[38] Here we see two familiar Pleiadian themes; notably the mythological connection between these stars and music in regard to the creation of the universe, and the idea that young boys are only allowed to stay within women's domain provided they had not reached adolescence or become initiated as men.

In addition to her role as the sacred cow and goddess of all things beautiful (as is typical of Taurus sun signs), Hathor is often referred to as the daughter of Ra, but some writers such as Wallis Budge and Barbara Walker entirely disagree.[39] Walker claims that it was Hathor in her incarnation as the great cow or as the Nile goose that gave birth to the Sun god and not the other way around.[40] This, she argues, is the true meaning behind the glyph depicting the Cow Goddess with the sun disc lying between her horns, for it symbolises the male logos spirit 'enclosed and protected, soon to be reborn.'[41] It is at this juncture that we enter the world of symbolism in an attempt to comprehend the meaning associated with various Egyptian images, insofar as they relate to the stars of the Pleiades and their associated themes identified in the opening chapter.

To begin with, the golden orb safely nestled in the goddess' horns combines two other archaic but interrelated signs — that of the 'Cosmic Egg' and the 'Tree of Life', whose setting sun sinking below the horizon is caught by the fork of its branches.[42] Perhaps the best-known image of the Tree of Life is the Norse Yggdrasil,[43] which featured in the movie *Conan the Barbarian* starring Arnold Schwarzenegger, to which the hero was chained. In ancient Egypt, not surprisingly, it took the form of the sycamore tree sacred to Hathor.[44] Common to both icons is the eternal serpent (sign of the constellation Draco) that entwines itself around the Cosmic Egg or the Tree of Life. Both reveal different images of our creation: that of the egg and sperm, which creates life, and the caduceus, an ancient symbol for the double helix of our DNA.

In *Symbols of Birth and Death*, Dorothy Cameron suggests the Ancients identified the horns of bulls with the female reproductive system, which they would have discovered during the burial process.[45] Within this context, the horned sun disc would have denoted the female reproductive organs, whereby the fallopian tubes represent the cow's horns, and the orb the womb. Several writers have associated the Tree of Life with the Tau cross, which astronomically represents the *axis mundi* or world axis, the significance of which is explored further on and in the concluding chapter.[46] The Egyptian version of the Tau cross found its expression in the sacred ankh, which combines the union of male and female principles, that of the phallic cross and yonic circle.[47] Consequently it came to represent life, and the wearing or holding of the sacred icon indicated that person had 'the power to give and take life.'[48] Other writers such as John Bowker in *World Religions* have noted similarities between the Egyptian ankh and the astrological glyph of Venus, whose sign denotes 'life, love, and sexuality.'[49]

Like all other variants of the cross and circle, the ankh speaks the initiatory language of the 'First Creation' that the ancient Egyptians referred to as Zep Tepi.[50] The Greek *tau* is in fact 'the oldest form of the letter' that signifies the fall from heaven and the separation of the sexes, says Blavatsky.[51] In the Druidic alphabet it was called tinne, which represented the holly named for the Teutonic goddess Holle.[52] Its 'red berries were seen as drops of her life-giving blood, in which lay the secret of the tree's immortality or year-round greenness,' says Walker.[53] In *Starlore of All Ages*, William Tyler Olcott sees a link between the Greek tau, and 'the sacred scarabaeus or tau beetle of Egypt.'[54] As he does not elaborate one can only wonder what he meant. Beyond depicting our creation, the amalgamation of cross and circle represent codified astronomical symbols. The circle of the cross represents the Cosmic Wheel, whose crossing of the ecliptics and the revolutions of our solar system are depicted in the hub's spokes at precise points of intersection. Some of these images, as in the Kalacakra or Hindu 'Wheel of Time', include our solar system's magical journey around the centre of our Milky Way galaxy. The cross and circle, wheel and hub, symbolise the circular aspect of time as reflected on our clock faces, which the Goddess of the seven stars held in her keep to distribute in accordance with fate and destiny as dictated by the Seven Fates or Seven Hathors.

The idea that the primeval universe was originally contained in the form of an egg or seed is fairly archaic and common to many cultures, says Barbara Walker. Its image lives on in the fairytale of 'The Goose that Laid the Golden Egg', for the goose is none other than Hathor whose sacred metal is gold.[55] Thus Hathor was often referred to as the 'Golden One', which not only implies that she gave birth to the Sun, but that she *is* the Sun, as in Aboriginal Australian cosmology. Her name *Hat Hor*, says Lewis Spence, associates her with the falcon-headed god Horus to whom she gave birth, for it means 'House of Horus.'[56] This reference implies that Hathor, like the Cosmic Egg, envelops and contains Horus within her womb and flies him up to the heavens. It brings to mind the story of the halcyon bird in Greek mythology that carries her dead mate on her back, wailing and mourning his passing.[57] Instead, we have the opposite portrayal; an embryonic bird-god soon to be born, while his mother lifts him up into the sky. Some Egyptian myths refer to Horus as Hathor's husband and consort and not her son, and so in this context she may be said to be carrying her mate like the legendary halcyon. It certainly gives added meaning to the term 'airborne', for that is precisely what happens to Horus in Hathor's mansion in the sky. This alternate meaning of her name, says Isidora Forrest, indicates she lived in the sky, as well as being the personification of it.[58]

Hathor's tears, the Egyptian Deluge

The reputation of the Pleiades as harbingers of rain in many world cultures was somewhat qualified in ancient Egypt, where they specifically ascribed to Hathor 'the tradition of a deluge or other race-destroying disaster.'[59] Consequently, her tears came to symbolise the greater cataclysmic flood that destroys humanity as opposed to the annual flooding of the Nile that enables life to flourish. The annual flooding was attributed to the heliacal rising of Sirius, the brightest star in the night skies and the celestial embodiment of Isis, Hathor's dark twin.[60] Hathor was celebrated during a three-day festival in her honour that began on the seventeenth day in November when the Pleiades rose after the Sun went down, and ended three days later when the stars culminated at midnight.[61] This period agrees with the Mosaic Deluge account, say Gertrude and James Jobes in *Outer Space*, which was said to have occurred on the 'seventeenth day of the second month of the Jewish Year.'[62]

According to Giuseppe Sesti in *The Glorious Constellations*, the month of November was named for the Goddess as Athar-aye, which he says 'is equivalent to Month of the Pleiades,'[63] and the same custom was common among the Chaldeans and the Israelites.[64] Although both Hathor and Isis were equally associated with water, the gemstones turquoise and *lapis lazuli* were especially sacred to Hathor because their colours represented the sky,[65] which she ruled, and her tears the 'primordial element of creation.'[66] As the so-called 'daughter' of Ra the Sun god who gave birth to Hathor through his eye, she was called the Eye of the Sun and her tears indicated wisdom.[67] Numerous legends recount how Hathor was sent by Ra to punish human beings on his behalf, for her powers were considerable. On one occasion when Hathor's father was particularly disturbed by the merciless slaughter she was causing by her unrestrained flooding, he devised a plan to make her 'drunk with the waters of the Nile.'[68] Legend tells how Ra ordered his subjects to brew a large quantity of beer and to colour it with red dye to make it look like blood. The brew was placed in her path on the ground and Hathor immediately set upon drinking the red beer, whereupon she became so inebriated that her killing frenzy subsided and she forgot all about slaying humanity![69]

But this is not just another tale about a drunken woman, says Isidora Forrest in *Isis Magic*, for Hathor is no ordinary woman but a goddess, and an unstoppable one at that. Defending her, Forrest points out that music, drinking and dance not only celebrate 'the abundant life Hathor provides' but that it also pacifies her 'fierce aspect.'[70] In this regard, as the goddess of death and destruction, Hathor bears a striking similarity to her Hindu counterpart, the bloodthirsty, blood-drinking Hindu deity Kali, whose frightening appearance signifies completion and, on the grander scale of things, the end of Kali Yuga, the Age of Sin and Sorrow. Kali's creative and destructive forces are renowned and celebrated in Tantric traditions as the goddess Shakti, whose rituals include blood offerings, caste-free sex and 'the drinking of alcohol in spiritually polluted places, such as cremation grounds.'[71]

Like Kali, whose other manifestation Durga was often depicted as a woman riding on a lion, or as a Lion Goddess,[72] Hathor's other aspect was Sekhmet (or Seshat, the Egyptian lion goddess or woman with a lion's head).[73] Lions are a well known solar symbol, and Durga riding the animal represents her taming the Sun, suggesting her supreme powers (like that of Kali and Hathor) to control time. This explains why Hathor is sometimes portrayed as having 'lion heads that looked forward and backward, symbolizing time.'[74] These images of Durga-Hathor survive in the major arcana tarot cards 'Strength' and the lion or sphinx driven 'Chariot', which is the seventh card of the

Major Arcana. It also explains the image of a woman taming a lion on the Round Zodiac of Dendera. This woman is none other than the goddess Hathor, whose zoological form is embodied in the immortal sentinel Sphinx on the Giza plains.

Hathor houses the clocks of time

Like Isis, who was much loved and worshipped, many temples were built and dedicated to Hathor. The most renowned of these is the Temple of Dendera situated some 648 kilometres south of Cairo. Although legend refers to Seven Hathors, there are just six goddesses carved into the pylons at the front entrance of the temple, alluding to the mystery and interplay between six and seven identified as one of several recognisable Pleiadian themes.[75] Once a place of holy annual pilgrimage by Hathor's followers, it is best known for its two zodiacs; the 'Square Zodiac' that remains within the temple's wall, and the more famous circular or 'Round Zodiac' once positioned above the roof chapel and now housed in the Louvre Museum in Paris. Both zodiacs depict familiar constellations of our present day zodiac, along with some less recognisable and unique Egyptian constellations. They include Taurt (or Taweret), the pregnant hippopotamus goddess that represents the constellation of Draco, together with a rather peculiar sole hind leg of a bull, which the hippopotamus rests its hands on that represents the stars of Ursa Major.[76] The ancient Egyptian name for this constellation was Maskheti.[77]

The depiction of a one-legged being might at first appear odd, but as Giorgio De Santillana and Hertha Von Dechend point out, it is in fact a universal motif for the *axis mundi* or the world axis often shown as a cosmic tree or mill.[78] The fact that ancient Egyptians placed the stars of Ursa Major in the sole hind of a bull, as opposed to any other animal, is highly significant and clearly suggests a joint relationship between that constellation and the Pleiades star cluster in Taurus.[79] Although we cannot be entirely certain, it is feasible that the bird perched aloft the heavenly bull on the astronomical ceiling of the tomb of Set I in the Valley of the Kings at Luxor is the Pleiades, as that joint image was commonly understood in Mesopotamia and ancient Greece.[80] Here in this same painting we also see the bull's tail (of Taurus) joined by a rope or some sort of cord to the bull's leg (Ursa Major), which affirms the affiliation that exists between these stars in many traditions. What possible meaning might these images be trying to convey?

One interpretation based on the insightful work of *Hamlet's Mill* suggests it clearly reveals scientific knowledge of the astronomical phenomenon of the precession of the equinoxes, which gives us our world ages. The celestial relationship between these stars is tied to the notion of the world axis or the cosmic mill, whose unhinging — or removal of the heavenly plug — destabilises the universe and causes a variety of cataclysms which destroy the current world age and pave the way for a new one. Seen from this perspective, the sole hind leg of Ursa Major, which the hippopotamus goddess leans on, represents this cosmic plug at the bottom of the Tree of Life, with Draco the Serpent entwined around the tree that is anchored by the heavenly bull. This appears to be a pictorial confirmation of Blavatsky's assertion that it is the stars of Ursa Major acting in concert with the Pleiades that govern the various cycles of time, including the cyclical destruction and reconstruction of the cosmos.[81] This mutual, co-dependent relationship is further emphasised by the presence of the hippopotamus goddess Taurt, who not only took the form of a water goddess (common to many Pleiadian mythologies) but like Hathor and Isis, wore horns containing the sun disc.[82] Within this context, her composite physical make-up and her name Taurt is especially revealing of this connection.

This notion of an astral plug holding back the rushing waters is not unfamiliar to Judaism. We saw in the opening chapter that a passage in the Talmud states that when God wished to send rain or floods he would simply remove a star from the Pleiades, and when he wished for the waters to abate, he simply replaced the star.[83] Cultural stories of the unhinging of the cosmic mill in world mythology, say the authors of *Hamlet's Mill*, reveal scientific knowledge of the astronomical phenomenon of the precession of the equinoxes. It is clear from the paintings on the ceiling of Set I in the Valley of the Kings and the visual images from both Dendera zodiacs that the ancient Egyptians were familiar with this concept. Some writers such as Graham Hancock view the Round Zodiac, in particular, as a giant astronomical clock. Whoever built it, he says, must have had 'a good knowledge of precession' because although it was constructed sometime during the first century BCE at the beginning of the Age of Pisces, it does not depict a Piscean sky as one would expect but an Aquarian one instead.[84] To understand the significance of this design is to have a basic appreciation of what it means, astronomically, to be living at the dawn of a New Age. Put simply, our current Piscean Age is soon to be replaced by Aquarius as the zodiacal sign that houses the Sun during the spring equinox of the Northern Hemisphere. Hancock explains the effect of these changes in *Heaven's Mirror*:

> This also means, when the New Age is born, that sunrise on the autumn equinox (21 September) will take place in the house of Leo (when it is now in Virgo), that the winter solstice (21 December) will be housed by Scorpio (now Sagittarius), and that the summer solstice (21 June) will be housed by Taurus (presently Gemini).[85]

How this plays out in the night skies above us means that the precessional cycle appears to move backwards (or precesses) through all the signs of the zodiac to reflect the change in ages and other seasonal markers as seen by an observer on Earth. The Ancients clearly understood precession and gave us the language of astrology to comprehend the mechanism of this vast Wheel of Time. To begin with, twelve signs of the zodiac allow for the grouping of six pairs of opposite signs, or for the grouping of four sets of triple signs, referred to as the four elements or triplicity of fire, water, earth and air.[86] Another triplicity involving opposite signs and their perceived astrological characteristics of mutability (Pisces, Virgo, Sagittarius and Gemini), cardinality (Cancer, Capricorn, Aries and Libra) or fixity (Leo, Aquarius, Taurus and Scorpio), form what are known as crosses.[87] So, for example, the opposite signs of Leo–Aquarius and Taurus–Scorpio form what is known as the 'Fixed Cross' because of their perceived astrological characteristics of inflexibility and immovability, which suggests permanence of a fixed nature.

These pairings are strictly adhered, says Hancock, 'so that, when the mechanism shifts, *everything shifts*.'[88] The fixed cross therefore acts as an intrinsic cosmic wheel-brace that turns the Wheel of Time. Its fulcrum is none other than the world axis or Tree of Life, represented by Taurt the hippopotamus goddess (Draco), or else Maskheti the sole hind leg of a bull (Ursa Major). Around the perimeter of the Round Zodiac are four female principle bearers standing in direct opposition to one another beneath the constellations of Leo, Aquarius, Scorpio and Taurus, with their arms raised above their heads, presumably supporting the sky. Given that the zodiac was housed in Hathor's temple it is reasonable to assume these figures represent the Egyptian goddess of time in her quaternary aspect. Within this context the four Hathors may be seen to correspond not only with the four cardinal points (north, south, east and west), but equally with the four stations of the Sun (the two equinoxes and two solstices that mark our annual journey).[89]

This quartering of the zodiac also represents the four points of the three zodiac crosses (fixed, mutable and cardinal). The remaining four sets of twin figures complete the twelve signs of the zodiac in combination with the four single Hathors. The twin figures serve to emphasise the astrological principle of zodiac opposites crucial to the turning of the heavenly wheel of precession, as well as the turning of the seasonal wheel (our year) as reflected in the night skies. This is because, from a technical and practical viewpoint, we can only ever see one full opposite half of the zodiac at any one time when viewing the stars.[90] So, for instance, when Taurus is high in the night skies, Scorpio cannot be seen, and when Leo is visible, Aquarius is not.[91] Even during those times when both opposites signs are visible in the night skies together, they are only partially so. For example, during the autumn months when Taurus is setting it is possible to see Scorpio beginning to rise, but they will never be seen together in their entirety, and the same applies to Leo and Aquarius. Thus in winter when Scorpio is high overhead, Taurus cannot be seen in the evening skies, but early in the morning it can be seen rising in the east only after Scorpio has set. When straight lines are drawn across the circle of the

Round Zodiac to join the four Hathors, there — in perfect alignment where the four quarters intersect — are the four signs of the Fixed Cross: Leo, Aquarius, Taurus and Scorpio. This quartering of the cosmos is reflected in the myriad designs of the cross in many diverse cultural traditions from the Hindu swastika to the Celtic cross, and from the Egyptian ankh to the astrological glyph for Earth.[92] The sexual symbolism remains equally pertinent, for this astronomical logo reflects the macrocosmic creation of the universe as well as the microcosmic creation of human beings. Hence, the popular spiritual adage, 'As above, so below'.

Other writers argue the design of the Round Zodiac merely indicates the Age of Taurus, the period from 4380 to 2200 BCE that marks the generally accepted period of the beginning of Egyptian civilisation and represents nothing more. Hancock disagrees with this view and insists this Aquarian configuration goes beyond the Age of Taurus, which therefore suggests a far more remote period of antiquity. He says that although the Taurean age was an already ancient epoch by the time of the first century BCE, the evidence suggests Dendera had links to even earlier times. One example is a temple inscription that claims the blueprint for Dendera was formulated during the age of the 'Followers of Horus' many thousands of years before the temple was built.[93] The reference to the precessional cycle as the 'Great Year of the Pleiades' in Egypt's sacred canon the *Hermetica*, signifies the central role ascribed to the Pleiades by the Ancients as the principal stars governing the various cycles of time.[94]

This recognition is further heightened by the fact that ancient Egyptians named these stars after Hathor, their Goddess of Time. To be aware of a cycle of 25,900 years suggests the ancient Egyptians must have observed its entire length — or at the very least a substantial amount of the cycle beyond two or three ages — in order to be conscious of the staggeringly slow turning wheel. This is because precession is an enormously difficult phenomenon to detect over thousands of generations, let alone in a single lifetime, where the stars would move just 1° every 72 years.[95] Certainly the ancient Egyptians maintained they recorded two such cycles, or a combined period of

51,736 years, thereby ascribing their civilisation a far greater antiquity than most Egyptologists have attributed them.[96]

Quite apart from revealing that the ancient Egyptians were familiar with the astronomical phenomenon of precession, the Round Zodiac of Dendera is an incredible timepiece that goes beyond mere measurements and calculations to our earliest beginnings. One of the most exciting discoveries of the Round Zodiac is the observation and analysis of John Lash featured in *The Atlantis Blueprint* by Rand Flem-Ath and Colin Wilson. In addition to the four axes of the Fixed Cross (A, B, C and D) in the Round Zodiac, Lash has identified a fifth *hidden* axis. Labelling this 'Axis E', it centres on Virgo the Grain Goddess, and in particular on the bright white star Spica or Alpha Virginis situated on the wheat ear that she holds in her left hand.[97]

Lash identifies three features of this virginal axis that indicate the Ancients knew about precession. Firstly, he observes that the axis bisects 'the altar mounted by four rams' heads' directly opposite the Virgin along the outer edge of the zodiac.[98] This alignment suggests some sort of connection between Virgo and the lunar cycle (the nakshatras of Hindu astronomy) whose decans run around the periphery of the zodiac. Secondly, he notes the axis crosses the tail of the lower fish in Pisces 'exactly where the spring equinox occurs today.'[99] Thirdly, the axis extends to the bottom of the Virgin's feet that 'marks the tail of Leo at a point that corresponds by precession to 10,500 BC.'[100] Effectively, Axis E 'marks the moment of precession when one full cycle ends and a new one begins' says Lash.[101] Spica, he concludes, appears to be the 'master key' to the Round Zodiac and is therefore referred to as the *precessional star.*[102] Besides confirming what many people have suspected about the extent of knowledge of ancient science, what could this zodiac possibly be telling us? Is there a message of hope for humanity?

The first thing that Lash noted was that although the zodiac depicts a number of constellations beside the traditional twelve signs, there are only two stars that are actually illustrated on it — Sirius in Canis Major and Spica in Virgo.[103] Sirius, we know, was the special star of Isis whose heliacal rising signalled the annual flooding of the

Nile. It is easily identified on the zodiac where it is positioned 'between the horns of the sacred cow on axis B' running from north to south.[104] Clearly this bull is not Taurus, but given that Isis is often described as Hathor's dark twin, the illustration makes sense in terms of the twin aspect, not to mention the Seven Cows or Seven Hathors. Spica, as we have already seen, marks the ear of wheat the Grain Goddess holds in her left hand. This image and gesture of Virgo was known to the ancient civilisations of Mesopotamia 'as early as the third millennium BC,' says Lash.[105] The Romans identified her with their Grain Goddess Ceres (Mother of the Corn) from which we obtain our modern word cereal.[106] It is interesting that in the relationship between these two stars, the entire constellation of Virgo, not just Spica, represents Isis, not just the Dog Star, Sirius. Here we see a pattern begin to emerge in the Round Zodiac that centres on the Virgin Isis. What can this possibly mean?

The answer lies in the arcane theories of Madame Blavatsky. Almost an entire century before Lash she wrote in *The Secret Doctrine* that one of the most esoteric cycles of the Hindu calendar involves the constellation of Virgo and the stars of the Pleiades.[107] The positioning of Virgo, in particular, signals the beginning and end of the Kali Yuga that presages the Age of Aquarius.[108] When we look to the Round Zodiac along the axis identified by Lash, we see the Pleiades perched upon Taurus the Bull on the Virgin's right shoulder to the left of the axis (from the observer's perspective), perhaps confirming this relationship. Furthermore, the Hindu calendar records that a lunar eclipse took place at the start of the last Kali Yuga in a point situated between Spica and the star Omega Virginis of the same constellation, thus confirming Lash's speculation of a lunar connection with Virgo.[109] Given Blavatsky's assertion that the ancient Egyptians derived their zodiac from ancient India, it makes sense that the Round Zodiac of Dendera would be identical in this respect.[110]

Yet what is the connection of the Virgin goddess with the changing tides of the ages? Firstly, we know that Hathor's tears cause the cataclysmic floods that destroy humanity as opposed to the beneficial annual flooding of the Nile heralded by Sirius–Isis. Secondly, as

Hathor's twin aspect, Isis is implicated in the Deluge. But if Hathor is in Taurus, then where do we look for Isis in the skies?

'Wherever we find the Tree of Life,' says Reginald Lewis in *The Thirteenth Stone*, 'we are bound to find the Queen of Heaven.'[111] And as we have already seen, when the tree is disturbed, the world axis is unhinged and the Deluge pours forth. Clearly, that queen is none other than the Virgin constellation. But not all is lost, for Virgo–Isis is the 'Mortal Woman' of *The Secret Doctrine* that seeds humanity after the flood. And it is not just humanity that is seeded but the Virgin gave birth to gods such as Isis who bore Horus, on which the Virgin Mary and the Christ Child are modelled. Thus we see Isis and Horus situated directly beneath Virgo on Axis E of the Round Zodiac. Blavatsky's reading of this icon would be that Virgo stands for the seeding of the third subdivision of the Fifth Root Race represented by Horus. Furthermore, the wheat she holds in her hand is not a mere pictogram but an actuality. Blavatsky claims that wheat is not a natural plant of Earth but was brought down from the heavens.[112] This is the true meaning behind the goddess' statement in *The Egyptian Book of the Dead* where she proclaims:

> I am the Queen of these regions. I was the first to reveal to mortals the mysteries of wheat and corn. I am she who rises in the constellation of the dog.[113]

The thirteenth stone is a reference to the Tree of Life that plays a significant role in the esoteric teachings of the Kabbalah and in prophecies of the Bible (of which the Virgin is its key player). Curiously, the Arabic name for Spica, Al Zimack translates as 'branch', which Lewis identifies as the all-important limb of the Cosmic Tree.[114] Spica is the sceptre, says Lewis, the branch from which 'Jesus comes out of his mother's womb, fertilizes her and becomes the father of the king of the Golden Age, his seed being wisdom.'[115] Now before some purists attack Lewis and say the Virgin or the Christ has nothing to do with the Kabbalah, they need to read his work to understand that the entire premise of his book is predicated on the belief that all religions are ultimately the same, that their fundamental beliefs and precepts derive from the one source. It is only that the players, characters and terminology change according to the various traditions (in much the same way that Monopoly remains essentially an 'English' board game, but the properties reflect the familiar and popular names of other recognisable and high-profile landmarks).

Within this revelatory system the 'fruit' of the tree is knowledge, says Lewis, and in the Bible bread denotes wisdom.[116] As wheat is an essential ingredient of bread, the Virgin's sheaf of wheat in her hand represents the particular wisdom which comes from the dawn of a New Age in the form of the thirteenth stone on the Tree of Knowledge.[117] Whichever interpretation is correct, there is no doubt that the mysteries surrounding the Round Zodiac of Dendera are equally bound to Hathor's stars — the Pleiades. If anything, the zodiac emphasises the central role of the goddess in the cycles of time in her many incarnations, whether she is designated the seven stars of Ursa Major, or the Pleiades, or that of the Virgin (Isis–Hathor). Significantly, the ancient Egyptians named the stars of the Pleiades after the Goddess Hathor, for her name Hat Hor, which refers to the House of Horus, is precisely where we get one of our foremost and well known time measurements, the 'hour.'[118] Thus in ancient Egyptian mythology Hathor is not only the goddess who personifies and keeps time but, like her Indian counterpart Kali (whose name derives from *kala* meaning time), she is one who transcends time and is therefore eternal.[119]

The Seven Fates

Hathor's talents and responsibilities were many and, like Isis, she was much revered. As the personification of the vigorous forces of nature that creates and destroys all things on a perpetual basis, she 'multiplied into seven fates that foretold the future, especially of children at birth.'[120] In this regard the Seven Hathors are analogous to other goddesses of fate such as the Greek Moirai, the Teutonic Norns, the Viking Valkyrie, the Celtic fey and the proverbial fairy godmother of modern day fairy tales.[121] This clairvoyant role of the Goddess accentuates her connections with time for her ability to see into the past, present and future. The Goddess' connection with fate derives from ancient beliefs that it was She who created the universe with her weaving tools and implements. The underlying premise of this belief is that the universe is non-chaotic and structured. What is more, its patterns of orderliness have purpose and direction as expressed by one's destiny, whose strands of fate are spun, measured, cut and woven by the supreme seamstress such as the Greek goddess Ariadne, the Celtic Arianrhod, the Roman Fortuna and all of her other incarnations.[122] In many cultural traditions the Fates are often referred to as weavers and, as Barbara Walker tells us, the word 'destiny' is actually derived from the Latin 'destinos' meaning 'that which is woven.'[123] Apart from being 'a warlike divinity,' Neith was regarded as a patron of weavers, which identifies her as a goddess of fate and so strengthens the claim that she may have been one of the Seven Hathors.[124] One of her many hieroglyphic signs has been identified as a weaving implement known as a shuttle, hence her epithet 'The Shuttle'.

In *The Temple of Man*, Schwaller de Lubicz explores the hidden symbolism behind this implement, and weaving in general. He notes Neith's special sign engraved on a cobra's chest in a Luxor temple,[125] and launches into an esoteric but intellectually challenging discussion of its significance. Taking the serpent's dual physicality (tongue, penis, vagina and so on) and relating it to the olfactory senses of human beings, he argues that the weaving language of warp and weft are metaphors for a 'physiological topography' of the human body, which reveals intricate, detailed biological and medical knowledge.[126] Referring to this process as 'olfactory intelligence,'[127] he notes that:

> With the sign of Neith being placed on the cobra's chest, across its trachea, there is an indication of a relationship between the function of weaving to make corporeal and the fixation of the spirit by respiration.[128]

Thus, 'the weaving of what animates (air) and what is animated (blood) is carried out by respiration.'[129] Essentially, de Lubicz was trying to articulate his theory on smell and sound as being the primary senses upon which we perceive and intellectualise our environment, hence the ancient Egyptian's predilection for burning incense in their temples. It is important, therefore, to note that it is not only the Goddess' thread which creates the matrices of the universe but also the sounds of her voice that cause objects to materialise. These sounds, classified as seven sacred vowels which represented the Supreme Deity in ancient Egypt, in turn gave rise to human languages and the alphabet.[130] Consequently, says Gertrude Jobes in her *Dictionary of Mythology, Folklore and Symbols*, the Hebrew verb to swear, means literally 'to come under the influence of seven things, thus seven lambs made up the oath gift between Abraham and Abimelech.'[131]

Beyond the world of communication the Goddess' sacred sounds also provide the necessary matrix upon which our physical world is built. This explains why mathematics, which forms the basis of our sciences, is sacrosanct to her in all its forms. The sacred geometry of Pythagorean traditions reveal the dimensions of her many shapes and forms. Physics enables us to understand the processes of her laws. And chemistry allows us to comprehend the elementary forms of the mother (*matter*) through the interdependent relationships of her progeny, their various networks and interactions with one another that enable their transmutation. This is the real meaning of alchemy and its true alchemists who sought to understand her scientific *mysteries* within its holistic context. Numerology, in particular, is especially sacred to the Goddess, for numbers express the glory of her numinous nature that leads and guides us to her well of spiritual insight.

Of all the sacred sciences, ancient Egypt gave her name to alchemy and chemistry from Kemennu, which Wallis Budge says in *Egyptian Mysteries* means 'Land of the Moon.'[132] Others suggest it means 'black' as in the colour of its soil, but there may be other associations as well.[133] For instance, black is the colour of the new moon. A more practical interpretation may be that it simply refers to the skin colour of its original inhabitants, given Egypt's location in Africa, 'Land of the Black People'. Idries Shah tells us in *The Sufis* that alchemy was often equated with 'black magic,' but that this 'misunderstanding exists only because of ignorance of the similarity of 'black' and 'wise' in Arabic by foreigners.'[134] In some cultures black is linked with death, which may account for the ancient Egyptians' fascination with the afterlife. Finally, black is often equated with that which is hidden, as in the occult. Perhaps it is in this sense that it was mistakenly connected with 'black magic.'

Many goddesses of fate often wore veils to emphasise the hidden or secret nature of things, as in one's destiny, or the face of divinity. Curiously, the etymology of the very word 'revelation', as Barbara Walker tells us in *The Woman's Dictionary of Symbols and Sacred Objects*, comes from the Latin *revelatio* which means, 'to draw back the veil.'[135] This notion of the hidden name or face of god is central to many great spiritual traditions, including Judaism, where the ineffable name was written as YHVH or referred to as the Tetragrammaton.[136] Concomitant to this belief is the idea that the future is equally unknown to human beings but not to divinity who may bestow clairvoyant and psychic skills to certain individuals to act as mediums or to prophesise. And as Walker points out, although some people have expressed a common desire to know or see the future, there are always those who fear what it might actually reveal. Thus it was commonly believed that a person might see their own death were they to take a furtive peek behind the Goddess' veil, and for this reason it was considered unlucky. This explains why some myths portrayed the hidden face of the Goddess as something to be feared and dreaded or even considered deadly, like that of the snake-headed Gorgon Medusa who was decapitated by the Greek hero Perseus.[137]

Sometimes the Goddess wore not one but *seven* veils like the dancing Salome, the priestess who enacted 'the sacred mystery of Ishtar's seven veils.'[138] These veils stood for 'the seven planetary spheres,' says Walker, thereby 'concealing the true face of celestial divinity.'[139] We see this imagery employed in the Seven of Cups tarot card of the Rider Waite deck, where one of the cups contains a veiled figure with arms outstretched to represent the inquirer's true spiritual identity in their quest of self-discovery. Isis, too, wore 'the same sevenfold covering' as the Seven Hathors, not so much as a goddess of fate but more to mark her divine aspect.[140] It is in this sense that Madame Blavatsky titled her great literary homage to the Egyptian deity *Isis Unveiled*.[141]

As in other mythologies, many of the gods and goddesses in ancient Egypt overlap with one another, although the fact that Hathor was referred to as a sevenfold deity implies that her other aspects are intrinsic to her divinity rather than coincidental. The fact that she was identified with a number of Egyptian goddesses beyond the seven nominated as comprising the Seven Sisters of the Pleiades is axiomatic to the pagan principle that 'All the Goddesses are One Goddess.'[142] With Isis she shared the honour of the title Great or Divine mother whose enduring image of motherhood lives on in the Christian icon of the Madonna and the Christ Child.[143] This is not to suggest that either Hathor or Isis were the supreme Mother Goddess, for that title goes to their mother Nut the Sky Goddess, or their grandmother Tefnut or Telfnut.[144] As the light and dark aspect of one another, Hathor and Isis represented the yin and yang energies of the cosmos (that is, in a broader context, not restricted to 'male' or 'female'). Both were renowned healers[145] who each had a place in the Boat of the Sun at creation.[146] The legends also refer to Isis and Nephthys as twins — not just Hathor and Isis — for they tell how the Sky Goddess Nut gave birth to two sets of twins — Isis, Osiris, Nephthys and Set, although they do not state which sets of twins were paired. So we have no way of knowing whether they were combinations of same-sex twins or one of each.

However that may be, there are references to both Isis and Nephthys and Isis and Hathor as carrion birds, namely kites, cleansing dead bodies before their entrance into the Underworld.[147] Certainly Isis and Nephthys mourned the death of their brother Osiris, also their lover and husband. This mortuary aspect carried over to the Vulture Goddess, Nekhbet. Sometimes this goddess also took the form of the cobra, which reveals she shared a serpent aspect with Neith, the Goddess of War. With Nekhbet and Nephthys, therefore, Hathor shared her fierce warrior aspect that can destroy worlds.[148] These serpent goddesses also featured the All-Seeing Eye, the emblem of

Maat, Hathor's aspect as the mother justice. With Nut her sister-mother, Hathor shared the sky that re-emphasises her status as the 'Mother of All.' Rachel Storm recounts a tale of how Nut assisted Ra, the Sun god in distancing himself from humanity after he had become 'disillusioned with their ways.'[149] On this occasion she took the form of a cow and 'raised the god upwards on her back.'[150] The higher she rose, 'the dizzier she became until she had to summon four gods to steady her legs.'[151] These gods then became 'the pillars of the sky.'[152] This story demonstrates how the various goddesses were identified with one another and the points at which they merge. Nut is Hathor is Nut is Isis is Nephthys, and so on, until the goddesses are multiplied many times. It is perhaps in her aspect as Maat, the goddess of truth and justice, that her Pleiadian aspects rise to the fore.

Maat, the Goddess of truth and justice

The notion of the Seven Sisters as female justices in India found their expression in Egyptian traditions. In more ancient times, they took the form of Seven Hathors and then specifically as the goddess Maat, the Egyptian goddess of truth and justice. Her emblems of justice were several and include her feather of truth, balancing scales, her All-Seeing Eye, her throne, the gates to the underworld and her faithful guardian, the dog. Rich in esoteric meaning, their decipherment reveals a wealth of spiritual wisdom and insight into the language of the goddess and in particular her affiliation with the Pleiades. An obvious link is the feather of truth with birds, which characterises the Seven Hathors as bird goddesses of the Pleiades. Many Egyptian goddesses were portrayed as birds or were featured with outstretched arms of wings, which revealed them to be bird goddesses.[153] Isis, in particular, was often linked with a number of birds of prey, including the vulture, owl and kite. Together with her twin sister Nephthys, they were illustrated in Egyptian art as carrion kites feeding on live prey and scavenging corpses.

In this respect they resemble the Hindu goddess, Kali the Black. Their screeching cries made them sound 'like wailing, lamenting women,' says Isidora Forrest in *Isis Magic*.[154] Like Hathor who carries her son Horus into the heavens, the shrill cries of the kites brings to mind the Greek legend of the halcyon and its hauntingly woeful mourning cries. Mourning aside, we should not forget Isis' power as a sacred healer, for after she collected the 72 dismembered pieces of her husband's body (Osiris as Orion), she set about making him whole once more.[155] Within this tale, Graham Hancock sees the Ancient's reference to 72 as the magical number of precession.[156]

The ancient Egyptians believed human beings possessed seven souls, one of which represented a bird.[157] The most worthy of these souls was the heart-soul that was weighed in the balances during the Inquest of the Dead (over which Maat presided as goddess of truth, justice and the underworld) and where the souls of the recently departed were judged.[158] When a person died they 'had to appear before 42 judges of the dead and declare whether they were innocent or guilty of numerous crimes' during their lifetime.[159] This would take the form of 42 declarations in the negative; for example, number 4 (I have not stolen), number 6 (I have not uttered falsehood), number 38 (I have not cursed the God) and so on.[160] This verbal affidavit in the Court of the Dead became known as the 42 Negative Confessions.

After making this declaration the soul of the dead would then be weighed on the scales against the feather of truth to determine their worthiness to enter heaven. 'To be light as a feather meant one was unburdened by sin and therefore exonerated in death,' says Walker.[161] As the embodiment of the higher pursuit of integrity, Maat was often drawn wearing a feather on her head and her followers pledged to live 'by Maat, in Maat and for Maat.'[162] In later times, Egyptian judges would wear a gold necklace with an image of Maat 'as their badge of office.'[163] Originally there were only seven women judges at the Death Tribunal. It is only in later texts that 42 judges somehow miraculously appear. At first it seems that there is some mistake or the court has multiplied rather expeditiously. How might we account for this change?

An obvious suggestion is the patriarchal takeover of Mother Goddess traditions. Whatever the reasons, there is no doubt who the chief magistrate is in these proceedings. Some writers have suggested that the story of the 42 judges simply came about to account for the number of confessions the deceased made and means nothing more. For instance, in *The Complete Gods and Goddesses of Ancient Egypt*, Richard Wilkinson points out that each judge represented a specific deity who was 'responsible for a particular crime, which ranged from murder and rapaciousness to religious transgressions such as blasphemy or damaging the image of a god.'[164] However that may be, a numerological analysis suggests this archaic ritual has a deeper connection to the Pleiades than is immediately apparent. As Barbara Lesko points out in *The Great Goddesses of Egypt*, 'The forty-two judges of the Dead were really only a multiple of seven.'[165] Thus we find that 7 x 7 equals 42. Furthermore, in numerology, multiple numbers like 42 are added together so that 4 plus 2 equals six. But when six is combined with the goddess Maat, we not only have the magical number seven but the interplay between six and seven that alludes to the story of the Lost Pleiad as well as the mystical Tetragrammaton and the great cycles of time.[166]

Although Johanna Lambert draws several key parallels between the Egyptian Hathors and the Hindu Krittika in *Wise Women of the Dreamtime*, she offers no further explanation for this observation. Madame Blavatsky, on the other hand, argues that these similarities are not coincidental at all for she sees in India the birthplace of ancient Egyptian culture.[167] Barbara Walker also notes other cultural crossovers, especially in regard to the relationship between the Sisters, sharp cutting objects and the administration of justice.[168] Hathor, she points out, was identified with several goddesses, including the Greek crone Hecate. This affiliation is especially revealing for, as we know, it is the midwife who traditionally cuts the umbilical cord to separate the child from its mother. On a symbolic level, this may be read as our cosmic severance from God or the universe on the physical plane. In many cultures, this cord is imbued with sacred meaning and the esoteric lore surrounding it is prolific. How it was perceived in one Aboriginal culture in Bandaiyan (Australia) is considered in the final chapter, but its figurative use in regard to law and justice in ancient Egypt relates back to the Seven Hathors in their role as judges, especially through Maat the personified whole.

Maat was often depicted with a magical cord, says Walker, that 'signified the binding force of matriarchal law.'[169] Even the greatest of Egyptian deities, the Sun god Ra acknowledges the supreme authority of the Seven Hathors where he states: 'Their law is the cord in Amentet.'[170] This notion of the binding powers of the law as symbolised by Maat's cord survives in contemporary times through the colloquial expression of couples 'tying the knot' when they marry. The saying actually derives from the old ceremonial custom of placing cords across their joined wrists, which were then tied together. As well as acknowledging the law of the Mother and the aspirations of truth and harmony in a marriage, this knot symbolised the 'knot of eternity' that binds two people to one another. On a deeper level, these knots may be equated with the strands of our DNA that are joined together to create new life, the ultimate binding of partnership.

Apart from the underlying symbolism of these marital and relationship ties, Barbara Walker points out that three interlaced rings or three joined triangles creating a triple vesica were often employed to invoke or denote the fates.[171] Even though they were seven in ancient Egypt does not mean that the triple deity was not of spiritual import and therefore the triangular shapes of pyramids and their physical layout at Giza would have taken on some symbolic significance in this regard. Certainly their numeric and physical configuration attracted the attentions of Robert Bauval and Adrian Gilbert. In *The Orion Mystery*, the authors argue that ancient Egyptian writings such as the Pyramid Texts suggest that the three large pyramids were purposely built to model the three stars of Orion's Belt, as that constellation represented their god Osiris or Sahu and his heavenly abode.[172] Orion as you may recall, was the legendary hunter and eternal amorous pursuer of the Seven Sisters in Greek and Aboriginal Australian mythology. And, as previously noted, to some Aboriginal people the stellar region surrounding Orion's Belt represents the celestial home of their supreme creator, the Rainbow Serpent.

Astronomer Edwin Krupp disagrees with Bauval and Gilbert's interpretation on the basis that the pyramids and other geographical features such as the Nile are not in the correct positions as the Giza starmap suggests.[173] However that may be, the triangular nature and symbolism of the pyramids cannot be ignored or dismissed, nor their potential involvement with the stars of Hathor. The association of magical cords, knots and weaving with women generally — and with the Pleiades specifically — essentially derives from those cultural traditions that perceive the universe as the creation of the mother goddess. In Native American traditions she is known as 'Spider Woman' or 'Grandmother Spider' who, like the Greek Ariadne, weaves the fate and destiny of human beings.[174] The web that she weaves is the very matrix (or *maatrix*) of our existence. Besides reflecting the patterns and orderliness of our lives, a spider's web provides the perfect natural graphic of a coordinates system that reflects modern computer-generated images of the curvature of time and space.

These representations of Maat and the Krittika as female judges endure today within the Western adversarial legal system. Statues of a blindfolded woman standing guard outside courts with sword and scales in both hands, otherwise known to the legal fraternity as 'Justice', is none other than the Goddess in her guise as wise woman and crone. Her sword is the sword of discernment that enables judges to distinguish right from wrong, and in this context the airy element of Maat's feather becomes the sword of truth that cuts through illusion like the razor of the Krittika. Her blindfold suggests that justice should be impartial and fair, neither influenced by 'fear, favour or affection' as in the judges' oath. The scales represent the weighing of rights and wrongs in the balancing of justice like Maat's scale, which weighed hearts in the Egyptian underworld. The portrayal of justice as a woman goes back to the ancient Egyptian and ancient Hindu notions of the seven mother judges of the world which indicates the longevity of this symbol.

These impressions also live on in the astrological sign of Libra the Scales and in tarot cards, namely the major arcana card of Justice and the minor arcana suit of swords (the air element), notably the Queen of Swords and the androgynous Page. In the Rider Waite tarot we see

Justice perched on her throne wearing a red cloak with a blue sword held upright in her right hand with the balancing scales in the left. In *The Tarot Revealed*, Paul Fenton-Smith says this posture indicates the goddess can take appropriate action when required but is always mindful of the consequences. Her white shoe protruding from under her cloak, he says, alerts us to the consequences of our decisions and the blue sword (a colour associated with the Pleiades) represents her 'mental clarity.'[175]

The mother judges' associates, Thoth and Anubis

Maat was assisted in her official judicial duties by her husband Thoth the Egyptian Moon god, and her faithful companion, the jackal-headed god, Anubis. During the weighing of the heart-soul in the inquest of the dead, Anubis held the balancing of the scales while Thoth recorded the verdict.[176] As the goddess' scribe, Thoth was considered the god of 'magic, spells, writing and record keeping.'[177] The Greeks identified him with Hermes, the messenger of the gods and the Romans with their counterpart, Mercury.[178] Said to be the author of Egypt's sacred canon, the *Hermetica*, he was therefore deemed the God of Wisdom.[179] Because of these literary achievements the god was credited with having invented magic and the alphabet, but Barbara Walker and Madame Blavatsky reject these claims for different reasons. Walker disagrees largely on the basis that writing was more often linked with goddesses than gods, as with Isis or Maat in ancient Egypt, or Carmenta in Rome, the Norns of Scandinavia or the Gulses in ancient Babylon.[180] As to the claim that Thoth invented magic, Blavatsky vehemently protests by exclaiming, as if 'the unveiling of the eternal and actual mysteries could be *invented*.'[181] Thoth could not possibly be the 'inventor' of anything, says Blavatsky, for it is a generic title rather than the name of any one individual.[182] At most, she says, they were priests and initiates, or 'serpents of wisdom'.[183]

Whatever the true situation, in recognition of these supposed achievements the Greeks referred to him as Hermes Trismegistos, that expresses his greatness three times: Hermes the 'Great, Great, Great' god.[184] In Latin this expression is written as Hermes Trimegistus but even his so-called Egyptian name was a Greek corruption of the Egyptian Djuhti or Djehuty.[185] Curiously, this word phonetically resembles the English word 'duty' with its connotations of moral or legal obligation, as would be expected of a judge's associate. Ultimately of course, it does not matter whether the Moon is male or female or which of the sexes created a particular invention before the other, for there is no competition in the world of spirit. In this sense, the emphasis placed on either sex is irrelevant in terms of understanding the *mysteries* — the principal goal of which is the unification and synthesisation of the two monads or the divided whole. This was the main purpose of the mystery school known as 'The Law of One' that took place in the Great Pyramid.

What is interesting about the gender-specific Writer Fates is their general identification with the stars of the Pleiades in several world cultures, and specifically to the Seven Hathors of ancient Egypt. It draws attention to Blavatsky's belief that the Pleiades, or Atlantides, are the prime determinants of human destiny and the underwriters of karmic law. This notion is further heightened by their perception as mother judges in ancient Egypt and India, where one's fate was ultimately sealed by the deliverance of her holy decree or judgment. The goddess' feathers were called the 'plumes of Maat', for the earliest writing implements were made from birds' quills called plumes, thereby sealing the identification of the Bird Goddess with providence.[186] Under her 'alternate name of Seshat', Maat came by another one of her many titles, that of 'Mistress of the House of Books.'[187] This epithet is especially intriguing in light of the channelled reference of Barbara Hand Clow to the 'Great Library' on the star Alcyone in *The Pleiadian Agenda*.[188] It appears to suggest a similar title for that goddess, if not as head librarian.

Although there are many similarities between the Greek and Egyptian Hermes, one major difference is that in the Greek legend Hermes is not the husband but the son of Maia. Not that it matters a great deal, says Blavatsky, for the practice of merging the roles of gods into husbands, fathers and sons and goddesses into wives, mothers and daughters is a widespread mythological theme.[189] It explains to some extent the synthesis of male gods into the Sun, the giver of life, and the merger of various titles of the goddesses into the generic name for Mother as in Maia, Maya, Maria, Mary and so forth. This might account for the linguistic similarities between the Greek *Maia*, the Hindu *Maya*, the Bundjalung *Maimai*, the Egyptian *Maat*, the Hawaiian *Makali'i* and the Maori *Matariki* (pronounced Maatariki). It also explains why several Egyptian goddesses are described alternatively as the wife, mother, daughter or sister of Ra the Sun god. Aside from this distinction, there are several notable similarities between the Greek and Egyptian legends of the Seven Sisters of the Pleiades. Both goddesses are described as exceedingly beautiful and both are coupled with a god, Hermes or Thoth, who is responsible for many scientific achievements and inventions including the alphabet and writing.[190]

In this regard they bore titles of the Messenger of the Gods, a status affirmed by their close relationship with the supreme deity, the Sun.[191] Hence in Greek starlore, Hermes is relegated to the closest planet to the Sun, that of Mercury.[192] Thoth, on the other hand, is spoken of as the Egyptian Moon god, often referred to as the companion of the Sun.[193] What is more, their physical characteristics and personal traits are exactly alike. For instance, mercury was the name given to the chemical element whose silvery liquid substance is shiny and swift like the Moon, whose movement in the night skies is the fastest of all astral objects. To complete the symbolism the celestial couriers, Hermes and Thoth, both employed avian imagery; Thoth is illustrated with the head of an ibis and Hermes or Mercury is often drawn wearing feathered boots or sandals.[194]

The latter image of Hermes or Mercury is strikingly similar to the specialised lawmen and women in Bandaiyan (Australia) known as kadaicha. The kadaicha or 'feather foot' as they are commonly known (because their special shoes are made of emu feathers), are Aboriginal messengers who dispense and administer justice when certain tribal laws are broken. Quite apart from Hermes' footwear, the emu provides yet another connection between Maat and some Aboriginal legends of the Pleiades. You may recall for instance, the story of the Magara or

'Emu Women' who are chased by the Dingo Men of Orion. Emus are large Australian birds that are related to the ostrich and it is interesting to note that Maat's 'feather of truth' is that of an ostrich.[195] The cultural interconnections and crossovers are astounding, but predictable, given our common heritage and origin as human beings.

The goddess' other faithful assistant in the ritual of the weighing of the heart was the black jackal-headed god Anubis, who was 'depicted either as a jackal or as a man with the head of a jackal.'[196] As previously noted, there are many similarities between the spiritual beliefs of the ancient Egyptians and those of Aboriginal Australians and Native Americans, especially in relation to the diverse canine species. All are endowed with similar characteristics — that of teacher, friend, companion, funeral attendant and caretaker. The Moon, in particular, appears as an ally to dog that guards the gates at the entrance to the underworld.[197] In ancient Egypt the Moon was male, as it is in Bandaiyan (Australia). We saw in the Australian Aboriginal chapter that there is a clear demarcation between men's and women's domains codified by dogs and birds to symbolise this gendered division. We also saw that there were two jackal gods in ancient Egypt — Anubis and Wewawet (or Upuaut). Both held similar titles such as Opener of the Body.[198] This title, says Walker, customarily denoted the firstborn child of the goddess.[199] Some accounts suggest that Isis is his mother,[200] or else her sister Nephthys through an illicit affair with her sister's husband and their brother, Osiris.[201]

In this respect, Anubis resembles the Hindu god, Skanda or Karttikeya, conceived in similar fashion between Agni and the Krittika. Another interpretation of this title may be a reference to the notion of the Goddess as the ultimate gatekeeper where allegory conceals the fact that 'the gate was another euphemism for female genitals.'[202] This was the real 'gate', says Barbara Walker, where life emerged and where souls passed through a 'sexual paradise' and the death of the 'phallic spirit.'[203] We see this metaphor played out in ancient Greece in the seven-gated city of Thebes where the Pleiades 'were represented by Seven Mantric Priestesses, who were linked to the Seven Pillars of Wisdom,' says Robert Graves in *The Greek Myths.*[204]

'This was an era when the Seven Hathors ruled over a rather grim calendar cycle which had the sacred king slain every seventh year.'[205]

Of all her many symbols of justice, perhaps the most enduring is the All-Seeing Eye of the Goddess. Like the Polynesian peoples of the Pacific, the ancient Egyptians perceived stars as the eyes of the gods.[206] We see traces of this belief in ancient Greek traditions that referred to the brightest star in Taurus as the eye of the bull. Egypt's sacred eye, the utchat, became one of its most holy icons and was depicted everywhere: on papyri, in temples, on tombs, coffins, and in the pyramids. It was identified with a number of Egyptian gods and goddesses such as Ra the Sun god, or Osiris, or his son Horus. The association of the utchat with male deities is a later historical development, says Walker. Originally it represented the ancient emblem of the uraeus, the Egyptian cobra glyph of the goddess in her serpentine aspect before merging with Maat and other Egyptian goddesses. As well as meaning 'truth', Maat's name was 'based on the verb to see,'[207] thus implying that 'no crime could be hidden' from her.[208] The famed blue lapis lazuli gemstone became linked with Maat and was called the 'stone of truth' for no one could hide from her All-Seeing Eye.[209]

This explains the prevalent belief in China that it could 'cure eye diseases,' says Walker.[210] The Chinese revered the gemstone as one of 'Seven Precious Things', and its blue colour signified 'water as the primordial element of creation.'[211] Once again, this illustrates the connection of the Pleiades with water and time as in the First Creation. Several writers have commented on the application of the All-Seeing Eye in the seal of the United States of America and on their dollar bill, pointing out that its designers 'were strongly influenced by Freemasonry, which adopted a number of Egyptian religious symbols,' including this.[212] What is more, the symbolism of law and order implied in the utchat panopticon — together with the Latin motto, *Novus Ordo Seclorum* (New World Order) — makes the political agenda of the designers quite clear.[213]

More often than not, the utchat was often drawn as just one eye, although on some occasions two eyes were drawn to indicate the Sun and Moon or Ra and Thoth respectively. It was particularly significant as to which eye was illustrated, says Bob Frissell, for both eyes represented two separate twelve-year spiritual training programs referred to as the 'Right and Left Eye of Horus' in the Egyptian Mystery School known as 'The Law of One.'[214] Drawing on the work of Drunvalo Melchizedek, he tells us the right eye training program taught initiates sacred geometry through the Flower of Life essence.[215] Its primary purpose was to integrate the left and right brains by 'convincing the left brain or the male side of us that there is one and only one spirit that moves through everything.'[216] Once that was achieved then integration would take place to allow the movement 'from polarity into unity consciousness.'[217] The left eye training involved working with 'various emotions, feeling, fears, and both positive and negative aspects of the chakra.'[218] One particular exercise in learning to overcome one's fear involved the unenviable task of swimming in an underground watery maze with crocodiles![219] Upon completion of the two separate twelve-year training periods, initiates 'would descend into the Great Pyramid for a three-and-a-half-day period of final initiation.'[220]

Mountains of fire, the Giza pyramids

The three pyramids of Giza near Cairo are the last remaining members of the Seven Wonders of the ancient world. Like the Sphinx beside them, they remain a riddle. The subjects of countless books, every part of their walls has been analysed and measured, their steps counted, their interiors explored, every nook and cranny probed and searched, even by intrepid robotic archaeologists. The objective of these examinations has been to extract some meaning or deeper understanding of their purpose and existence, including information on who built them and how. Numerous explanations and suggestions have been proposed, some plausible, some stimulating, others fantastical. Whatever the answers, there is no doubt that the pyramids 'have certainly kept the world wondering all these years,' say John and Elizabeth Romer, and that their symbolism remains highly charged.[221] Their distinctive triangular shape, exact precision and sheer size are simply awe inspiring. Much has been made of their symbolic value as cosmic mountains, pharaohs' mountains or mountains of the star gods. This has great significance in a country like Egypt that has mostly a flat, featureless terrain, at least where the pyramids of Giza are located. When we consider a possible connection with the Seven Sisters of the Pleiades, whose pyramids, hillocks and towers are essential landing and launching platforms for their various sojourns on Earth, then their meaning takes on an extraterrestrial dimension. For there can be no doubt that the Sisters' presence was equally felt on the Giza plateau as it was elsewhere.

The largest and most famous of the three pyramids is the Great Pyramid of Khufu, named for its alleged builder whom the Greeks called Cheops (hence its other name, Cheops' Pyramid).[222] Some writers disagree with this historical account. Blavatsky, for instance, boldly states that Cheops never built the Great Pyramid, and argues that it 'was built ages before him.'[223] If anything, she says, the pharaoh 'desecrated' the pyramid 'by giving it another use.'[224] Its real purpose, she points out, was to house the *mysteries*, which were revealed only to initiates in special ceremonies conducted within its hallowed walls.[225] Just exactly what these rituals involved is looked at in the following section, but the fact that there are seven chambers in the Great Pyramid is enormously significant for a number of reasons.[226] As previously noted in the opening chapter, seven is considered the ideal number of initiation and its association with the Pleiades, in particular, makes it even more so. Of these seven chambers, the so-called King's and Queen's chambers played the most crucial roles in these ceremonies.

More than a hundred and twenty-three years ago the former Royal Astronomer of Scotland, Charles Piazzi Smyth, wrote that these seven chambers were built to commemorate the seven stars of the Pleiades.[227] Even more astonishing is the revelation that the precessional cycle or 'Great Year of the Pleiades' was named for an extremely rare astronomical alignment involving the previous Pole Star Thuban (Alpha Draconis) in the constellation of Draco the Serpent and Alcyone of the Pleiades. Not only does this extraordinary configuration occur just once during the entire cycle of 26,000 years, but it can be viewed from the entrance passage to the Great Pyramid.[228] Such an incredibly rare alignment clearly reveals the importance attached to this star cluster by the ancient Egyptians. Piazzi Smyth writes that the astronomer-priests recorded the last alignment in 2170 BCE during the vernal equinox, when these stars crossed the meridian at midnight.[229] If the heavens are mirrored on the Earth plane, surely this must provide the most telling clue as to the identity of the legendary Seven Builders? Could this be what Robert Graves meant when he alluded to Alcyone as the 'mystical' leader of the Pleiades in *The White Goddess*? What does this tell us about the antiquity of Egyptian civilisation or of the pyramids?

The prevailing orthodox archaeological view is that the Egyptian civilisation arose a mere 4,000 years ago during the Age of Taurus and a much shorter time is ascribed to the alleged age of the pyramids. In recent years much more daring estimates have been suggested to indicate a far more remote period of antiquity for the Egyptian civilisation (and by association the Giza pyramids) than previously held. An increasingly popular date favoured by some authors, including the Sleeping Prophet Edgar Cayce, is that of 10,500 BCE.[230] Although this timeframe 'accords so well with the geology of the Sphinx,' Graham Hancock points out this does not necessarily mean the pyramids 'were actually built in 10,500 BC.'[231] They may have been built in 'piecemeal' fashion over many thousands of years.[232] Or alternatively, this date may represent the creation of the mound — the sacred isle of Egyptian cosmology that provided the architectural blueprint for their foundation.

Madame Blavatsky suggests an even older date of 31,105 years ago for the building of the pyramids.[233] Although she does not elaborate on how she arrived at this figure, it is possible she may have done so by not counting backwards from our current Piscean age to the Age of Taurus, but rather in the opposite direction, thereby completing an entire cycle of precession including Taurus and Pisces. Curiously, by counting in this manner, Blavatsky has effectively drawn time not as a circle but in a conical shape reflecting a tube torus in a pyramid. Whether this was a deliberate move on her part is not known, but this particular geometric configuration, as Frissell points out, 'is the primal shape of the Universe.'[234] It is unique primarily because of its ability 'to move in on itself,' which no other shape can do.[235]

Whatever the correct date might be for the erection of the Giza pyramids or of the beginning of ancient Egyptian civilisation, it is a simple fact that the Pleiades has greatly inspired and influenced the architects and builders of the Great Pyramid. This astronomical alignment involving the stars of the Pleiades and Draco is uniquely precessional. It not only reveals the marvels of an ancient civilisation ahead of our own but it marks the Pleiades unquestionably as the absolute sovereigns of human fate and destiny, as the Ancients have always maintained. In *The Great Pyramid*, Piazzi Smyth suggests that the starlight of Thuban the Dragon Star, shining down the descending passage of the Great Pyramid, represented humanity's 'descent into materialism, and for those who continued on this path, death and damnation.'[236] By contrast, the divine light of the Pleiades 'refer(s) to the path of the ascending passage and is thus in direct opposition to the Dragon Star's path of the descending passage.'[237] Could this be the true meaning and hidden message of the Bible in its reference to the 'sweet influence' of the stars of the Pleiades?

Pleiadian initiation in the Great Pyramid

Throughout many cultures in the world the Pleiades have played a central role in initiation rites and ceremonies, and it seems this may also be true of ancient Egypt. When archaeologists first discovered the inner sanctums of the Great Pyramid they claimed it was built as a mausoleum to house dead pharaohs, despite the fact that no mummies were found inside except for a solitary empty stone coffin in the King's chamber. Several writers, including Staniland Wake, Madame Blavatsky and Bob Frissell, disagree with this rather absurd hypothesis and argue that it was used as a place of initiation instead, of which the porphyry sarcophagus was its crowning glory. As Frissell points out, one of the more glaring anomalies is that this stone coffer is larger than the entrance. This suggests it 'had to have been put there while the place was being built,' which is 'not the custom for burials at all.'[238] According to Frissell, when the container was opened for the first time archaeologists found within it 'an unusual white crystalline powder.'[239] This mysterious substance, he claims, is the by-product of a deep meditative state excreted 'from the pituitary gland, which crystallises into a powder.'[240] For there to have been such a noticeable amount, says Frissell, it must have accumulated over a very long time, which suggests this 'baptismal font' had many graduates.[241]

Blavatsky and Frissell provide similar accounts of the alleged initiation rituals that took place inside the Great Pyramid, although Frissell's is more detailed on the use of black and white spiralling light.[242] In simple terms, an initiate representing the energising rays of the solar god would be led into the King's chamber and directed to lie overnight within the sarcophagus of the mother's womb to ritualise fertilisation to ensure the Earth's fruitfulness. The novice's re-emergence from the tomb the following morning signified their resurrection from their ritualised death. In Egypt the womb of the great mother took the shape of the sarcophagus but it changed its shape and form in other countries, says Blavatsky. The only requirement for the mysteries was that it remain 'a vessel, a symbolic navis or boat-shaped vehicle, and a container, symbolically, of germs or the germ of life.'[243]

Given their central role in the Great Pyramid's construction and their general role in initiation ceremonies, it is logical to assume the Pleiades may have played a key role in these rituals. Frissell's reference to the specific use of a black and white spiral light during the initiation procedures inside the Great Pyramid is especially revealing within the context of the Draco–Pleiades alignment. Unlike the other writers, he suggests the initiation rites first commenced under the pyramid in the subterranean chambers before the postulants made their way up to the other levels.[244] This was done to ensure their first encounter with the 'black light spiral,' which penetrates into the Earth's centre as well as extending up into the Halls of Amenti. Then, making their way into the King's chamber, the postulants lay down inside the sarcophagus to experience the healing energies of the 'white light spiral.'[245] The tomb was positioned to allow this spiralling white light to penetrate into the pineal gland, which enabled the initiate to experience different levels of consciousness. Deep in trance the postulants would then 'find their way back into their bodies' through the use of a shamanic umbilical cord by applying techniques and principles of sacred geometry.[246] Upon their ritualised resurrection, the new initiates would enter the Queen's chamber to rest after the ordeal. This ritual in the Great Pyramid, says Frissell, represented the completion of the Law of One training that took twenty-four years to complete. It represented the integration of two separate twelve-year training periods that began with the Left Eye 'emotional body' and ended with the Right Eye 'unity consciousness' school.[247]

Lending weight to the discussion in *Heaven's Mirror*, Graham Hancock says scholars 'have failed to consider the possibility' that all of the pyramids may have been used for initiation purposes 'in order to gain foreknowledge of the afterlife realm.'[248] Introducing modern concepts to the debate, he suggests the Great Pyramid may have been built as a 'three-dimensional model' of the Egyptian underworld, a 'sort of simulated Netherworld.'[249] In other words, a huge, hi-tech computer playground arcade full of virtual reality games. The analogy is an appealing one but this was no game and, as Madame Blavatsky suggests, it was much more than this. The cycles of initiation that took place within the Great Pyramid were miniature reproductions of the great cosmic changes that occurred during the various world ages in the cycle of the Great Year of the Pleiades.[250] Just as the 'heavenly bodies' returned to 'the same relative positions as they occupied at its outset' at the close of the Great Year, so too did the initiates regain their 'state of divine purity and knowledge' as when they set out on their 'cycle of terrestrial incarnation.'[251]

Furthermore, the notion of black light is not unknown to science or religion. One of the leading mystical Kabbalistic texts, the *Zohar*, makes several references to 'Black Fire', which represents 'Absolute Light Wisdom.'[252] Other spiritual traditions refer to the 'Black Sun', the central bulge of our galaxy as the source of all wisdom, and scientists believe that the core of all galaxies may contain black holes, which are not the absence of light but the containment of it. This 'central sun' is of great esoteric significance to Kabbalists and other occult philosophies.[253] Of course, in reality there is no such thing as a 'sun' at the centre of our galaxy and these references must be read on a purely symbolic level according to these teachings.

Unveiling the Pleiades

The search for the Seven Sisters of the Pleiades in ancient Egypt has been an engaging one. On the surface it appears very little has been written about them, but a deeper investigation reveals otherwise, for there is no doubt that the Stars of Hathor were much revered and that they played a significant role in the lives of ancient Egyptians. One of the major difficulties confronting researchers in this field is the fact that there were numerous periods of Egyptian history, making it difficult to isolate one particular epoch as representative of an entire civilisation. Add to this the difficulty of deciphering hieroglyphics and the many interpretations that ensue, and the task becomes incredibly complex. But not insurmountable, for mythical language is a language unto itself that becomes all the more clear when one learns to read and understand the underlying metaphors, symbols and archetypes. The fact that many of the early scientists, Egyptologists, archaeologists and the like were not as concerned or as knowledgeable about astronomy may have led to enormous cultural deficiencies in some of their interpretations of ancient Egyptian beliefs and lifestyles. It is possible, therefore, that some, if not many, were simply unaware of the archaic significance of the Pleiades in ancient Egypt.

Another historical factor to take into account is the process of colonisation, whereby people, culture and the environment are significantly altered or distorted in all its manifestations, including the outright destruction, displacement and tampering with matriarchal images and traditions by opposing patriarchal forces. Much remains to be 'uncovered' in order to be 'discovered.' Perhaps part of the delay is — as Madame Blavatsky suggests — that apart from the initiates 'no one has understood the mystic writing,' or perhaps time has been reckoned for so long in that part of the world that the Pleiades, like the legendary Lost Sister, have simply faded into antiquity in the desert sands.[254] Even so, I still draw hope and inspiration from the wise theosophist whose invaluable words of encouragement reverberate in my mind. For she maintains the *key* to the mystery remains and the answers are still there, where they may be found written 'on the timeworn granite pages of cave-temples, on sphinxes, propylons and obelisks.'[255] These relics and remnants, she says, 'have stood there for untold ages, and neither the rude assault of time, nor the still ruder assault of Christian hands, has succeeded in obliterating their records.'[256] But, 'Who will unseal them,' she asks, 'Who of our modern, materialistic dwarfs and unbelieving Sadducees will dare to lift the Veil of Isis?'[257]

Who indeed?

Notes

1.   Lamy, *Egyptian Mysteries*, p. 82. The quote is made up of two separate quotes — one from Lucy Lamy and the other from Serge Sauneron who is quoted in her book.
2.   Blavatsky, *The Secret Doctrine*, vol. 2, p. 631.
3.   Hancock, *Heaven's Mirror*, p. 66. See also *Myths and Legends of Ancient Egypt* by Lewis Spence, p. 147. For an esoteric treatment of the Seven Builders see the chapter on 'The Builders' at pp. 182–91 in *The Sufis* by Idries Shah. See also *Keeper of Genesis* by Robert Bauval and Graham Hancock, pp. 200–202, for further information on Egypt's Seven Sages.
4.   Allen, *Star Names*, p. 399. See also *Isis Magic* by Isidora Forrest, p. 132.
5.   Tyler Olcott, *Starlore of All Ages*, p. 410.
6.   Jobes and Jobes, *Outer Space*, p. 337 and Watterson, *Gods of Ancient Egypt*, p. 120.
7.   Forrest, *Isis Magic*, p. 38. See also *The Book of the Dead* by E. A. Wallis Budge at pp. 644–45.
8.   Wallis Budge quoted in Barbara Walker's *The Woman's Dictionary of Symbols and Sacred Objects*, p. 76.
9.   Ibid.

10. Storm, *Egyptian Mythology*, p. 48.
11. Ibid.
12. Hart, *A Dictionary of Egyptian Gods and Goddesses*, p. 79.
13. Spence, *Myths and Legends of Ancient Egypt*, p. 169.
14. Forrest, *Isis Magic*, p. 132. The actual island of Philae is now submerged under the waters of the Nile. Due to the building of the Aswan Dam the island of Philae would become periodically submerged under water for six months of every year. The temple was removed piece by piece between 1972 and 1980 to the nearby island of Agilkia where it was reassembled and re-erected.
15. For a colour photograph of the Seven Cows and Heavenly Bull on the Tomb of Nefertari see *Egyptian Mysteries* by Lucie Lamy at p. 83. See also *House of Eternity* by John McDonald at pp. 54–56 and 80–81.
16. Spence, *Myths and Legends of Ancient Egypt*, p. 169.
17. Richard Allen writes in *Star Names* that the stars of the Pleiades were also identified with the goddess Neith and were called *Chu*. See p. 399.
18. For an excellent treatment of the notion of seven primary Egyptian goddesses see Barbara Lesko's *The Great Goddesses of Egypt*. Although she does not identify these seven goddesses with the Pleiades, nonetheless, it supports my speculative ideas surrounding the Seven Sisters in ancient Egypt. We seem to agree on five out of the seven goddesses as being principal deities, except that she includes Mut whereas I suggest that Maat was more significant. Furthermore, she includes Wadjet the Cobra Goddess, whereas I have chosen Nephthys as one of Nut and Geb's set of quadruplets, who is as powerful and significant as her other siblings — Isis, Osiris and Set.
19. Spence, *Myths and Legends of Ancient Egypt*, p. 169.
20. Hart, *A Dictionary of Egyptian Gods and Goddesses*, p. 79.
21. Spence, *Myths and Legends of Ancient Egypt*, p. 163.
22. Walker, *The Woman's Dictionary of Symbols and Sacred Objects*, p. 368.
23. Ibid.
24. Ibid, p. 368.
25. Ibid, p. 233.
26. Ibid, p. 344.
27. Ibid, p. 343.
28. Ibid, p. 90.
29. Ibid.
30. Storm, *Egyptian Mythology*, p. 36.
31. Walker, *The Woman's Dictionary of Symbols and Sacred Objects*, p. 470.
32. Storm, *Egyptian Mythology*, p. 36.
33. Forrest, *Isis Magic*, p. 41.
34. Ibid, p. 39.
35. Watterson, *Gods of Ancient Egypt*, p. 118.
36. Ibid.

37. Walker, *The Woman's Dictionary of Symbols and Sacred Objects*, p. 105.
38. Watterson, *Gods of Ancient Egypt*, p. 105.
39. Some writers spell Ra as Re. I have followed writers such as E. A. Wallis Budge who refer to the Sun god as Ra in *The Book of the Dead* at pp. 164–67.
40. Walker, *The Woman's Dictionary of Symbols and Sacred Objects*, p. 402.
41. Ibid, p. 207.
42. Ibid, p. 5.
43. Ibid, p. 460.
44. Ibid, p. 470.
45. Cameron, *Symbols of Birth and Death in the Neolithic Era*, pp. 4–5.
46. Walker, *The Woman's Dictionary of Symbols and Sacred Objects*, p. 22.
47. Blavatsky, *The Secret Doctrine*, vol. 1, p. 5 and vol. 2, pp. 546–47.
48. Bowker, *World Religions*, p. 12.
49. Ibid, p. 225.
50. Hancock, *Heaven's Mirror*, p. 66.
51. Blavatsky, *The Secret Doctrine*, vol. 1, p. 5.
52. Walker, *The Woman's Dictionary of Symbols and Sacred Objects*, p. 466.
53. Ibid.
54. Tyler Olcott, *Starlore of All Ages*, p. 410.
55. Walker, *The Woman's Dictionary of Symbols and Sacred Objects*, pp. 402 and 466.
56. Spence, *Myths and Legends of Ancient Egypt*, p. 164.
57. Graves, *The Greek Myths*, p. 163.
58. Forrest, *Isis Magic*, p. 38.
59. Jobes and Jobes, *Outer Space*, p. 337.
60. Temple, *The Orion Mystery*, pp. 152, 124 and 350.
61. Jobes and Jobes, *Outer Space*, pp. 336–37.
62. Ibid, p. 337.
63. Sesti, *The Glorious Constellations*, p. 447.
64. Ibid.
65. Mackenzie, *Legends of China and Japan*, p. 182.
66. Walker, *The Woman's Dictionary of Symbols and Sacred Objects*, p. 515.
67. Forrest, *Isis Magic*, p. 39.
68. Jobes and Jobes, *Outer Space*, p. 337.
69. Forrest, *Isis Magic*, p. 39.
70. Ibid.
71. Bowker, *World Religions*, p. 24.
72. Ibid, p. 23.
73. Walker, *The Woman's Dictionary of Symbols and Sacred Objects*, p. 276.
74. Bowker, *World Religions*, p. 24.
75. For a photograph of the entrance to Hathor's temple at Dendera, see p. 94 of Barbara Lesko's *Great Goddesses of Egypt*, or p. 58 of *Gods and Symbols of Ancient Egypt* by Manfred Lurker.

76. De Santillana and Von Dechend, *Hamlet's Mill*, pp. 247 and 405. See also *Beyond the Blue Horizon* by Edwin Krupp at p. 227.
77. Ibid.
78. De Santillana and Von Dechend, *Hamlet's Mill*, p. 127.
79. The authors of *Hamlet's Mill* agree — see footnote #12 at p. 416.
80. For a colour photograph of the astronomical ceiling of Set I, see *Heaven's Mirror* by Hancock at p. 89.
81. Blavatsky, *The Secret Doctrine*, vol. 2, p. 549.
82. Walker, *The Woman's Dictionary of Symbols and Sacred Objects*, p. 378.
83. De Santillana and Von Dechend, *Hamlet's Mill*, p. 386.
84. Hancock, *Heaven's Mirror*, pp. 60–1.
85. Ibid.
86. See Lewis, *The Astrology Encyclopedia* at p. 510.
87. Ibid, p. 401. Zodiac 'crosses' are formed by four pairs of 'opposite' signs (i.e. signs that are 180° from one another and are opposite 'elements'). One quarter formed by the cross represents a square, meaning the signs are sequentially 90° from one another as you move around the cross in either direction. See the section on 'aspects' at pp. 40–41 for further clarification.
88. Hancock, *Heaven's Mirror*, p. 61.
89. The Seven Hathors that appear on the walls of Queen Nefertari's tomb in the Valley of the Queens at Luxor 'are accompanied by the four rudders representing the cardinal points of the sky . . .' says George Hart in *A Dictionary of Egyptian Gods and Goddesses* at p. 80.
90. To understand the changing night skies from a seasonal perspective, see *Stars of the Southern Skies* by Patrick Moore. Although written from a Southern Hemisphere point of view, the astronomical principles remain the same for the Northern Hemisphere.
91. Ibid, p. 153. See, for example, the Greek mythological explanation for explaining why Orion (near Taurus and the Pleiades) cannot be seen in the night skies together with Scorpio.
92. See Barbara Walker's discussion of 'Four-Way Motifs' in *The Woman's Dictionary of Symbols and Sacred Objects* at pp. 46–64.
93. Hancock, *Heaven's Mirror*, p. 61.
94. Allen, *Star Names*, p. 393.
95. Schwaller de Lubicz, *The Temple of Man*, vol. 1, p. 487.
96. Blavatsky, *The Secret Doctrine*, vol. 1, p. 435.
97. Flem-Ath and Wilson, *Atlantis Blueprint*, pp. 170–71.
98. Ibid, p. 171.
99. Ibid.
100. Ibid.
101. Ibid.
102. Ibid, p. 172.
103. Ibid, p. 170.

104. Ibid.
105. Ibid.
106. Walker, *The Woman's Dictionary of Symbols and Sacred Objects*, p. 295.
107. Blavatsky, *The Secret Doctrine*, vol. 2, p. 435.
108. Ibid, vol. 1, p. 663.
109. Ibid, vol. 2, p. 435.
110. Ibid.
111. Lewis, *The Thirteenth Stone*, p. 256.
112. Blavatsky, *The Secret Doctrine*, vol. 2, pp. 364 and 373–74.
113. From *The Egyptian Book of the Dead* (Book 1, chapter XIV) as quoted in Blavatsky, *The Secret Doctrine*, vol. 2, p. 374.
114. Lewis, *The Thirteenth Stone*, p. 28. According to Hindu beliefs, there are seven branches of the Tree of Knowledge. See Blavatsky, *The Secret Doctrine*, vol. 1, p. 168.
115. Lewis, *The Thirteenth Stone*, p. 107.
116. Ibid.
117. Ibid, pp. 25 and 107.
118. Ibid, p. 243.
119. Eliade, *Images and Symbols*, pp. 64–65.
120. Watterson, *Gods of Ancient Egypt*, p. 120.
121. Walker, *The Woman's Dictionary of Symbols and Sacred Objects*, p. 36.
122. Ibid, pp. 16–17.
123. Ibid, p. 158.
124. Lurker, The *Gods and Symbols of Ancient Egypt*, p. 85.
125. Schwaller de Lubicz, *The Temple of Man*, vol. 1, p. 465.
126. Ibid, p. 466.
127. Ibid, p. 467.
128. Ibid, p. 466.
129. Ibid.
130. Jobes, *Mythology, Folklore and Symbols*, Part 2, p. 1422.
131. Ibid.
132. Wallis Budge, *Egyptian Mysteries*, p. 20.
133. Shah, *The Sufis*, p. 187.
134. Ibid.
135. Walker, *The Woman's Dictionary of Symbols and Sacred Objects*, p. 161.
136. Ibid, p. 196.
137. Ibid, p.161.
138. Ibid.
139. Ibid.
140. Fenton-Smith, *The Tarot Revealed*, p. 66.
141. Blavatsky, *Isis Unveiled*.
142. Forrest, *Isis Magic*, p. 4.
143. Walker, *The Woman's Dictionary of Symbols and Sacred Objects*, p. 207.

144. Ibid, p. 53.
145. Blavatsky, *Isis Unveiled*, vol. 1, p. 532. See also *Isis Magic* by Isidora Forrest, pp. 74–79.
146. Wallis Budge, *Egyptian Ideas of the Afterlife*, p. 106.
147. Forrest, *Isis Magic*, p. 27.
148. Ibid, p. 37.
149. Storm, *The Encyclopedia of Eastern Mythology*, p. 60.
150. Ibid.
151. Ibid.
152. Ibid.
153. Ibid. See Isidora Forrest's discussion of the work of Marija Gimbutas, author of *The Language of the Goddess* and its application to the Egyptian context at pp. 24–28. Although she only talks about Isis as the 'Bird of Prey Goddess,' nonetheless her discussion equally pertains to the other Egyptian goddesses.
154. Forrest, *Isis Magic*, p. 26.
155. Ibid, p. 20.
156. Hancock, *Heaven's Mirror*, pp. 50–52, 113 and 153.
157. Walker, *The Woman's Dictionary of Symbols and Sacred Objects*, p. 135.
158. Ibid, p. 317.
159. Storm, *Egyptian Mythology*, p. 48.
160. Hancock, *Heaven's Mirror*, p. 71. For a complete list of the 42 Negative Confessions see *The Book of the Dead* by E. A. Wallis Budge at pp. 572–84.
161. Ibid.
162. Storm, *Egyptian Mythology*, p. 48.
163. Harris and Pemberton, *The British Museum Illustrated Encyclopaedia of Ancient Egypt*, p. 90.
164. See the complete list of gods who presided over the Death Tribunal and the list of 42 crimes they oversaw in Wilkinson, *The Complete Gods and Goddesses of Ancient Egypt*, at p. 84.
165. Lesko, *The Great Goddesses of Egypt*, p. 88.
166. Tetragrammaton refers to the hidden name of God in the Jewish religion. It is supposed to consist of four Hebrew letters *yod, he, vau* and *he*, which together spell YHWH, 'usually called the Ineffable Name of God,' says Barbara Walker in *The Woman's Dictionary of Symbols and Sacred Objects* at p. 223.
167. Blavatsky, *The Secret Doctrine*, vol. 2, p. 435.
168. Walker, *The Woman's Dictionary of Symbols and Sacred Objects*, p. 151.
169. Ibid, p. 130.
170. Ibid. Note the use of the pronoun 'their'.
171. Ibid, p. 41.
172. See *The Orion Mystery* by Bauval and Gilbert.
173. Krupp, *Skywatchers, Shamans and Kings*, p. 290.

174. Walker, *The Woman's Dictionary of Symbols and Sacred Objects*, pp. 51 and 419–20.
175. Fenton-Smith, *Mastering the Tarot*, p. 37.
176. Storm, *Egyptian Mythology*, p. 48.
177. Walker, *The Woman's Dictionary of Symbols and Sacred Objects*, p. 403.
178. Ibid, p. 214.
179. Hancock, *Heaven's Mirror*, pp. 72 and 77.
180. Walker, *The Woman's Dictionary of Symbols and Sacred Objects*, p. 117.
181. Blavatsky, *The Secret Doctrine*, vol. 3.2, p. 211.
182. Ibid.
183. Ibid, p. 210.
184. Walker, *The Woman's Dictionary of Symbols and Sacred Objects*, p. 204.
185. Watterson, *Gods of Ancient Egypt*, p. 182. See also *Gods and Goddesses of Ancient Egypt* by Richard Wilkinson at p. 215.
186. Walker, *The Woman's Dictionary of Symbols and Sacred Objects*, p. 135.
187. Ibid, p. 403.
188. Hand Clow, *The Pleiadian Agenda*, p. xx.
189. Blavatsky, *The Secret Doctrine*, vol. 1, p. 396.
190. Walker, *The Woman's Dictionary of Symbols and Sacred Objects*, p. 204.
191. Burnham et al., *Practical Skywatching*, p. 108.
192. Walker, *The Woman's Dictionary of Symbols and Sacred Objects*, p. 92.
193. Ibid, p. 404.
194. Ibid.
195. Spence, *Myths and Legends of Ancient Egypt*, p. 109.
196. Storm, *Egyptian Mythology*, p. 17.
197. Sams and Carson, *Medicine Cards*, p. 97.
198. Bauval and Gilbert, *The Orion Mystery*, p. 58.
199. Walker, *The Woman's Dictionary of Symbols and Sacred Objects*, p. 393.
200. Ibid.
201. Storm, *Egyptian Mythology*, p. 17.
202. Walker, *The Woman's Dictionary of Symbols and Sacred Objects*, p. 136.
203. Ibid.
204. Graves, *The Greek Myths*, vol. 1, pp. 3–4 and 45.
205. Ibid.
206. For an esoteric reading of 'The Eye of Heaven' in Polynesian societies see Hancock, *Heaven's Mirror*, p. 245.
207. Walker, *The Woman's Dictionary of Symbols and Sacred Objects*, p. 112.
208. Ibid, p. 201.
209. Ibid, p. 515.
210. Ibid.
211. Ibid.
212. Ibid, p. 201. See also *The Biggest Secret* by David Icke at pp. 359–61.
213. Campbell, *The Power of Myth*, p. 26.

214. Frissell, *Nothing in This Book*, p. 75.
215. Ibid, pp. 73–93.
216. Ibid, p. 74.
217. Ibid.
218. Ibid, p. 123.
219. Ibid, pp. 124–5.
220. Ibid, p. 124.
221. Romer and Romer, *The Seven Wonders of the World*, p. 176.
222. Spence, *Myths and Legends of Ancient Egypt*, p. 25.
223. Blavatsky, *Collected Writings*, vol. IV, p. 287. See also p. 2 of *The High Country Theosophist*, vol. 7, no. 8, Boulder, Colorado, August, 1992. The editor says that Madame Blavatsky recorded these comments in her own copy of *The Origin and Significance of the Great Pyramid* at p. 85. The newsletter states that a copy of this book with her personal inscription is in the Adyar archives, although it does not state its exact location. Presumably this may be the world headquarters of the Theosophical Foundation. See the following website for further information: <www.theosophy-nw.org/theosnw/world/med/eg-vonk.htm>
224. Ibid.
225. Blavatsky, *The Secret Doctrine*, vol. 1, p. 314 and vol. 2, p. 462.
226. These seven chambers within the Great Pyramid may be identified as: the King's and Queen's chambers, the antechamber, the construction chamber, the grand gallery, the subterranean chamber and the lesser subterranean chamber.
227. Piazzi Smyth, *The Great Pyramid*, p. 374.
228. Ibid, p. 375.
229. Ibid.
230. Hancock, *Heaven's Mirror*, p. 99.
231. Ibid.
232. Ibid.
233. Blavatsky, *The Secret Doctrine*, vol. 1, p. 435.
234. Frissell, *Nothing in This Book*, p. 79.
235. Ibid.
236. Piazzi-Smyth, *The Great Pyramid*, pp. 374–75.
237. Frissell, *Nothing in This Book*, p. 79. See also the article 'Prophecy and the Great Pyramid' by Jason Jeffrey in the magazine *New Dawn* no. 4 (Jan.–Feb. 1997). The article agrees with Piazzi-Smyth and Frissell but goes into further detail about the Scored Lines in the Descending Passage of the Great Pyramid.
238. Ibid, p. 128.
239. Ibid.
240. Ibid.
241. Ibid.

242. Ibid, p. 127.
243. Blavatsky, *The Secret Doctrine*, vol. 2, p. 462.
244. Frissell, *Nothing in This Book*, p. 127.
245. Ibid.
246. Ibid, pp. 127–28.
247. Ibid, p. 125.
248. Hancock, *Heaven's Mirror*, p. 81.
249. Ibid.
250. Blavatsky, *The Secret Doctrine*, vol. 1, p. 314.
251. Ibid.
252. Ibid, vol. 2, p. 162.
253. Ibid, p. 240. See pp. 239–51 for full discussion. See vol. 1 at p. 337 for a discussion on 'Darkness is Light' and p. 443 on 'The Mystery of Blackness.'
254. Blavatsky, *Isis Unveiled*, vol. 1, p. 573.
255. Ibid.
256. Ibid.
257. Ibid.

CHAPTER SEVEN

Matariki

Seven Little Eyes of Heaven

I haere mai koe it te ao o Puanga
I te Huihui o Matariki
I a Parearau, I a Poutu-te-rangi.
Ka mutu, e tama, nga whetu homai kai ki Aotea.

You came hither from the realm of Rigel,
from the Assembly of the Pleiades,
from Jupiter,
and from Altair
These alone, O child, are the stars, which provide food at Aotea.

— Traditional Maori lullaby[1]

The story of how the Polynesians peopled the Pacific is a fascinating tale that represents one of the greatest journeys ever undertaken by the human race. Their exploration of the largest ocean in the world was a magnificent and 'unique maritime achievement' unparalleled in human history.[2] Even the accomplishments of the skilful Phoenician sailors pale in comparison to these brilliant navigators and mariners, for the Phoenicians (unlike the Polynesians) sailed without ever losing sight of land. These people put out to sea with their animals, plants, families, hopes and dreams as they bravely traversed the unknown. Like other immigrants throughout the course of history, they took with them their spiritual and cultural traditions to flourish in the new lands but always with the memories of their beloved homeland they left behind. With their spirit ancestors to accompany them and their family of stars, te whanau marama[3] to shed light and guide them during te paki o ruhi — the long, long journey, these intrepid travellers were never alone.[4]

Their heavens were comprised of many levels to reflect their hierarchal societies and on the earthly plane their chiefs were paramount. Like the ancient Egyptians, they perceived the stars as 'eyes' which reflected the various mythical heroes, chiefs, gods and goddesses. It also conveyed an occult double meaning of the ability to see a spiritual vision, to see beyond the restrictions and materialism of this world and to gaze into other realms unbounded, a land of truth, integrity and justice.[5] The 'high born' stars[6] of the Pleiades embodied these ideals and paved the way for the New Year in many Polynesian cultures.[7] Among the Maori people of Aotearoa (New Zealand) these stars are known as 'Matariki', which means the 'little eyes of heaven.'[8] Although the Maori people, like many other Polynesian nations, depended on many of the night stars to guide their waka (canoes), Matariki and her six daughters played a major role. Their importance in heralding the rains that brought abundance of food were commemorated in songs and festivals where women would lovingly greet them with much aroha (love) in their hearts.

Aroha for Matariki

Elsdon Best, an ethnologist with the Dominion Museum in Wellington, New Zealand produced a number of publications on Maori history and starlore over a period of twenty years or so. In *The Astronomical Knowledge of the Maori*, he writes that the stars were much revered by Maori people, especially those of the Pleiades. Their heliacal rising in the north-east just before dawn in late May signalled the beginning of the Maori New Year, just as their disappearance in the evening skies the following mid-April marked its end. Their re-appearance in the early morning skies 'was a notable event,' says Best.[9] Women, in particular, would greet the Pleiades with song and dance.[10] They were songs of joy and sorrow but always expressed with outward displays of love and affection for these stars. Margaret Orbell provides an example of one song of lament in *The Illustrated Encyopaedia of Maori Myth and Legend*, but prefaces it with the qualification that although they sang in remembrance of the dead, 'the singer's tears were joyful too, because the New Year had begun.'[11] She writes:

Tirohia atu nei wheturangitia matariki,
Te Whitu o te tau e whakamoe mai ra,
He homai ana rongo kia komai atu au —
mataorehu o roto i au!

See where Matariki are risen over the horizon,
The seven of the year winking up there.
They come with their message so I can rejoice.
Here I am full of sorrow, full of sadness within![12]

The combination of joy and sorrow was symbolised by Matariki's disappearance and reappearance, says Orbell. Her departure in the western evening skies at the onset of darkness signified an association 'with death and sorrow,' while her return in the early dawn skies was linked 'with light, life and wellbeing.'[13] Other songs celebrated the abundance of food supplies that Matariki had brought and in times of scarcity other stars and planets were invoked to provide plenitude. The stars of the Pleiades were a particular favourite, as attested by the numerous references to its connection with kai (food). For instance, the saying *Matariki ahunga nui* refers to the star cluster as the 'provider of plentiful food-supplies.'[14] Another expression, *Nga kai a Matariki, nana i ao aka runga*, refers to food supplies that are 'scooped up' by her.[15] In Aotearoa, as in other regions of the Pacific, the Pleiades were celebrated in a festival known as the First Fruits that coincided with the New Year because of its relationship with the First Creation. Thus offerings of the first harvest of the kumara (sweet potato) were made to Matariki during the New Year festival.[16]

Yet another old saying is *Kakitea a Matariki, kua maoka te hinu,* 'When Matariki is seen, then game is preserved'.[17] This was because the stars indicated the time when food needed to be 'procured and preserved in fat' and placed in suitable containers. As with other cultures around the world, the Pleiades signified the various seasons and cycles, especially the agricultural and food calendar, which explains why the name Ao-kai was sometimes applied to the cluster.[18] Another saying related to fishing was 'When Matariki is seen by the eye of man, then the Korokoro (lamprey) is caught.'[19] In some parts of Taranaki during midwinter, people would prepare a hangi (earth oven) while waiting to watch the stars rise at dawn. When Matariki appeared in the early morning skies, the people would rise with the stars, weep and begin to recite the names of family members who had passed away since the stars had last set. Concerned that Matariki and her daughters were cold and weak, the people would remove the cover of the oven so the smell of food would rise and strengthen the stars.[20] Besides their relationship with food, the movements of the stars were closely observed to forecast the weather. When they were seen standing 'wide apart' then it predicted a season of warmth and abundance; hence the expression *paki o Matariki* denotes fine weather, but when they appeared close together then they indicated a cold season.[21]

Te Huihui o Matariki (The Assembly of the Pleiades)

Several commentators have remarked on the linguistic similarities between the ancient Egyptians and Polynesian peoples. Curiously, they share the same word for the Sun in Ra, the Egyptian Sun God, and Rangi, the Sky Father in Maori cosmology. Unlike Egypt, which had a Moon god, the Moon is female in Aotearoa and her name is Marama, although the woman in the Moon is known as Rona.[22] Stars are called whetu in Maori and planets are known as whetu ao; but because stars are considered the younger members of te whanau marama, they are also known as ra rikiki meaning 'little suns.'[23]

Sometimes the term huihui or assembly is applied to a constellation or groups of stars as in *Te Huihui o Matariki* or the 'Assembly of the Pleiades.'[24] They are a group of women who came down to Earth from the heavens, leaving Poutini (an unidentified star) 'on high.'[25] Best records that a Maori elder of Ngati-Awa gave the individual names of Matariki's six daughters as Tupua-nuku, Tupua-rangi, Waiti, Waita, Waipuna-a-rangi and Ururangi.[26] In her account of Polynesian astronomy in *The Morning Star Rises*, Maud Makemson gives the meaning of their Maori names in English and their translation as: Earth fairy (Tupua-nuku), Sky fairy (Tupua-rangi), Resembling water (Waiti), Water-dashing down or Bail water (Waita), Celestial spring of water (Waipuna-a-rangi) and Pierce the sky or Sky magic (Ururangi).[27] She also notes the linguistic association of these names and the Pleiades with water, in particular 'with the crop-giving rain.'[28]

Best suggests that the reference to just six daughters could possibly mean that 'Matariki is the name of a single star of the group, in which case we have the name of seven.'[29] These seven goddesses of the Pleiades are similar to the Seven Hathors of Egyptian mythology who were one and seven goddesses at the same time. Remarkably, Makemson tells us that an old Polynesian name for the Pleiades was *Nei Auti*, which means 'that woman.'[30] An ancient chant from the Cook Islands reveals more about her nature. She is the woman 'who descended into the pit of the setting sun in the west and who, after travelling around the tunuku (back of the Earth), emerges again in the east scatheless after her great adventure and with her six eyes sparkling on the face of dawn.'[31] Native Hawaiians, whose language is similar to Maori, refer to the Pleiades as Makali'i, which means a 'cluster of little eyes.'[32] Like the Maori people, they too refer to the Assembly of the Pleiades as in Na-huihui-a-Makal'i.[33] The Tahitians call the Pleiades Matari'i, the 'k' being dropped in that language, and at Horne Island and other parts of Polynesia they are known as Mataliki.[34] Samoans call these stars Mataali'i or simply Li'i in its abbreviated form and, like the Tahitians, the Tongans refer to the star cluster as Matari'i.[35] In the entire

Pacific as elsewhere, the Pleiades were highly venerated in song, ceremonies and festivals.

Te Ikaroa (Tane's basket of stars)

Edwin Krupp tells us in *Skywatchers, Shamans and Kings* that the multi-layered universe was a widespread feature of Polynesian cosmology that reflected a hierarchical structure. In some Pacific cultures it took on a dome shape; thus it resembled a coconut to the Cook Islanders of Mangaia.[36] Maori cosmology is no exception, for they speak of nga rangi tuhaha or many heavens, although the precise number differs among the various iwi (tribes).[37] Some say there are ten, twelve — even as many as twenty heavens. Whatever the number, these heavens are all perceived as being 'distinct realms or worlds.'[38] The lowest and nearest of these multiple heavens represents Ranginui the sky parent, or Great Rangi as he is sometimes called.[39] Like the Egyptian sky goddesses Nut and Hathor, Rangi's body is adorned with stars and planets placed there by his son, Tane. It was Tane who separated Rangi the Sky Father from his wife Papa Tunuku the Earth Mother, and as a result of this action brought Ao Marama or 'light' into the world.[40] And from his basket, Tane distributed te whanau marama, the Shining Ones.

Like all mythology there are variations to suit local and regional conditions. Best provides two examples of the legend, one from the Bay of Plenty area and the other a Takitumu version. In the first story, Tane visits Tangotango, one of the upper heavenly guardians. While there he makes a passing remark about the brightness of the star children. Understanding this to mean that the young man would like them, Tangotango asks him, for what purpose would he use them? Tane replies that he wishes to relieve the world of darkness by bringing light to the earth children. The old man first gives him Hinatore (phosphorescent light) but she proves too feeble 'and darkness held fast.'[41] Tane then obtains a basket of stars, which he distributes along his father's body but still the light is dim. Next he places the Moon upon his father and the light becomes stronger. 'Then Tane placed the sun on high, and bright light entered the world.'[42] In this way, Ao Marama, 'the realm of light' came into being.[43] In the Takitumu version, Tane sends his brother Kewa to act as his intermediary to far-off Maunganui to collect 'the children of *Te Ahuru*, that they might be taken to dwell with their grandparent Rangi.'[44] However, because their mother was a 'supernormal being', Kewa had to seek leave of Tangotango and the other heavenly guardians of whanau atua (supernatural offspring). Needless to say, he was granted permission, and so the Shining Ones were taken to decorate the Sky Father.[45] A. W. Reed tells a similar tale in his book *Maori Legends*. In this story, Tane goes to Maunganui himself where he finds Uru's star children playing at the foot of the Great Mountain.

> Tane begged Uru to give some of the Shining Lights to fasten on the mantle of the sky. Uru rose to his feet and gave a great shout. The Shining Ones heard and came rolling up the slope to their father. Uru placed a basket in front of Tane. He lunged his arms into the glowing mass of lights and piled the Shining Ones into the basket.

> Tane placed five glowing lights in the shape of a cross on the breast of Rangi and sprinkled the dark blue robe with the Children of Light. The basket he hung in the wide heavens. It is the basket of the Milky Way. Sometimes Uru's children tumble and fall swiftly towards the earth, but for the most part they remain like fireflies on the mantle of the night sky.[46]

Best says there was not one but three separate baskets for the Sun, Moon and stars.[47] The basket which held the Sun was called Rauru-Rangi, 'that of the moon was Te Kauhanga, while that of the stars was Te Ikaroa (the Milky Way).' One star, however, was not placed in any of them. Instead, it hung outside the basket where 'it still remains outside the Milky Way.'[48] This star is none other than Canopus (Eta Carinae) in the constellation of Carina the Keel. In the Southern Hemisphere toward the southern region there are very few bright stars shining at magnitude 1 or 2; therefore Canopus, the second brightest star in the night skies, does appear to stand alone and apart from all the others.[49] The Maori explanation for its solitary nature — that it was left outside Tane's basket of stars — makes complete sense within this context. Maori people refer to Atuahi (Canopus) as *tapu* (sacred) for he dwells alone as all tapu persons do.[50] Orbell says he 'clung to the outside of the basket' to ensure he 'would be the first-born' to ensure his high-ranking status.[51] This was because 'a man of high rank might be honoured' in poetry and oratory 'by being spoken of as Atutahi.'[52]

Having placed Te Whanau Marama into the basket of Te Ikaroa, we are told that Te Ikaroa (Milky Way) and Tama-rereti (Scorpio) were put in charge of ra ririki (little suns), or whanau ririki (star family). Occasionally some of the younger ones tend to 'stray' from their elders and family. These wayward stars are called matakokiri, otherwise known as meteors.[53] Native Hawaiians tell a similar story where a basket of stars was used to decorate the Sky Father's body. The central hero of this story is not Tane but Kane instead, and — like the Maori hero — he too brought light into the world after separating his parents, Papa the Earth Mother from Wakea the Sky Father.[54] Some accounts tell how he enlisted the aid of his brothers Ku, Kanaloa and Lono to accomplish the task, who were saddened that their father was unadorned. After distributing the stars on his body, they then set the brightest star Sirius over Tahiti, Spica over the island of Samoa and the rest of the stars in the constellation of Corona Borealis, the Northern Crown.[55]

This notion of a hanging sky is an archaic one familiar to ancient Egyptian and Indian cosmology. The three baskets feature in the tale of another sky hero figure, Whiro, who tried to reach the highest heavens in search of the 'three baskets of knowledge' which Tane obtained.[56] Best casually remarks that the three baskets may be compared to the three sacred texts of Hinduism but does not expand on the theme. What could he have meant by this? Given that stars play an important role in many agricultural calendars the association of baskets with food becomes compelling. Bread, in particular, is a metaphor for wisdom, hence the consumption of the consecrated bread or Eucharistic host in the Catholic mass known as the Holy Communion.[57] Within this context, baskets (like arks, ships and canoes) symbolise the germ of life, the seed from which all things grow. The knowledge sought becomes the mythical quest for the Holy Grail or other similar object. Furthermore, the choice of baskets as the symbolic container of knowledge placed in the heavens is intriguing. Baskets, as we know, are made from weeds, reeds or light pliable woods woven into shape primarily by women. Like the Ngarrindjeri women of South Australia who weave their history into their baskets, Maori women did the same.[58] Interestingly, the omphalos stone at Delphi in Greece with its engraved lattice patterns resembles a woven basket.[59] Apart from their mathematic application in architecture and geodesics as symbols of measuring glyphs, parallels and meridians, what else could they possibly represent?

The answers, says Reginald Lewis in *The Thirteenth Stone*, are to be found in the esoteric Kabbalist teachings of the Tree of Life. Its 'fruits' of knowledge are referred to as 'stones' and their strategic placement on the Tree is part of a complex web of secretive ritualised knowledge that lies at the heart of prophecies and the Christian Bible.[60] The navel stone, which lies at the base of the Tree, represents the 'all-important stone' whose function 'is to hold back the Great Flood.'[61] This thirteenth stone 'is one and the same as the Holy Grail' says Lewis, and is otherwise known as the Philosopher's Stone.[62] The Tree's branches, laden with forbidden fruit (forbidden only to the uninitiated), are held together by an invisible web or net, hence the reference to Jesus Christ as the 'fisher of men' and the role of the Fisher King in the story of the Holy Grail. This web, net or basket is none other than that woven by the Spider Goddess, which Lewis says explains the symbolism behind the constellation Coma Berenices (Berenices Hair) the golden tresses of Berenice, the beautiful Queen and wife of the Egyptian King Ptolemy III. Coma in Greek means 'hair', from which the English word comb is derived.[63] 'Its position on the Tree,' says Lewis, 'symbolises genital hair, the Tree represents Samson-like power and strength.'[64] This explains why the Mungingee (Pleiades) in the South Australian legend plucked their genital hairs as part of a trial of ordeals not unlike the Twelve Labours of Heracles (the thirteenth being revelation of the mysteries).[65] Lewis claims that Coma Berenices and Virgo hold the key to understanding the mysteries hidden in the Tree of Life.

Te Paki o Ruhi (the long voyage)

Many have speculated on the origins of the Polynesian peoples. One theory popularised during the 1950s and 1960s, largely due to Thor Heyerdahl's famous *Kon-Tiki* expeditions, was that they came from the Americas. However, in all fairness to the intrepid explorer, it must be said that Heyerdahl believed there were many migrations from different streams, including the Americas.[66] The current prevailing view is that they came from southern Asia around 3,000 BCE in their canoes. Multidisciplinary approaches in archaeology, biology and other sciences have provided ample evidence to support this view. For instance, biologists have established that certain animals and food plants, with the exception of 'the south American sweet potato' or kumara, 'were all conveyed aboard the immigrant's canoes ultimately from the Southeast Asian archipelagos.'[67] Although this finding does not preclude the possibility of physical contact with the Americas, genetic research has now proved this 'theory' beyond the shadow of a doubt.[68] According to Will Kyselka in *An Ocean in Mind*, the peopling of the Pacific gradually occurred over a period of 5,000 years. They first landed in Samoa and Tonga some 3,000 years ago and a thousand years later arrived in Tahiti before branching out to the Tuamotus, Marquesas and Hawaiian islands. After another thousand years they travelled to New Zealand.[69] As they spread across the Pacific, Tahiti became 'the main centre of Polynesian culture', which served as the platform for later exploration and migration to other islands.[70]

Another contested debate focused on whether the settling of Polynesia was 'accidental' by drifting canoes, or 'intentional' by deliberate, planned voyages. Kyselka says that so-called 'drift voyaging cannot account for that extensive contact, since wind and wave move parallel to the equator, not across it', therefore 'drift is simply not possible.'[71] He goes on to argue, 'Some form of locomotion, a sail or paddle, is needed in such purposeful voyaging.'[72] We now know that not only was this voyaging deliberate and planned, but also skilful and scientific, evoking praise and admiration from a wide range of disciplines. As David Lewis remarks in *We, the Navigators*, Polynesian explorers were simply 'outstanding navigators.'[73] Other writers extolling their virtues include British astronomers Heather Couper and Nigel Henbest who marvel at the extraordinary distances involved in which the Polynesians sailed 'out of sight of land.'[74] For instance, 'the longest non-stop voyage was from Tahiti to Hawaii,' a distance of about 4,000 kilometres,[75] and yet they 'had no problem in finding Hawaii at the journey's end.'[76] They conclude by noting the Polynesians had developed a navigational technique 'that was very different from anything that the Europeans or the Chinese had invented.'[77] In fact, they invented one of the most sophisticated navigation systems in the world, a feat all the more wondrous given the vastness of the Pacific Ocean and the fact that they did so without modern instruments such as the sextant, telescope or compass.

'Way of the finder' or 'navigator'?

In the last forty years or so, there has been a burgeoning interest in the customary navigation methods of Polynesians. A large part of this has been spurred on by the formation of the Polynesian Voyaging Society (PVS) based in Hawaii in 1973. Their intention was to build a large voyaging canoe 'to test the feasibility of making long-distance navigated voyages in a voyaging canoe guided solely by traditional navigation.'[78] Such a canoe was built, and duly christened *Hokule'a*, which is the Hawaiian name for the bright orange star Arcturus 'that passes directly over the island of Hawaii.'[79] Its name means 'canoe guiding star.'[80] The Hokule'a put to sea in 1976 and completed a modern test voyage, sailing successfully from Hawaii to Tahiti in thirty-one days and 'arriving on 4 June 1976' using traditional navigation methods.[81] The Micronesian navigator Mau Piailug from Satawal provided the navigational expertise to a crew of seventeen, including young Native Hawaiian sailor and navigation student Nainoa Thompson.[82] Thompson went on to navigate the second successful round trip between Hawaii and Tahiti in 1980.[83] In doing so, he became 'the first Hawaiian and the first Polynesian to practise the art of wayfinding on long distance ocean voyages since voyaging ended in Polynesia in the fourteenth century.'[84]

These experimental voyages reveal a wealth of information about Indigenous sciences, especially navigation. Through real-life experiences they demonstrate how navigators relied on a combination of factors, not just the stars. They include the Moon, the Sun (our biggest star), the wind, ocean currents and swells, the colouration and temperature of the water, birds, clouds and even phosphorescence. While the stars played an important role, it was a combination of techniques that contributed to the overall sophisticated navigation system that some writers refer to as *wayfinding*, as in 'a way for finding islands.'[85] As Thompson himself came to realise and appreciate, 'the principles of wayfinding are simple; the practicalities are very complex.'[86] So complex in fact, that it took a lifetime to learn these skills, including an early childhood education followed by specialisation in these matters. As Herb Kane, the celebrated Native Hawaiian artist and writer tells us in *Voyage: The Discovery of Hawaii*, Polynesian navigators 'were also priests in the sense that they could invoke spiritual help and conduct the rituals of their profession.'[87] Thus, a 'brotherhood of experts', variously called *tohunga* or *kahuna* were 'trained to acute powers of observation and memory.'[88]

The term wayfinding is not a traditional Polynesian term at all. Members of the Polynesian Voyaging Society invented it to describe Indigenous navigation methods and techniques that were not reliant on instruments. 'Non-instrument navigation' sounded too 'condescending' says Kyselka, and 'even land finding too incomplete.'[89] While the philosophical spirit behind this rationalisation may be appreciated, the term wayfinding detracts from a more rigorous scientific form to something softer and therefore less scientific. I prefer to simply call it by what it is — navigation. After all, in the words of Kyselka himself, navigation is 'directing a vessel over the sea to an intended destination.'[90] Whether instruments are used or not should not be the defining criteria. Early European explorers, including the Phoenicians and the Vikings who were formidable sailors, sailed without instruments and yet no one would dare suggest that they did not navigate the seas. If anything, the so-called wayfinding system of the Polynesians is even more scientific because of its holistic, encompassing nature. As Lewis points out, their methods were not compartmentalised into divisions such as steering or deviating from a course, or fixing a position other than 'during their initial training.'[91] Their entire navigation system was a unified one; the sum input 'from such disparate sources as stars, swells and birds being processed through training and practice into a confident awareness of precisely where they were at any one time, where they were going, and how best to get there.'[92]

Lewis is suggesting that Polynesian people did not separate their 'navigating arts from their social roots and the psychological and spiritual values of which they are an expression.'[93] So Polynesian navigators 'were not merely in tune with their environment as Western seafarers might be, they were literally a part of it.'[94] It is in this sense perhaps that those who coined the word 'wayfinding' believed they had an all-embracing, all-encompassing term which acknowledged both the spiritual dimension as well as the scientific component of the Polynesian navigation system. Ultimately though, it does them a great disservice. Firstly, we are not talking about an 'art' at all but a fully fledged, complex and sophisticated body of knowledge that can match any so-called 'science' of the Western world. And secondly, just because a technique is grounded in spirituality that does not necessarily make it any less scientific, and vice versa. In other words, one need not be necessarily devoid of the other, but each can enhance and inform the other dimensions. In short, there can be spirituality within science and science within spirituality. Indeed, Albert Einstein, arguably one of the greatest scientific minds of the twentieth century, is renowned for his acceptance of this truism as evidenced by his writings, quotations and famous phrases where he acknowledges a higher power in the workings of the universe.

Tangata Whenua (people of the land)

Maori people trace their whakapapa (tribal lineage) from the names of the waka or canoes their people travelled in when migrating to the 'Land of the Long White Cloud' (Aotearoa). Within this context, the word *canoe* is somewhat misleading, says Lewis, for it conjures up images of tiny craft hollowed out of tree trunks. In fact, some of these so-called 'canoes' 'were longer than Cook's *Endeavour*,' he argues, and deserve to be called 'ships' in their own right.[95] As Makemson points out in *The Morning Star Rises*, Cook particularly admired the great double-hulled canoes of the Tongans. Writing in his journal, he praised their 'ingenuity' and 'workmanship', which 'exceed everything of this kind we saw in this sea.'[96] Be that as it may, one legend having 'its origin in ancient chants' describes the migration of a 'Great Fleet' of seven canoes that travelled to Aotearoa around 1350 CE.[97] The chants name these seven canoes as Aotea, Arawa, Kurahaupo, Tokomaru, Takitumu, Tainui, and Mataatua.[98] Consistent with the account given by Kyselka, most of the canoes migrated from eastern Polynesia, in particular from the region of Tahiti. The largest of these canoes, the Tainui — which made several landings before its final resting-place at Kawhia on the West Coast of Aotearoa — came from the Tahitian island of Raiatea.[99]

Leon Paoa Sterling, a Native Hawaiian crewmember of the Hokule'a, confirms the legend of the Great Fleet migration in *Voices of Wisdom*. He tells of a 'special place' in Rarotonga in the Cook Islands marked by stones to commemorate the seven canoes that left on their voyage to New Zealand.[100] At one time, science would have referred to this legend of the Great Fleet of seven canoes as a 'myth.' However, this has now been scientifically affirmed.[101] Geneticists have since traced the origin of the Maori people to Taiwan and New Guinea. In a fantastic tale sounding more like a science fiction novel, a group of women left Taiwan in canoes for the southern Pacific Ocean where they landed on the island of New Guinea. Stocking up on provisions, including a few carefully chosen male specimens, these Taiwanese Amazonians set sail to populate the Polynesian New World. Thus modern day Polynesians owe their ancestry to two very different racial strands. Scientists have estimated the numbers required to travel and populate and have arrived at the magical number of seven canoes!

Perhaps nowhere nearer to being solved is the precise location of the fabled homeland of the Polynesian peoples — *Hawaiki*.[102] Throughout many of these migratory tales there are countless references to this mysterious land. Because the 'k' is dropped in the Hawaiian language it becomes Hawaii and therefore some people have suggested it is the legendary homeland. In a paper presented in 1980 at Brigham Young University at Hawaii, Vernice Wineera tells how 'a small group of Maori people from Rotorua set out to visit the Big Island of Hawaii in 1978 in the belief that Ka'u was the place of their beginning, the Hawaiki of the old chants and legends.'[103] Wineera is dismissive of the claim, on the basis that the region is historically subject to severe drought and famine and therefore would not have been able to supply provisions for long ocean voyages. This argument fails to take into account climatic changes over long periods of time and conditions may have been very different in the past. Neither does it consider the possibility of the travellers obtaining supplies from elsewhere. In any event, although many Maori people trace their origins to eastern Polynesia or Tahiti, there were many voyages throughout Polynesia. The settling of Aotearoa, therefore, does not necessarily preclude the possibility of ancestors of *some* Maori people having come from other regions within the Pacific, including Hawaii.

I am very much a strong advocate in supporting a people's-own process of *self*-identification. Therefore, if these people's cultural traditions tell them that that is where their ancestors are from —

and even if science is unable to support their stories — then I am more inclined to accept their version. Science may prove their 'myth' some day, or it may never do so. What is more important is that these people, for whatever reason, feel some sort of spiritual, cultural and emotional connection with Hawaii and basically that is all that matters. As Wineera notes, like many other Indigenous peoples, 'The first question one Maori asks another when meeting for the first time is not "Who are you?" but "Where are you from?"'[104] People reply by listing their tribal name, area and pa (village) 'and possibly canoe.' Thus, he says, 'By asking for origin, identity is established.'[105] Wineera adds, 'It is the same in the larger perspective — the Maori quotes his or her legendary home as place of origin, and thereby reveals their identity in all its cultural heritage. He (and she) is secure in this knowledge.'[106]

Regardless of where the true Hawaiki is located, there were many voyages to and from Aotearoa and other areas of Polynesia, including the fabled Hawaiki. Several Maori legends speak of two-way travels between Aotearoa and Hawaiki, beginning with the early Polynesian explorers Maui and Kupe. Margaret Orbell says there are many examples and reports of people sailing back to Hawaiki from Aotearoa. The most well known of these was an expedition led by Pahiko from the east coast following an ignominious defeat. As they headed east on these return journeys 'the ship's course was naturally directed towards the rising sun.'[107] Similarly, Kyselka tells us there was much 'repeated voyaging between Hawaii and Tahiti' as recorded in their 'chants, legends, songs, dances and stories.'[108] Lewis says that Rarotonga proved a logical staging post between Tahiti and New Zealand because of its linguistic and cultural linkages to both countries.[109] Consequently the Cook Islanders share the same names of canoes with Aotearoa 'that staged in Ngatangia lagoon on their way south.'[110] Even the Tainui canoe, 'the biggest ever built in the Tuamotus, which sailed away to unknown lands, turns up again in New Zealand legend and, moreover, with the same captain.'[111] Even more captivating is the revelation that an old name for Raiatea, where the Tainui was built and sailed from 'was Havaiki, or Hava'i.'[112] Makemson says that this ancient name is referred to as the birthplace of the lands, calling to mind the great voyages of exploration and colonisation which set out from this centre in the golden age of Polynesia during the tenth to fourteenth centuries.[113]

Lewis recounts an old tale from Tahiti that provides a mythical explanation for the close cultural ties between the Cook Islands, Tahiti and Aotearoa. This tells of a time long ago when the lands were one and Rarotonga was joined 'to the southern end of Raiatea.'[114] One day, some visiting priests from Rarotonga who brought gifts and tributes to place at the 'Raiatean high altar at Opoa' were murdered.[115] Angry at this sacrilege, the gods took immediate action by carrying Rarotonga away. No one knew for certain where it had been taken but some people thought it might have been removed somewhere to the south.[116] Many other stories are told of drifting islands throughout the Pacific that were once joined long ago and there are as many tales of sunken lands. A common metaphor in Polynesia to describe the ending of friendships and intimate relationships is to say that 'the islands have drifted apart' or that 'the bridge between them has sunk.'[117]

Tatai arorangi o Tohunga kokorangi *(Maori priestly astronomers and navigators)*

On these long sailing voyages the waka carried 'one or two expert star-gazers, men versed in the lore of tatai aorangi.'[118] These were the navigator priests who assisted the captains in their navigation of the seas. For instance, the two experts on the Takitumu waka from eastern Polynesia to New Zealand were Puhi-whanake and Whatuira.[119] During the night the navigators would scan the heavens to give directional orders and forecast weather conditions. Their assistants or apprentices would tend to navigation during the day while the experts rested. Not everyone could become a tatai aorangi — only a limited number of the tohunga class — thus adepts in starlore were known as the tohunga kokorungi.[120] In *Maori Legends*, Reed says there are several grades of tohunga, the highest of these being the *to*hunga ahurewha and the lowest the tohunga kehua.[121] There are many specialists among the tohunga, such as the tohunga tatai-arorangi who interprets the stars, the tohunga makutu who exacts revenge and the tohunga matakite who foretells future events.[122]

One writer suggests that the term tatai aroangi means 'to study the heavens for guidance in navigation,' but Best says it was used 'to denote the personified form of astronomical knowledge.'[123] Thus a *tangata tati arorangi* is a person who undertakes such studies, effectively an astronomer,[124] and the expression *Ko Tatai-arorangi he kai arataki i te ra* means 'a conductor or guide of the sun.'[125] These expert astronomers were taught skylore from an early age and in some Polynesian cultures attended special astronomy schools to formalise their education.[126] In terms of archaeological evidence, stone slabs can still be seen on some islands today that were 'used to teach navigational skills', says Anthony Aveni in *Ancient Astronomers*.[127] Referred to as 'navigation stones' by archaeoastronomers, students would place themselves between two upright stone slabs, which simulated a canoe, face 'in one of the cardinal directions' and memorise the individual stars and constellations as they faced the desired destination.[128] Locals in the Gilbert Islands refer to these navigation stones simply as 'stone canoes' or 'the stone for voyaging.'[129]

Although Kane and Best refer to a priestly brotherhood of navigators, this is not to say that women were never included. Lewis records an origin myth from the Carolinas, which tells how navigation long ago was once 'the gift of women.'[130] Furthermore, even though it was customary to hand down these teachings 'from father to son,' it was permissible to train a daughter in circumstances where there were no male offspring.[131] Lewis tells the story of a famous woman navigator Paintapu, whose navigation skills were considered suspect on one occasion by the chief of an expedition, much to his expense. One of Paintapu's seemingly odd techniques was to lie in the bottom of the canoe and gaze up at the stars to give tactical directions. Convinced she was a sorcerer, the chief 'had her unceremoniously thrown overboard.'[132] The last canoe in the fleet rescued her, whereupon she guided them home. 'None of the others were ever heard of again,' says Lewis.[133] The notable absence of any mention or discussion of women in much of the literature available is not lost on Lewis, who notes that the 'powerful role of women' in Polynesian navigation 'has been virtually ignored, except when, like the navigator Paintapu, they are glimpsed through the mists of time.'[134] History is all the poorer, says Lewis, for 'it was women, ashore or afloat, who held the ocean in their keeping, sharing a common affinity with the moon and with fruitfulness.'[135]

Steering by the stars

The stars relied on during the voyage of the Takitumu waka were: Atutahi (Canopus), Tautoru (Orion's Belt), Puanga (Rigel), Karewa (unidentified), Takura (Sirius), Tawhera (Venus as Morning Star), Meremere (Venus as Evening Star), Matariki (Pleiades), Tama-rereti (Tail of Scorpio) and Te Ikaroa (the Galaxy).[136] Technically speaking, Te Ikaroa does not count as it is not an individual star as such but refers to the Milky Way, or Tane's basket of stars. Orion's belt, on the other hand, is made up of three stars and Scorpio has about seven to eight bright stars that make up its tail, whereas Matariki are visible as a small cluster. So the total amount of stars relied on the Takitumu voyage to Aotearoa exceeded fifteen.

If, however, we count Orion's Belt, Scorpio's tail and the Pleiades as one star, then it becomes more manageable in terms of thinking and talking about seven stars in all. Specific directions were given to 'carefully keep the prow of the vessel laid on Venus during the night; during the daytime follow behind Tama-nui-te-ra (the Sun).'[137] At the time of writing his monograph, very little was known about the specific technicalities of Polynesian navigation so Elsdon Best was perplexed. How could Maori navigators follow specific stars, especially when the majority of these rise in the east and set in the west? What is more, Aotearoa is on a south-west course from eastern Polynesia, yet many of the stars relied on would have risen in the east.[138] He became curious: at what point did the steersman 'commence to steer by it' and at what point were they utilised?[139]

The answer to Best's queries lies in the special techniques of the Polynesian navigation systems that they developed. As previously noted, the stars were just one of the many things that contributed to the entire process. Certainly, even in Best's era, he would have known that the position of stars gave an indication of latitude. For instance, the further south one travels from the Northern Hemisphere, the lower Polaris (the Pole Star) appears on the horizon. When you cross the equator from the Northern to the Southern Hemisphere, the Pole Star disappears entirely from sight. Couper and Henbest explain the reason why this occurs in *The Stars: From Superstition to Supernovae*: 'As the Earth rotates, a steady procession of stars seems to move across the sky, with some stars passing right overhead.'[140] Therefore, 'If you change your latitude, moving north or south around the curve of the Earth, you look out on to slightly different areas in space, and different stars will pass through the zenith.'[141] It was only a matter of time before the Polynesian voyagers discovered that certain bright stars passed directly overhead that marked specific islands. For instance, Sirius, the brightest of all stars, passes directly over Tahiti and therefore became 'a special star for that island.'[142] Similarly, the bright orange star Arcturus or *Hoku'ele* passes directly over Hawaii[143] and the white-blue star of Spica passes directly over Samoa.[144] Consequently:

> A Polynesian crew sailing from Tahiti to Hawaii would keep watching the bright orange star Arcturus. They knew that this star, which they called *Ana-tahu-a-ta'ata-metua-te-tupu-mavae*, passed directly overhead at their destination. As they sailed northwards, Sirius would gradually slip downwards, crossing the sky further from the overhead point every night. But Arcturus would be rising higher as the weeks went by. When the navigator saw Arcturus passing overhead, he knew that they had reached the correct latitude for Hawaii. Now he turned the great canoe west. As the canoe covered the last short stretch to the Hawaiian Islands, the

navigator kept a look out for the tell-tale signs of land — birds, dolphins and towering clouds — that would guide them to a precise landfall. To return to Tahiti, they sailed for the bright beacon Sirius, whose Hawaiian name, *Hoku-ho'okele-wa'a*, in fact means 'canoe guiding star.'[145]

What Best did not realise was that in addition to being able to tell latitude by the stars, an indication of longitude can be ascertained by those stars rising in the east and setting in the west. The Pleiades and Orion follow this pattern so they are very good indicators of an east-west alignment. In addition to tracing an east-westerly route, the stars of the Pleiades draw a curved arch in the sky from the north-east to the north-west. Thus stars can act as a virtual compass, with individual stars rising and setting at precise compass points such as north-north west, north-west, south-south west, and so on. Stars such as Arcturus, which indicate latitude (north-south), or the Pleiades that indicate longitude (east-west) are known as 'horizon' or 'guiding' stars.[146] These are stars that 'are low in the sky that have either just risen or are about to set.'[147] They are by far 'the most accurate direction indicators,' says Lewis.[148] To get to a particular location, 'you steer toward whichever star rises or sets' in that direction or, in more technical terms, your 'bearing.'[149] You follow the bearing (or azimuth) 'of its guiding star, at rise if the course be an easterly one, at set if it be westerly.'[150] Horizon stars, unlike circumpolar stars which never set (they simply encircle the poles) rise in the east to their zenith (ultimate height) and then proceed to set low in the night skies with some stars actually disappearing below the horizon. Horizon stars, therefore, as Lewis points out, 'can only be used to steer by for a certain time.'[151]

Is this what Best was alluding to when he expressed his curiosity as to *what exact point in time* did the navigator steer by a particular horizon star? The answer lies in what Polynesian sailors describe in English terms as the 'star path', or the 'carrier.'[152] The star path is simply the succession of guiding stars toward the bow of the canoe that are either rising or setting, and by which one steers. When the star is low in the skies it is used as a guide and when it rises overhead 'it is discarded and the course is reset' by the next star in the series, and so on until dawn.[153] The Tahitians call the star path *avei'a*, the Tongans *kaveinga* and in Tikopia it is called *kavenga*.[154] Lewis tells us that a star path is always named after the lead star so that it is not always necessary to know the name of each individual star of a star path.[155] Generally, ten stars provide a good star path 'for a night's sailing, which occupies roughly twelve hours in the tropics.'[156] So the Takitumu voyage relied on only seven stars; excluding Te Ikaroa the Milky Way galaxy in which the stars are placed, counting Venus just once as both the Morning and Evening star and Matariki as one star, even though it is actually a cluster of stars. In addition to these stars, the Sun was used to guide sailors on their voyage across the ocean.

Kupe the fisherman and sailor

Although it was Maui 'who fished up' the North and South Islands of Aotearoa, it was Kupe who 'discovered' them while looking for the fish of his ancestor, Te-Ika-a-Maui. It was during his search that the new land obtained its name. Legend tells that while searching for signs of land, his wife Kuramarotini spotted the snow-covered alps of the South Island and, thinking they were clouds, called out 'He ao' which means 'A cloud.' Drawing closer, she exclaimed 'He Aotea, he Aotearoa', meaning 'A white cloud, a long white cloud.'[157] Best tells us Kupe's sailing directions were 'fairly explicit.'[158] Essentially, they were to 'keep the sun, moon, or Venus just to the right of the bow of the vessel, and steer nearly south-west.'[159] Kupe's voyage was made during November or December, says Best, who notes that 'the true course from Rarotonga to Auckland is about S. 56 W., or SW by W.'[160] After discovering Aotearoa, Kupe returned to Hawaiki. On his return he was asked if there were any people in Aotearoa. He indicated that he only saw kokako (crow) and tiwaiwaka (the fantail) which, says Anthony Alpers in *Maori Myths and Legends*, was 'more polite than saying *no*.'[161] When asked if he 'intended to return', he said 'E hoki kupe?' meaning, 'Will Kupe return?'[162] Now those who 'wish to indicate indirectly but firmly that they will never return to a place' have immortalised the phrase ever since, says Orbell.[163]

Some writers suggest that the exploration and peopling of Aotearoa only happened in Kupe's wake, but Lewis suggests a far more basic reason — that of the annual migration of the long-tailed cuckoo. He says the Tahitians must have observed its flight path and because it was a land bird it would have signalled that it was 'making for land in the southwest,' the direction taken by the birds every year.[164] This alone provided them with proof that an unknown land lay in that direction. All the mariners would have to do is follow the star path the birds had taken. As easy as that sounds, we must not forget that Aotearoa lay some 1,630 miles beyond Rarotonga. Still, Lewis argues that the idea is not as far-fetched as it may seem given an ancient Tuamotuan sacred chant or *fangu*, whose words were 'sacrilege to alter':[165]

> Mine is the migrating bird
> Winging over perilous regions of the ocean,
> Ever tracing out the age-old-path
> of the Wandering Waves.[166]

What this chant reveals is that birds played a major role in navigation, as in ancient Greece, where the release of doves signalled the commencement of the sailing season.[167] In Greek mythology these doves symbolised the Seven Sisters of the Pleiades in anthropomorphic form and in ancient Egypt they took the form of various Bird Goddesses. Not surprisingly, a number of birds are featured in several Pacific star constellations and as navigational tools. A common Polynesian term for a bird is *manu*, and the shape of its body resembling a cross (formed by its outstretched wings in flight) provided a visual compass for these navigators. Native Hawaiian navigator Nainoa Thompson explains this compass division: 'The four houses of Manu, midway between the four cardinal directions, can be seen as the points of the beak, tail, and outstretched wing-tips of a bird.'[168] Thus, 'on early voyages to Tahiti, the Hokule'a sailed in the direction of Manu Malanai, with its wings and Manu Ko'olao and Manu Kona, and its tail pointed back at Manu Ho'olua.'[169] Apart from its practical application in the science of navigation, the metaphoric symbolism of the bird as a solar symbol and as an ark, boat, seed or canoe is perhaps more revealing. To this end, the Polynesian manu may have its linguistic roots in India as the name of the Vedic leader, Manu of the Seven Sages, the eponymous ancestor of humanity who sailed in the ark with his six sons to escape the Deluge.[170] Neither can we ignore the glaring similarities between the ancient Greek legend of the Phoenix and its counterparts in the Egyptian Firebird and the Maori Sunbird.

Te Manu i te Ra (Bird of the Sun)

In his monograph on Maori starlore, Best provides several examples of different Maori names or phrases that are used in relation to the Sun, including Te Manu i te Ra, which refers to the 'Bird of the Sun.'[171] He lists two other terms for the Sun that were rarely used, komaru and mamaru, and notes that all three are 'applied to a canoe sail' although he failed 'to see any connection between the two.'[172] In fact, several connections can be made on a number of bases including linguistics, anthropology and mythology. Furthermore, much of the evidence appears to suggest that there may be strong cultural and spiritual affiliations between Polynesia and ancient Egypt. The first clue lies in the traditional Polynesian metaphor of the bird for the canoe. Within this context the symbolism behind the Tuamotuan sacred chant becomes abundantly clear — the migrating 'bird' winging over the perilous waves of the ocean is none other than the explorer's canoe. This means that Te Manu i te Ra is not only a Bird of the Sun but may be interpreted as the Maori Sun god's canoe as in Te Waka i te Ra. It brings to mind ancient Egyptian images of their Sun god *Ra* travelling in a sailboat 'across the body of the sky goddess from one horizon to the other.'[173] Is this pure coincidence? Best doesn't think so and points out there are more linguistic examples that suggest the Sun had 'many manifestations' in ancient Egypt and Polynesia.[174] For example, in ancient Egypt the setting Sun was called *Ra-tum* and in eastern Polynesia it is known as *Ra tumu*.[175] Another Egyptian term *Kau*, he says, is used in the Maori language to refer to 'the movements of the heavenly bodies.'[176] In addition to these similarities between the two languages, profound parallels in their mythologies also suggest a possible common source.[177]

In ancient Egypt and Mesopotamia the Sun god was often depicted as the winged sun disc that some writers suggest is a symbol of the legendary Phoenix, which rose from the ashes. Also known to the Ancients as the 'Firebird', it was none other than the sun 'who flew on wings through heaven and was constantly immolated and reborn from the fires of sunset and sunrise,' says Barbara Walker in *The Woman's Dictionary of Symbols and Sacred Objects*.[178] The symbolism behind these avian images of the Sun is easy to deduce, says Edwin Krupp — the Sun's movements through the sky appear to imitate that of a bird. Therefore, in the minds of the Ancients, it looked as if the Sun 'could fly on feathered wings.'[179] The Assyrians and Persians had much the same idea, for they too 'equipped their winged sun disk with tail feathers.'[180] Similarly, Maori people claim that not only is there a bird in the Sun but that Te Manu i te Ra is the Sun itself.[181] The Egyptian Firebird was revered, for it carried 'the life-giving essence' known as *hike* from the Isle of Fire.[182] This was a mystical land beyond the known world, a 'place of everlasting light . . . where the gods were born or revived' and from where 'they were sent into the world.'[183] Many temples were built to honour the Sunbird, the most famous being the Temple of the Phoenix in the ancient Egyptian city of Heliopolis. Built on 'a sacred hill or mound' that marked the spot where the First Sunrise occurred, the temple was dedicated to Atum the father of the Egyptian gods who later became identified with the Sun god Ra.[184]

Within the grounds of the temple stood the Benben stone. Krupp says this stone served to mark the spot where the first mound rose out of the watery abyss and from which the Bennu bird of creation took flight. Robert Bauval and Adrian Gilbert say in *The Orion Mystery* that it represented a meteorite that fell from the sky 'near

Memphis some time in the third millennium BC, perhaps during the Second or Third Dynasty.'[185] Furthermore, this stone symbolised the 'legendary cosmic bird of regeneration, rebirth and calendrical cycles.'[186] This meteorite, say the authors, was massive and estimated to weigh anywhere between six to fifteen tons, thus the 'frightful spectacle of its fiery fall would have been very impressive.'[187] Coupled with the 'loud detonations caused by shock waves,' this visual image would have left an indelible print in the minds of ancient Egyptians who witnessed its entry into the Earth's atmosphere.[188] This meteoric 'firebird,' say the authors, 'would have evoked the notion of a returning phoenix crashing in from the east.'[189] Upon inspecting its landing site, ancient Egyptians 'would have seen that the firebird had disappeared, leaving only a black, pyramid-shaped object or cosmic egg.'[190] This sacred stone was then transported and placed within Atum's temple to be worshipped and revered by his followers.

Curiously, Heliopolis is now located in the modern Cairo suburb of Matariya that phonetically sounds like the Maori Matariki of the Pleiades.[191] Maori mythology tells a similar tale of a Sunbird who lived in a faraway land on the top of a mountain, which even death 'could not reach.'[192] Best suggests that Mount Hikurangi, as it is called, formed part of the fabled homeland or else a 'mythical place.'[193] Orbell says it is both: 'as well as existing in mythology, the name Hikurangi was given in reality to a number of prominent hills and mountains in different parts of the country.'[194] Like the biblical landing place of Noah's Ark, Hikurangi was 'a place of refuge for people threatened by a great flood.'[195] According to Ngati Porou of the east coast, their sacred mountain Hikurangi was 'the first land fished up by Maui, and that his waka was still up there on the summit, turned to stone.'[196] They say, 'at dawn the sun lights up this high peak while all around is in darkness,' for 'this event repeats the first occasion on which the sun shone upon Hikurangi as it rose above the waters.'[197] Does this sound remarkably familiar to Atlantean traditions of a great flood whose living relic is now seen in present day Mount Atlas, or of similar cataclysms? Is there any link between this story and tales of a much older, lost continent, the great continents of Pan and Mu, that some believed once existed in the Pacific Ocean?

Te Ra o Tainui (the sail of Tainui)

Of all the migratory waka, the Tainui alone is celebrated and commemorated in Maori starlore as having a direct relationship with the Pleiades. Like the former constellation, *Argo Navis* (the Ship Argo) now comprised four smaller constellations — Carina (the Keel), Puppis (the Stern), Vela (the Sail) and Pyxis (the Compass) — the Maori constellation of the Sail of Tainui does not refer to just that particular component but to the entire canoe.[198] It also comprised a large group of stars that form its individual components. Thus the Pleiades form its bow, the three stars of Orion's Belt make up its stern, the Hyades represents its sails, the pointer stars of the Southern Cross make up its cable, whereas the cross itself, Puanga a Tama-rereti is the anchor.[199] It makes sense that the Pleiades should form the bow of this starry vessel for they were the favourite guiding stars of Polynesian sailors.[200] And not just sailors either but also the gods it seems, for Best notes that according to one Maori elder, the stars of the Pleiades 'are seen on the breast of their forebear Rangi, seen paddling their canoe.'[201]

Orbell traces the journey of the legendary Tainui from first creation, to its migration to Aotearoa, to final resting place on the shores of Kawhia Harbour on the west coast of the North Island.[202] The famous canoe was named for the stillborn child of Tinirau and his wife Hine-kura. When he was born, Tainui was said to be 'incomplete', for although he was 'perfect from head to chest' he 'had no waist or legs.'[203] Where he was buried, a tree later grew and 'many generations later' when it was 'fully grown, it became the hull of the Tainui.'[204] After 'appropriate rituals' were performed, the tree was felled, the waka built and prepared for its maiden journey.[205] On its voyage to Aotearoa, Hoturoa captained the Tainui and as such his place 'was at the stern, the other high ranking men were immediately in front, and the tohunga sat at the bow beside the tuahu.'[206]

There is some disagreement, however, over the true identity of the navigator who accompanied Hoturoa. Some accounts say it was Ngatoro-i-rangi, even though the captain of Te Arawa, Tama-te-kapua, kidnapped him. Others say it was Raka-taura, but there are conflicting accounts that say he 'was deliberately left behind in Hawaiki because he was lazy or a thief, or because his son had been killed and the people feared his retaliation.'[207] Some say he was replaced by Riu-ki-uta, whose sacred chants called to 'the taniwha in the ocean and the birds of the air to bear the vessel forward.'[208] In addition to disagreement over who really navigated the Tainui, there are differing accounts as to the directions taken by the waka after it entered Hauraki Gulf on the east coast of the North Island. Some say it then went to the far north to Muriwhenua, then on to Tamaki; others say that it visited Tamaki first and then sailed 'right round Te Tai Tokerau (Northland) and down the west coast.'[209] The general belief is that the Tainui 'reached the west coast' by being carried 'across the Tamaki isthmus.'[210]

Coasting south from Manuka, the Tainui encountered high waves that Kupe is said to have left behind on the west coast 'to obstruct the passage of later expeditions, but Hoturoa calmed the waves with a ritual chant.'[211] On the way to its final resting-place at Kawhia, 'further landmarks were established and names given to the places they passed.'[212] One such place was Whitianga, where it is claimed that one of the sails of the Tainui 'was left hanging on a cliff, which became known as *Te Ra o Tainui*.'[213] The canoe now lies buried near Maketu in the harbour at Kawhia. Says Orbell:

> In the harbour there, beside the waters at Maketu, two tapu stone pillars, Hani and Puna, mark the positions of the prow and sternpost of Tainui; the prow belongs to Rakataura and the sternpost to Hoturoa. The ship itself lies buried, turned to stone, between them. The skids employed in pulling it ashore took root and turned into a tree, the Tainui, which formerly grew only in a small area between Kawhia Harbour and the Mokau River.[214]

In 1997 I visited Kawhia (pronounced Karfia)[215] with my Maori partner whose iwi (tribe) is Waikato, one of several Maori tribes whose ancestors came to Aotearoa on *Te Waka o Tainui*. On my visit there, I saw the harbour where the Tainui sailed into and walked along the beach where it landed. I also saw the schoolyard where, according to Maori history, it is buried. Later that night back in Te Awamutu, I dreamt I was on top of one of the mountains at Kawhia. It was very windy and the mountain was devoid of trees. In the dream, an old Maori woman came walking towards me. Her hair was grey and her face wrinkled and weathered like a rusty sailor from years spent at sea. On her chin she wore the customary *moku* (tattoo). She spoke to me and said that it was the custom of her people to bury their canoe in the sands of the new lands where they decided to settle, and to bury the sails separate from the vessel. She said that she would show me where her people had taken the sails of the Tainui and took me to the very spot on the mountains where they lay interred. She told me that the burial grounds of the sails were tapu and that this information was privileged. I recall feeling extremely humble and especially honoured that she had entrusted this information to me.

The next day, I recounted my dreams to friends Lesley Morgan and Marama Tane. Morgan is a pakeha woman and Tane a direct descendant of those who sailed on the Tainui. She confirmed it was a widespread Maori cultural practice to bury the sails and canoe separately upon settlement, but although it is common knowledge where the waka is buried, only Maori tohunga knew where the sails were laid to rest. Imagine my surprise and sheer delight when I stumbled upon the star-legend of Te Ra o Tainui four years later as part of my research for this book!

Not surprisingly, the Tainui waka and its relationship to the stars of the Pleiades played an important symbolic role in Maori cultural and historical identity. Orbell tells us that 'after the Waikato War of the mid-1860s', many Maori descendants of the Tainui 'had their lands seized by the Government, but that they continued to maintain their own political, social and religious organization the King Movement (Te Kingitanga)'.[216]

The Tainui waka and the stars of Matariki were important symbols of this movement and were depicted on one of the King's flags — the flag of Mahuta.[217] In more recent times, the Tainui has been commemorated once again in the night skies, this time in Hawaii. Although it is not a traditional constellation, Native Hawaiian Wayfinder Nainoa Thompson has given the name *Ke Ka o Makali'i* (the Canoe Bailer of Makali'i) to a line of stars to assist him with navigation using Indigenous methods.[218] The new star path honours the memory of Noa, one of the greatest Hawaiian navigators and explorers. Although five principal stars make up this star path (Capella, Castor, Pollux, Procyon and Sirius), the Pleiades and the Maori constellation Te Waka o Tainui are important stars 'in and around Ke Ka o Makali' that ensures the memory of their voyaging canoes is forever immortalised in the night skies.[219]

The great star of Matariki (a symbol of unity)

Like the Pawnee legend of Alcyone the 'Broken Chest Star', a common Polynesian tale tells how Matariki once formed a single star, although in this story Matariki is a man and a very handsome one at that. The memory of this legend may be encoded in the design of King Tawhiao's Coat of Arms, *Te Paki o Matariki*, which shows six small stars and one large star.[220] Gertrude and James Jobes record the story in *Outer Space*, as does Edwin Krupp in *Beyond the Blue Horizon*. Essentially, it involves a mini 'Star Wars' conflict involving Tane (the guardian of the Four Pillars of heaven) and the stars Sirius and Aldebaran, who sided together against Matariki, at that time the most brilliant star in the night skies. Tane's jealousy was aroused by Matariki's boasting of his celestial splendour and he enlisted the aid of the two stars to extinguish Matariki's lights.

They hatched a plan to ambush Matariki at night by creeping up on him but before they reached him, he saw the trio and fled to the Milky Way to hide. Tane, however, spotted him and hurled Aldebaran at the star, which broke into six pieces. Afterwards, Tane struck a truce with Aldebaran.[221] Although this legend refers to Matariki as a male star that broke into six pieces, not seven, it does not preclude it from the common themes identified at the beginning of the book, for it survives in the Maori tale of Matariki and her six daughters. It stands as a symbol of unity as well as a demonstration of emanation of the deity expressed in the principle of the One in the Many and Many in the One. The fact that six pieces emerged from the one implies the interplay between six and seven that alludes to the mystery of the Lost Pleiad. In this sense, it resembles the Babylonian 'Sevenfold One', a name suggestive of a spiritual entity rather than an astronomical one.[222]

Ultimately, whether there is any astronomical truth to the story is not important. What matters more is that human beings not only have something to strive for, but that they have something tangible to symbolise this concept of unity, such as Matariki herself. For all of Te Whanau Marama, including Matariki, have much to teach us. The Maori perception of stars as family members mirrors the spiritual beliefs of other Indigenous peoples, which acknowledges the kindredness of all beings with the cosmos. It ensured that *Te Paki o Ruhi*, the long, long journey 'that lasted for so many centuries' was not a lonely voyage.[223] For as they bravely sailed across the wide, blue expanse of the Pacific with their loved ones beside them 'combating the wrath of Paeweranui' their Ancestors travelled alongside their waka to guide and accompany them on their journey.[224]

> For with Hine-korako and Kahukura to guide them,
> with Tutara-kauika and Ruamano to guard them,
> with pale Hina and her younger relatives to illumine
> their path, wherefore should fears assail them?[225]

Is it any wonder that Maori and other Polynesian peoples would greet te Whanau Marama, the Star People, with songs and aroha (love) in their hearts? For the 'Shining Ones' guided and protected them along te ara moana (the sea road), while during the day they followed te ara whanui a Tane, 'the golden path of the setting sun.'[226] And so at the end of the day, at the end of the long, long voyage — as Wineera states, it matters not where the 'real' Hawaiki is, it is enough to know the name.[227]

I ahu mai tatou i Hawaiki roa, Hawaiki pamamao.

Notes

1. Best, *The Astronomical Knowledge of the Maori*, pp. 7 and 26.
2. Lewis, *We, the Navigators*, p. 3.
3. Best, *The Astronomical Knowledge of the Maori*, p. 5.
4. Ibid, p. 80.
5. Hancock, *Heaven's Mirror*, p. 245.
6. Makemson, *The Morning Star Rises*, p. 76. They were perceived as 'high born' stars as they held the highest rank of all the stars, presumably because they commenced the New Year.
7. Best, *The Astronomical Knowledge of the Maori*, p. 34.
8. Ibid, p. 30. In the Maori language, *ra* means sun, *riki* is star and *mata* means eyes. *Ra ririki* means 'little suns.' For an esoteric reading of 'The Eye of Heaven' in Polynesian societies see Hancock, *Heaven's Mirror*, p. 245.
9. Best, *The Astronomical Knowledge of the Maori*, p. 54.
10. Ibid.
11. Orbell, *The Illustrated Encyclopaedia of Maori Myth and Legend*, p. 113.
12. Ibid.
13. Ibid.
14. Best, *The Astronomical Knowledge of the Maori*, p. 53.
15. Ibid.
16. Ibid, pp. 32 and 34.
17. Ibid, p. 53.
18. Ibid.
19. Ibid.
20. Orbell, *The Illustrated Encyclopaedia of Maori Myth and Legend*, p. 113.
21. Best, *The Astronomical Knowledge of the Maori*, p. 53.
22. Ibid, pp. 23–24.
23. Ibid, p. 28.
24. Ibid, p. 30.
25. Ibid, p. 53.
26. Ibid, p. 52.
27. Makemson, *The Morning Star Rises*, pp. 232 and 266.
28. Ibid, p. 267.
29. Best, *The Astronomical Knowledge of the Maori*, p. 52.
30. Makemson, *The Morning Star Rises*, pp. 104–105.
31. Kirch and Green, *Hawaiki, Ancestral Polynesia*, p. 264.
32. Kyselka, *An Ocean in Mind*, pp. 9 and 48.
33. Ibid.
34. Best, *The Astronomical Knowledge of the Maori*, p. 55. Horne Island is part of the Wallis and Futuna Islands located in the South Pacific Islands in Oceania, which is about two-thirds of the way from Hawaii and New Zealand but closer to the former islands.
35. Ibid, p. 54.
36. Krupp, *Skywatchers, Shamans and Kings*, p. 169.

37. Best, *The Astronomical Knowledge of the Maori*, p. 8.
38. Ibid.
39. Ibid.
40. Ibid, p. 13.
41. Ibid.
42. Ibid.
43. Ibid.
44. Ibid, p. 14.
45. Ibid.
46. Reed, *Maori Legends*, p. 44.
47. Best, *The Astronomical Knowledge of the Maori*, p. 17.
48. Ibid, p. 14.
49. Levy, *Skywatching*, p. 150.
50. Best, *The Astronomical Knowledge of the Maori*, p. 42.
51. Orbell, *The Illustrated Encyclopaedia of Maori Myth and Legend*, p. 32.
52. Ibid.
53. Best, *The Astronomical Knowledge of the Maori*, p. 46.
54. Kyselka, *An Ocean in Mind*, p. 9.
55. Ibid.
56. Best, *The Astronomical Knowledge of the Maori*, p. 17.
57. Lewis, *The Thirteenth Stone*, p. 30. For a brief explanation of the Catholic sacrament of Holy Communion see *World Religions* by Paul Oliver at p. 36.
58. Bell, *Ngarrindjeri Wurruwarrin*, p. 542.
59. Hancock, *Heaven's Mirror*, p. 252.
60. Lewis, *The Thirteenth Stone*, p. 22.
61. Ibid, p. 23.
62. Ibid.
63. Hoad (ed.), *The Concise Oxford Dictionary of English Etymology*, p. 85.
64. Lewis, *The Thirteenth Stone*, p. 23.
65. See David Unaipon's story of the *Mungingee* (Pleiades) in *Legendary Tales of the Australian Aborigines*, pp. 145–49.
66. Heyerdahl, *Sea Routes to Polynesia*, pp. 34–35. See also *The Kon-Tiki Expedition* by the same author.
67. Lewis, *We, the Navigators*, p. 7.
68. 'Maori Origins', Australian Broadcasting Corporation TV Science Program *Catalyst*, broadcast Thursday 27 March 2003. See the following website for further information:
    <www.abc.net.au/catalyst/stories/s823810.htm>.
69. Kyselka, *An Ocean in Mind*, p. 14.
70. Couper and Henbest, *The Stars*, p. 22.
71. Kyselka, *An Ocean in Mind*, p. 14.
72. Ibid.
73. Ibid, p. 3.
74. Couper and Henbest, *The Stars*, p. 22.

75. Ibid.
76. Ibid.
77. Ibid.
78. See the website of the Polynesian Voyaging Society at: <http://pvs.hawaii.org/navigate/stars.html>.
79. Kyselka, *An Ocean in Mind*, pp. 15–16.
80. Couper and Henbest, *The Stars*, p. 23.
81. Lewis, *We, the Navigators*, pp. 313–14.
82. Ibid.
83. Ibid, p. 337.
84. See the Public Broadcasting Service website on the documentary 'Wayfinders: A Pacific Odyssey' at: <www.pbs.org/wayfinders/wayfinding.html>. See under 'Ask the experts'.
85. Kyselka, *An Ocean in Mind*, p. x.
86. See footnote #84.
87. Kane, *Voyage: The Discovery of Hawaii*, p. 105.
88. Ibid.
89. Kyselka, *An Ocean in Mind*, p. x.
90. Ibid.
91. Lewis, *We, the Navigators*, p. 48.
92. Ibid.
93. Ibid.
94. Ibid.
95. Ibid, p. 53.
96. Makemson, *The Morning Star Rises*, p. 39.
97. Vernice Wineera, 'The Story Behind the Legend of the Seven Maori Canoes and the Descending Maori Chiefs,' paper presented to *First Annual MPHS Conference*, Brigham Young University, Hawaii, 1–2 August 1980, p. 1. Available on the internet at: <www.mormonpacific.org/Proceedings/Wineera80.htm>.
98. Ibid.
99. Lewis, *We, the Navigators*, p. 20.
100. Leon Paoa Sterling, *Voices of Wisdom*, p. 199.
101. 'Maori Origins', Australian Broadcasting Corporation TV Science Program *Catalyst*, broadcast Thursday 27 March 2003. See the following website for further information: <www.abc.net.au/catalyst/stories/s823810.htm>. See also the research findings of geneticist Professor Bryan Sykes in *The Seven Daughters of Eve* in his individual chapters on 'The Puzzle of the Pacific' and 'The Greatest Voyagers'.
102. Orbell, *The Illustrated Encyclopaedia of Maori Myth and Legend*, pp. 50–1.

103. Vernice Wineera, 'The Story Behind the Legend of the Seven Maori Canoes and the Descending Maori Chiefs,' paper presented to *First Annual MPHS Conference*, Brigham Young University, Hawaii, 1–2 August 1980, pp. 1 and 3. Available on the internet at: <www.mormonpacific.org/Proceedings/Wineera80.htm>.

104. Ibid.

105. Ibid.

106. Ibid.

107. Orbell, *The Illustrated Encyclopaedia of Maori Myth and Legend*, p. 52.

108. Kyselka, *An Ocean in Mind*, p. 14.

109. Lewis, *We, the Navigators*, p. 374.

110. Ibid, p. 20

111. Ibid.

112. Ibid, pp. 3 and 374.

113. Makemson, *The Morning Star Rises*, p. 13.

114. Lewis, *We, the Navigators*, p. 310.

115. Ibid.

116. Ibid, pp. 310–11.

117. Ibid, p. 310.

118. Best, *The Astronomical Knowledge of the Maori*, p. 34.

119. Ibid.

120. Ibid, p. 31.

121. Reed, *Maori Legends*, p. 45.

122. Ibid.

123. Best, *The Astronomical Knowledge of the Maori*, p. 30.

124. Ibid, pp. 30–31.

125. Ibid, p. 30.

126. See Makemson's chapter on 'The Astronomical School' at pp. 272–78 in *The Morning Star Rises* for further information on these schools in Polynesian societies.

127. Aveni, *Ancient Astronomers*, p. 152.

128. Ibid.

129. Ibid.

130. Lewis, *We, the Navigators*, pp. 20 and 48.

131. Ibid, p. 285.

132. Ibid.

133. Ibid.

134. Ibid, p. 48.

135. Ibid.

136. Best, *The Astronomical Knowledge of the Maori*, p. 35.

137. Ibid, pp. 35–36.

138. Ibid, p. 36.

139. Ibid.

140. Couper and Henbest, *The Stars*, pp. 22–23.
141. Ibid, p. 23.
142. Ibid.
143. Ibid.
144. Kyselka, *An Ocean in Mind*, p. 9.
145. Couper and Henbest, *The Stars*, p. 23.
146. Lewis, *We, the Navigators*, p. 82.
147. Ibid.
148. Ibid.
149. Ibid.
150. Ibid.
151. Ibid, p. 83.
152. Ibid, p. 84.
153. Ibid.
154. Ibid.
155. Ibid, p. 98.
156. Ibid, p. 83.
157. See the story of Kupe's discovery of New Zealand at the website of Tai Tokerau Maori Tourism, Northland, New Zealand at: <www.taitokerau.com/features/maori/oral/ngatoki.htm>.
158. Best, *The Astronomical Knowledge of the Maori*, p. 36.
159. Ibid.
160. Ibid.
161. Alpers, *Maori Myths and Tribal Legends*, p. 139.
162. Ibid.
163. Orbell, *The Illustrated Encyclopaedia of Maori Myth and Legend*, p. 94.
164. Lewis, *We, the Navigators*, p. 21.
165. Ibid.
166. Ibid.
167. Allen, *Star Names*, pp. 395–96.
168. See the website of the Polynesian Voyaging Society at: <http://pvs.hawaii.org/navigate/stars.html>.
169. Ibid.
170. Doniger O'Flaherty, *The Rig Veda*, pp. 100–101 and Blavatsky, *The Secret Doctrine*, vol. 2, p. 139.
171. Best, *The Astronomical Knowledge of the Maori*, p. 17.
172. Ibid, p. 9.
173. Krupp, *Beyond the Blue Horizon*, p. 47.
174. Best, *The Astronomical Knowledge of the Maori*, p. 16.
175. Ibid, p. 16.
176. Ibid.
177. See, for example, chapter VIII 'Egyptian Links with Polynesia' in *South Seas* by Donald McKenzie, pp. 99–116.

178. Walker, *The Woman's Dictionary of Symbols and Sacred Objects*, p. 407.
179. Krupp, *Beyond the Blue Horizon*, p. 47.
180. Ibid.
181. Orbell, *The Illustrated Encyclopaedia of Maori Myth and Legend*, p. 166.
182. Bauval and Gilbert, *The Orion Mystery*, p. 198.
183. Ibid. Note: The authors are actually quoting Rundle Clark in this passage.
184. Ibid, pp. 16–17.
185. Ibid, p. 204.
186. Ibid, p. 17.
187. Ibid, p. 204.
188. Ibid.
189. Ibid.
190. Ibid.
191. Krupp, *Skywatchers, Shamans and Kings*, p. 223.
192. Best, *The Astronomical Knowledge of the Maori*, pp. 17 and 66.
193. Ibid.
194. Orbell, *The Illustrated Encyclopaedia of Maori Myth and Legend*, p. 53.
195. Ibid.
196. Ibid.
197. Ibid.
198. Levy, *Skywatching*, p. 150.
199. Best, *The Astronomical Knowledge of the Maori*, p. 60.
200. Ibid, p. 37.
201. Ibid, p. 53.
202. Orbell, *The Illustrated Encyclopaedia of Maori Myth and Legend*, pp. 169–71.
203. Ibid, p. 169.
204. Ibid.
205. Ibid.
206. Ibid.
207. Ibid.
208. Ibid.
209. Ibid.
210. Ibid, p. 170.
211. Ibid.
212. Ibid.
213. Ibid.
214. Ibid.
215. See *The Reed Dictionary of Modern Maori* by P. M. Reed at p. 7 on the pronunciation of the Maori language. He says *wh* 'is usually pronounced like *f*.' However, 'in some districts it is spoken like an *h* (e.g. in Hokianga) and in others like a *w* (e.g. in Taranaki), in others again like *wh* in when.'

216. Orbell, *The Illustrated Encyclopaedia of Maori Myth and Legend*, p. 170.
217. Ibid.
218. See the website of the Polynesian Voyaging Society at:
    <http://pvs.hawaii.org/navigate/stars.html>.
219. Ibid.
220. Orbell, *The Illustrated Encyclopaedia of Maori Myth and Legend*, p. 112.
221. See the story in Jobes and Jobes, *Outer Space*, pp. 340–41 and Krupp,
    Beyond the Blue Horizon, p. 246.
222. Krupp, *Beyond the Blue Horizon*, p. 242.
223. Best, *The Astronomical Knowledge of the Maori*, p. 80.
224. Ibid.
225. Ibid.
226. Ibid.
227. Vernice Wineera, 'The Story Behind the Legend of the Seven Maori
    Canoes and the Descending Maori Chiefs,' paper presented to *First
    Annual MPHS Conference*, Brigham Young University, Hawaii, August
    1–2, 1980, p. 5. Available on the internet at:
    <www.mormonpacific.org/Proceedings/Wineera80.htm>.

CHAPTER EIGHT

Subaru
The Story of the Lost Sister

Tengasema i ka yo
Sumaru bosha narabu yo
Umigasema i ka Ebi kagomu

Is the sky too small for you, Pleiades?
I see you are packed in.
Is the ocean too small for you, Shrimp?
I see you're hunched over.

> — Japanese 'Song of the Pleiades' from
> Uoshima Island in Ehime Prefecture.[1]

In the Land of the Rising Sun the legend of the Seven Sisters of the Pleiades lives on in the hearts and minds of the Ainu, the Indigenous people of Japan and in those who came after them, the Wajin[2] (foreigners) or Japanese. Like the Pawnee Native peoples of the American plains, they saw within this star cluster a celestial expression of unity, hence their name Subaru for the Pleiades.[3] Five separate Japanese companies adopted this concept in 1953, when they merged to form a huge transportation conglomerate and chose the star cluster as the official logo for their car-manufacturing outlet. Here, as elsewhere around the world, the stars of the Pleiades governed agricultural calendars and marked certain festivities. The most famous of these is the annual Feast of Lanterns, which celebrates their gentle heavenly glow. The Japanese, and indeed the world love affair with this beautiful star cluster, continues on the 'Big Island' of Hawaii atop the 4,200 metre summit of Mauna Kea where the National Astronomical Observatory of Japan has installed its very latest technological masterpiece, the Subaru Telescope. Here in science, as in mythology, we find yet again these faithful celestial companions continuing to guide and steer humanity in our exploration of the cosmos. And from automobiles to telescopes, the application of the Subaru name has ensured that the legend of the Seven Sisters of the Pleiades will continue to exist in this part of the world during the new millennium and beyond.

At sixes and sevens

Most people would be familiar with the Subaru car logo with its six stars — five small stars and one larger star framed in an ellipse. Although only six stars are depicted this does not mean that the Japanese people only ever saw that amount of stars in the cluster. According to Japanese astronomy experts, Steve Renshaw and Saori Ihara, the Japanese frequently saw seven, not six stars in the Pleiades. What is more, they point out that even today some Japanese people still refer to the Pleiades as *Shichi Fukujin*, 'which are often seen at temples and in miniature at souvenir shops throughout Japan.'[4] These 'Seven Happy Gods' as they are more popularly known, are frequently portrayed sailing together in an ark much like the Seven Sages who fled the rising waters in India to propagate the seeding of the human race. In this septenary context Benten, the Japanese goddess of the sea and the only female in the Japanese ark, may represent the 'mortal woman' of Madame Blavatsky's *Secret Doctrine*, a female progenitor like Manu Vaivasatva.[5] That it bears a striking similarity to this tale is not surprising, given that the seven deities can be traced back to the early history of Japan through its Buddhist roots, largely through the influence of China and Korea. Ultimately, Buddhism — like Hinduism — originated on the Indian subcontinent and although it is the younger of the two great religions there are clear linguistic and cultural links between them, thereby making the paper trail complete. Quite apart from their mythological references in these spiritual traditions, the stars of the Pleiades also feature in Japan's national religion — Shintoism. Although it has no written records as such, some of its legendary stories appear in two ancient collections from the eighth century, the *Kojiki* and the *Nihongi*. According to most historians, the process of compiling the Kojiki began around 682 CE and ended around 712 CE, whereas the Nihongi 'was published by imperial order' in 720 CE.[6] In *Nihongi: Chronicles of Japan*, Aston tells us the only stars mentioned in these two anthologies 'are Venus, the Pleiades and the Weaver Star (Alpha Lyrae).'[7] In Western astronomy this star is better known as the blue-white giant Vega.[8]

The Ainu name for the Pleiades is *Iwannociw*, which refers to six stars in the cluster, 'nociw' being the operative word for 'star.'[9] Although theirs is a moral instructional tale of six idle girls, Tomio Sueko in *Ainu no Hoshi* (Ainu Stars) says originally there were seven daughters instead and that the cluster was therefore previously known as *Arwannociw*, which means seven stars.[10] Quite clearly this reveals that there were once seven visible Pleiades stars in Ainu history. But rather than attempt to explain why one of them disappeared from sight like the Aboriginal Australian and Greek myths of the Lost Pleiad, the Ainu simply stopped referring to them as seven lazy sisters and spoke of six idle girls instead. But even though the Japanese and Ainu legends do not specifically refer to a lost or missing sister within this group of siblings, clearly the Ainu tradition embraces this familiar theme because of the previous reference to seven and not six sisters in their mythology.

There may well be a tale of the Lost Pleiad among the Ainu, which needs to be further explored. However that may be, the Subaru car logo has left an indelible mark in the psyche and collective unconscious of Japan and its people. By this I mean that the legend of the Lost Pleiad has to some extent now become forever immortalised by one of their most popular and successful car manufacturers. This is a perfect illustration of how mythologies created by human beings may eventually become incorporated or assimilated into a culture, even if by default. In this way the Subaru design and the seven gods of fortune raise the issue of the legendary lost or missing sister and provide the opportunity to explore that theme in more detail in this chapter.

Seven Happy Gods sail the seas

As Wajin left their homelands in Korea and China to cross the Sea of Japan to colonise its islands sometime around the fourth century, and in successive waves of immigration, they brought with them their stories, their languages, their customs and their religions.[11] Like so many other human migration stories they encountered Indigenous peoples within the new lands whom Maori respectfully call *Tangata Whenua* — the 'People of the Land'. Sadly, as history tells us, many of these encounters resulted in the substantial loss of life, language and culture of the colonised. Even where Indigenous peoples managed to survive the brutal onslaught and impact of dispossession, they now find in contemporary times that their descendants continue to struggle for recognition of their unique cultural identity and heritage. Japan is no exception. The Ainu have only in recent years become officially recognised by the Japanese government following the famous Nibutani Dam case (Japan's administrative *Mabo*) and to some extent acknowledged for their Indigeneity.[12] Their stories and histories are just beginning to be heard. And while Shintoism may be regarded as an indigenous Japanese religion, in truth this title belongs to Ainu spirituality. Be that as it may, there is no doubt that these diverse spiritual and cultural traditions have all shaped the soul of Japan and its people.[13] Difference has long been the site of clash and contestations, and within this social and cultural milieu commonalities are often overlooked, even in the subtle ways in which we influence one another. This may happen even without realising the points where we have crossed over and shared ways of being.

And so it is that into these cultural waters the famous seven sailed. They brought with them prosperity and abundance through Benten, goddess of the sea, longevity through Jurojin, wisdom through Fukurokuju, compassion through Hotei, wealth through Bishamon, knowledge of farming, trading and fishing through Daikoku and Ebisu respectively.[14] Practising Buddhists still evoke the individual attributes of these seven deities in prayer and worship, even today. Elsewhere, in souvenir shops and the like, their religious and esoteric symbolism is a forgotten fatality of the past. For who would think that these seven figures might represent the encoded history of humanity's origins or the containment of ancient astronomies? Their combination of the sexes, male and female (through the one goddess) clearly represents both the seven stars of the Pleiades and those of Ursa Major (the Great Bear). And this amalgam reveals ancient knowledge of the obliquity of the ecliptic and the joint, cooperative relationship of these stars in the turning of the mill (precession) that produces the world's ages.[15]

Benten (Benzai-ten)

This Japanese sea goddess also presents as a mixture of several goddesses from around the world. As the daughter of Ryu-wo the sea king (the Japanese Poseidon), like her Greek counterpart Aphrodite she emerged from the ocean.[16] As patron of the sea she resembles Pleione, mother of the Pleiades, dubbed the 'Sailing Queen' by the ancient Greeks, to whom sailors would beseech protection on their sea journeys.[17] Not surprisingly, as Juliet Piggott points out in *Japanese Mythology*, numerous shrines built in Benten's honour are found either on island locations or else by the sea.[18] Like Maia, eldest of the Pleiades and Venus, the Love Goddess, Benten exemplifies feminine ideals and virtues such as beauty and deportment.[19] Famed for her jealousy, she resembles Hera, Queen of the Heavens, while her love of music — in particular playing the biwa (a mandolin or lute instrument) — reminds us of Hathor, the Egyptian goddess of music, joy and good times.[20]

'As Zeniarai Benten (coin-washing Benten), she increases the wealth of those who toss coins into her well,' says Gertrude Jobes in her *Dictionary of Mythology, Folklore and Symbols*.[21] This connection to wells and her relationship with the sea identifies Benten as a Pleiadian water girl, thereby equating these spheres with the waters of life. 'Springs, fountains, ponds, wells were always female symbols in archaic religions,' says Barbara Walker in *The Women's Encyclopedia of Myths and Secrets*.[22] In the pagan springs of northern Europe they were 'often considered water-passages to the underground womb,' and were 'associated with Mother Hel, whose name also gave rise to *holy* and *healing*.'[23] Furthermore, the universal 'correspondence between water and mother' identifies Benten with the Mother Goddess, the creator of all life.[24] Anciently called *Ichiki-Shima-hime*, Japanese legends refer to Benten's two sisters, thereby evoking images of the Triple Deity, or the Goddess in her triple aspect as Crone, Mother and Maiden.[25] Within this tale of the Seven Gods of Fortune we see many other Pleiadian themes and similarities between the various world legends of this special star cluster.

Legend tells that Benten was married to fellow crewmember Bishamon, the God of Wealth of the Shichi Fukujin, the Japanese equivalent of the Seven Rishis of India who married the Seven Sisters of the Pleiades. According to the Hindu legend of the Krittika, one of the Sisters went to live with her husband in the constellation of Ursa Major and Benten's berth on the Japanese ark may reflect aspects of this story. As we have already seen, this sister Arundhati — or Amba as she was also known — became the star Alcor or Zeta Ursae Majoris, the other half of Mizar the binary star in the middle of the Big Dipper's handle.[26] Yet another account tells of her marriage to the Dragon King of Enoshima Island, which explains better her association with dragons and the sea, and in particular her depiction as 'either riding or being escorted by a dragon.'[27] Says Juliet Piggott:

> There is a tale that before Enoshima appeared, a dragon lived on the mainland near the strip of sand, which now connects it at low tide to the island. The dragon used to eat children in the village now known as Koshigoe, very near Enoshima and Kamakura. It was in this same village that Nichiren escaped decapitation through a miracle in the thirteenth century. In the sixth century an earthquake caused Enoshima to erupt, with Benten appearing in the sky above it. She then stepped on to the island and married the dragon. This caused the dragon's appetite for human flesh to cease.[28]

In another version of the story, Benten married 'one of the dragon kings after his constant wooing.'[29] That the Seven Sisters of the Pleiades, who are renowned for their legendary beauty, should constantly attract suitors and amorous admirers is hardly surprising. Thus we see in the Bundjalung story of the Maimai, the boys from Orion used honey as a ruse as part of their wooing rituals to entice the young maidens, but to no avail for their love went unrequited.[30] To commemorate this tale a shrine to the goddess was built on the island's peninsula which, like Japan's famous torii shrine at Miajima, becomes engorged by the rising tide.[31] Curiously, a cave on the island 'is also reputed to have been the home of the dragon or serpent husband of Benten,'[32] and legend says 'there is an underground tunnel running from the cave to the base of Mount Fuji.'[33] Here we see yet again two other familiar Pleiadian motifs — that of a hill or mountain which the Sisters climb, and the secret subterranean passage connecting the two. That Mount Fuji-San (Fujiyama to those outside Japan) is implicated is enormously interesting, because she is considered a mountain goddess.[34] Some believe her name derives from the older Ainu language and their fire goddess, and given that Mount Fuji is a volcano this makes perfect sense. The goddess' name — at least the shortened version — is *Apemerukoyanmat*, which means 'Fire-sparks-Rise-Woman'.[35] She is sometimes called *Kamui huchi* for Elderly Goddess, or *Ape huchi* for 'Grandmother Fire.'[36] In Ainu, an elder is called *fuchi*; hence the Japanese derivation.[37]

The rising island of Enoshima bears a striking resemblance to Devils Tower in the Sioux legend of the Seven Star Girls, where they seek refuge on a tree that rises into the heavens, taking them to safety.[38] This island rising from the waters is also similar to ancient Egyptian and Maori mythologies; namely the primeval mound of the First Creation associated with the benben stone and the bennu bird,[39] and that of Mount Hikurangi in the Pacific, the legendary home of the Bird of the Sun.[40] Whether this Dragon King is another aspect or incarnation of Bishamon is not known. Nonetheless, despite her mythological bigamy and her jealous temperament, Benten symbolises romance, marital bliss and personal happiness.[41] Single men and women would therefore pray to her to help them find suitable spouses, while married couples sought her spiritual guidance and wisdom on marital issues. Piggott recounts a beautiful, traditional romantic tale involving the Japanese goddess of love to rival that of medieval European troubadours with their songs and stories of courtly love.

In this story from Kyoto a young man discovers an exquisite love poem near Benten-of-the-Birth-Water temple in the grounds of Amadera Temple.[42] Curious about the identity of the unknown writer, Baishu resolves 'to discover her identity and to marry her.'[43] Because he found the poem near Benten's temple, he beseeches her to assist him in finding his true love. Significantly, toward the end of his prayer vigil on the seventh day he experiences a spiritual encounter with his beloved and two intermediaries, a younger and older man who were 'angelic beings' sent by Benten.[44] Walking home soon after this unearthly experience, Baishu meets a young woman that he recognises from his apparition, and whom he believes is his future wife. She tells him that the goddess Benten had appeared to her with the exact same message. They soon become lovers and live together. But in his coupled state of bliss, Baishu is unaware that other people cannot see her! One day while out walking the streets of Kyoto the servant of a wealthy master summons him and he agrees to meet with this man, even though he is a stranger. At their meeting, the master tells the young man that he recognises him as his future son-in-law and explains how this came about. Keen to find a good husband for his daughter, the old man had scattered her poems on the grounds of all temples dedicated to the goddess in Kyoto. Benten then appeared to him in a dream, 'saying that a good husband had been found for his daughter and that he would meet him during the coming winter.'[45]

The goddess reappeared to him once again closer to the time and informed the master of this young man's appearance. She also told him of the day he would pass by the master's house 'and that it would be in order to send for him.'[46] Baishu was just about to tell the father that he was already spoken for when the old man opened a door leading into the next room. Much to his astonishment there stood 'the waiting bride, no other than the young woman Benten had already made his wife' and whom he first saw in the goddess' temple.[47] The happy couple were reunited in 'an earthly wedding' and, needless to say, lived happily ever after.[48] This story is different, says Piggott, in that unlike tales of ghosts who visit their partners after death, here 'the order was unusually reversed for the girl's spirit came before she was married.'[49] Near where this story has its origins, Lake Biwa is named in honour of Benten after her favourite instrument. Much emphasis is placed on the geographical contours of the lake that resemble the shape of a biwa.[50] Several other locations throughout Japan are connected with the goddess, including a group of seven islands collectively referred to as Bentenjima in the southern part of Lake Hamana in Shizuoka Prefecture in central Honshu.[51] A shrine dedicated to Benten dates from the Edo period (1600–1868).[52]

Finally, Benten's depiction in Japanese art and in temples as having eight arms may be a zoological image of her arachnoid aspect as the Spider Goddess that creates the world.[53] It also suggests an association with weaving and therefore implicates her as a goddess of fate. Sometimes in these images two of her eight arms 'are folded in prayer' says Piggott, 'and in that form she resembles Kwannon (Kwan Yin), the goddess of Mercy.'[54] In *The Woman's Dictionary of Symbols and Sacred Objects*, Barbara Walker notes that the Hindu goddess Kali is often represented as having 'six arms and two legs,' which may be related to 'the eight-legged spider.'[55] In some instances the goddess is drawn with an upright sword in her hand, which evokes images of the Krittika, the Mother Judges of India and precursors to the female justices who stand guard outside courts in the Western legal adversary system.[56]

Hackin and others confirm in *The Mythologies of the East* that Benten and her husband Bishamon are of Hindu origin.[57] In addition to being the God of War he is also the God of Happiness, perhaps vicariously through Benten his wife. Like Ares, Mars and other war gods, Bishamon is often 'represented in full armour, holding in his right hand a small pagoda and a lance in his left.'[58] In this regard he resembles the Hindu war god Karttikeya, son of the Pleiades, known also as Skanda, thereby strengthening his alliance with the Sisters. As the sole female in the ark, Benten may represent the second eldest of the Greek maidens, Alcyone, as Robert Graves' mystic leader,[59] or the eldest sister Maia as the Great and Bountiful Mother whom the Hindus called Maya, representing the illusion of Time.[60] Or else she may represent the Egyptian goddess Hathor who symbolised Seven Cows or seven goddesses of fate.[61]

Like the Japanese, the Ainu celebrated a goddess of the sea or of the waters, says Donald Philippi in *Songs of Gods, Songs of Humans*. She went by two names, either as *Petru-ush-mat*, which means 'Woman Dwelling in the Watering Place, or else *Wakka-ush-kamui*, the 'Goddess Dwelling in the Water.'[62] As Goddess of the River Rapids she was called *Chiwash-kor-kamui*, who reputedly saved humanity from starvation during a particularly long famine.[63] 'By observing the proper hunting and fishing rituals' as dictated to them by the River Goddess and Petruushmat in song and dance, the Ainu people are able to avoid errors of the past by ensuring the natural cycles of food are not exhausted such that they cannot be replaced.[64] The story of the famine and the role of the goddesses in restoring plenty are recorded in the *Song of the Goddess of the Waters*, as told by Ainu informants Hiramura Kanunmore and Nitani Kunimatsu.[65]

Because Ainu cultures were primarily orally based like many other Indigenous cultures, their history, sciences and other knowledge bases were transmitted via epic songs known as *Kamui Yukar* (Songs of the Gods) or else through their stories or folk tales, which they call *Uwepekere*.[66] Kamui is the Ainu word for God or Goddess. Like the Polynesians, the Ainu heavens were multiple, some say there were six, others say there were five heavens.[67] Curiously, the second heaven is called *Rangi kando*, which sounds suspiciously Polynesian.[68] Ainu believe in a supreme creator of which the Sun Goddess is his representative and holds 'the most important office.'[69] Some believe the Japanese Sun Goddess Amaterasu is modelled after the much older Ainu deity. As in Aboriginal Australia (Bandaiyan) and ancient Egypt, the Sun is female and the Moon male.

In addition to their sea goddess, and in view of a sole female voyager in the Japanese ark, the Ainu curiously refer to an 'Old Boat' goddess by the name of *Onne-chip-kamui*, who went up to the heavens after her 'death.'[70] Although the song and author do not reveal her starry identity, clearly the song alludes to a star constellation, possibly an ark. Curiously, an entire ship constellation in the night skies of Europe was once called Argo Navis, which now consists of its constituent parts such as the stern, keel, compass and sails.[71] Another interesting observation is the Ainu reference to one of the sea goddesses as *Wakka-ush-kamui*. In several Polynesian languages, *waka* means canoe or ship and one of these, the Tainui, forms a constellation known as Te Ra o Tainui or the 'Sails of Tainui', formed in part by the stars of the Pleiades.[72] Furthermore, the Polynesian *waka* is a linguistic cognate of water[73] as in Waitangi (Sorrowful Waters) and in the Maori names for three of the Pleiades sisters, including Waiti (Water), Waita (Water-dashing down) and Waipuna-a-rangi (Celestial spring of water).[74] Beyond the linguistic similarities between Ainu and Polynesians, the dance of males resembles Torres Strait Islanders, both in style and tempo and in their use of bow and arrows. It is interesting to note also that the facial tattoos of Ainu women closely resemble the *moko* of Maori women.[75] The linguistic and cultural affiliations between Ainu and Polynesian peoples is not surprising, given the recent genetic findings that confirm the ancestors of Polynesians originated from southern Asia in Taiwan.[76]

Diversity within and without

Although Subaru is recognised as the official Japanese name for the Pleiades, the star cluster is known throughout the country by a variety of names that represent a number of different perspectives. Hence the emphasis behind Takeshi Uchida's *Dialects of the Stars and Cultures* is directed toward the many cultural and regional differences between the Japanese — let alone that which exists between Ainu and Japanese — thereby exploding the myth of homogeneity within Japanese culture.[77] To demonstrate the array of star names, Uchida lists over 60 different names for the cluster alone. Thus in *Starlore of Japan*, Kouichi Kitao observes that in Ishikawa Prefecture another name for the Pleiades was *Shibari*.[78] In their article on the Pleiades, Steve Renshaw and Saori

Ihara suggest that cultural adaptation is the key to this burgeoning diversity. For even though 'much of the myth of Subaru came to Japan from China,' nonetheless over time this starlore was adapted to local prefectures and towns based upon seasonal needs.[79] This included regional information on when to plant produce, when to go fishing and many other activities of daily life. Local religious customs influenced these mythologies, including 'individual differences in the perception of what the stars look like.'[80]

As previously noted, the term 'Subaru' means 'united' or 'getting together.'[81] According to Renshaw and Ihari, the Chinese character (Kanji) for Subaru also has connotations of being 'bright'; thus the 'bright' Subaru stars seem to 'get together in one place.'[82] In some regions of Japan — especially agricultural areas — Subaru becomes *Suharu*, which refers to a 'tied group' of things, usually 'shocks of grain tied in a bundle.'[83] This may explain why the Song of the Pleiades in the opening quote to this chapter refers to these stars as *Sumaru*, says Kitao. 'This song asks why Sumaru is clumped together so close in one place even though the sky is so broad, and why shrimp bend their backs even though the sea is so spacious.'[84]

In some areas of Japan the Pleiades were seen as 'seeds', says Renshaw and Ihara, 'and their rising with the sun in spring signalled a time to plant seeds.'[85] This demonstrates that the Pleiades would have played a significant role in their agricultural calendar, as in other parts of the world. It is also possible that their reference to 'seeds' might have an older association, namely through the Shichi Fukujin and the 'Ship of Renewal' that carries the seeds of life, as Marija Gimbutas calls these vessels.[86] Curiously, the stars of Ursa Major, specifically those of the Big Dipper, are seen as a ship's 'rudder' in Japan, which suggests a common relationship between these stars through the Shichi Fukujin.[87] Hence they are known as the 'Rudder Stars.'[88]

These maritime images and symbols continue with both the stars of the Pleiades and their sister stars the Hyades, with much emphasis on the V-shaped asterism of this cluster. In Ishikawa Prefecture, for instance, the V-shape of the Hyades reminds the local people of a temple bell. This image is easily accounted for, says Kitao. 'In Japan, a Buddhist temple stands in nearly every neighbourhood, and people regularly heard the solemn deep sound of its big bell.'[89] As Japan is an island country, it is not surprising that some people saw the Pleiades as a fishing net made of stones and bamboo. Accordingly, they used the rise and setting of these stars to tell them 'when to cast their own nets into the sea.'[90] Once again, this is a common if obscure Pleiadian theme. Kitao records that in Tottori Prefecture, the Hyades were seen as a triangular fishnet called *sashidamo* that was 'used to catch small fish like sardines or sand lances.'[91] And in Iwate Prefecture they were referred to as 'Carrier Basket' or *mokko*.[92]

An esoteric reading of nets, baskets and other related objects suggests a scientific coding of our creation and the correlation of these stars with goddesses of fate, as discussed in the opening and closing chapters. In Japanese starlore, as elsewhere, the stars of the Pleiades and the Hyades are connected with rains and with food. Kitao tells us that in Aomori Prefecture the V-shaped asterism of the Hyades resembled 'a fork-shaped hook that is attached to two lures and used especially for squid fishing,' hence its name *yamade* or 'squid lure hook.'[93] This connects with stories from Iwate Prefecture of the carrier basket of the Hyades, as 'the basket mentioned in this story was handmade of wood and twine, and used by squid fishermen in the region in their work.'[94] Says Kitao, 'The fisherman telling this story readily associates the rise of Hyades and a good catch.'[95] These maritime symbols connect the Hyades and the Pleiades with 'aqua-cosmogony' — the 'life source, and the birth of human, animal, and plant life,' says Gimbutas in *The Language of the Goddess*.[96]

Other patterns in the sky that people attributed to the stars of the Pleiades in Japan include a cup of sake, a strainer and a set of paintbrushes, and even 'a person's elbow joint'![97] In Koiki Prefecture, for instance, Ihara's grandmother refers to the Pleiades as *Hokiboshi*, which literally means 'brush star' or 'brush stars.'[98] *Boshi* is Japanese for star or stars. Next to the term Subaru, say Renshaw and Ihara, this is probably one of the more widely used Japanese titles for the Pleiades — the 'Brush Daub.'[99] As Kitao points out, the variety of appellations for the Hyades and the Pleiades show that 'most local star names were derived from the names of articles found in people's everyday life.'[100] Furthermore, they reveal that 'these stars were just as much a part of the lives of the people who named them as the items they represented.'[101]

The Japanese Feast of Lanterns and the First Fruits Festival

Despite these regional and cultural differences, the stars of the Pleiades played an important role in the agricultural calendar of Japanese and Ainu people, where they marked the seasons and their activities such as hunting, farming and ritual events. Two foremost events in Japan that can be traced back to celebrations involving this star cluster are the famous Lantern festival and the First Fruits ceremonies. As we have seen in the Hindu chapter, all Asian lantern festivals owe their origins to the Pleiades through Diwali, the Feast of Lamps.[102]

According to William Tyler Olcott in *Starlore of All Ages*, this event 'commemorated at this same season of the year some great calamity which was supposed to have overwhelmed the race of man, in the far distant past, when these seven little stars were prominent in the heavens.'[103]

Like the Europeans, the Japanese celebrated a harvest festival, *Toshigo no Matsuri*. Although there appears to be no written evidence linking the stars of the Pleiades with this festival, the experience of another diverse group of people separate to the Japanese, the Ryukyuu Islanders, suggests that they were involved.[104] A Shinto ceremonial ritual also makes reference to the First Fruits of this harvest, which in Polynesia was announced by the heliacal rising of the Pleiades.[105] In *History of Japanese Religion*, Masaharu Anesaki tells us this First Fruits festival (*Nii-name*) was 'celebrated at midnight in the late autumn' and is still 'observed annually even today.'[106] He says the ceremony that accompanies the festival 'is the most solemn and mysterious of all the Shinto ceremonies.'[107] Various deities were invoked during this ritual, including the Japanese Sun Goddess, to whom the following prayer was offered:

More especially do I humbly declare in the mighty presence of the
Great Heaven-shining Deity who dwells in Ise. Because the Great
Deity has bestowed on him (the sovereign) the lands of the four
quarters over which her glance extends as where the walls of Heaven
rise, as far as where the bounds of Earth stand up, as far as the blue
sky extends, as far as where the white clouds settle down; by the
blue-sea plain, as far as the prows of ships can reach without letting
dry their poles and oars; by land, as far as the hoofs of horses can go,
with tightened baggage-cords, treading continuously widening the
narrow regions and making the steep regions level, in drawing
together, as it were, the distant regions by throwing over them (a net
of) many ropes — therefore let the first fruits for the Sovereign
Deity be piled up in her mighty presence like a range of hills, leaving
the remainder for him (the sovereign) tranquilly to partake of.[108]

This prayer is an intriguing one in terms of our cosmology and
the structure of the universe. It speaks to several familiar Pleiadian
themes of nets and ropes and celestial seas. And while the Sun Goddess
may not be identified as one of the Seven Sisters in Japan, she clearly
is in some world mythologies. Furthermore, there may be a vicarious
relationship between the Sun Goddess and the Pleiades by virtue of
their leading role in other First Fruits ceremonies around the world. In
the southernmost point of Japan, the Yaeyama Islands off Okinawa,
where the greatest number of stars can be seen (including the Northern
Star and the Southern Cross), the Pleiades are called *Murubushi*, or
'cluster of stars.'[109] On several of the islands and villages of Yaeyama
there are many stone viewing spots where people can observe the
stars, including an observatory tower at Taketomi. The female priests
of the village shrine (*ogan*) would gather at these stone markers and
fix the days for Shinto rituals that related to farming and agriculture.

The Murubushi can be seen in the east at the start of November,
and gradually rise higher each day after sunset until by February the
cluster is almost directly overhead. At the start of May the cluster can
be seen low in the western sky at sunset. The six-month period when
the stars of Murubushi can be seen in the evening sky signals the
seasonal time for planting and cultivating wheat and barley. When

the Murubushi first appear in the night sky in November, light rains provide an opportune time for the local people to sow their plant seeds. This light rainfall is ideal for these seeds, which would otherwise be washed away in heavy showers. From the beginning of this busy season for farming, the intermittent yet ongoing Shinto rituals begin with the season festival. After the barley, the rice crop, millet, barnyard millet and 'shin' (another form of millet) seeds are sown ('scattered') in that order. And when the Murubushi is in the western sky at sunset — towards the end of May — the final stage, harvesting, occurs. In May the stars of Murubushi are now in the east and can no longer be seen as they disappear into the morning brightness, and after this in mid-May, in the earliest rainy season in Japan, the rain begins. The Murubushi can be seen again at daybreak at the end of June. This period marks the end of the rainy season. In June, in the earliest harvesting festival in Japan, the Shinto rituals also come to an end.

Seven lazy stars

As previously noted, the Ainu originally referred to the Pleiades as Seven Sisters, who were renowned for their laziness. It is only later when one of the stars faded from view that they became known as six idle girls. At first glance, their 'laziness' may be perceived in human terms as lack of virtue, because of a fixation and preoccupation with their own beauty, and in this regard they present as classic comic celestial drama queens and divas. But in actuality, their so-called 'laziness' has more to do with their presence in the night skies for a substantial amount of the year, as in the Yaeyama Islands off Okinawa where they signal the calendar year, and in particular the planting and harvesting seasons. Also, myths of this genre that describe its actors in a negative light are actually very important lessons to humanity on how *not* to behave, and are therefore as essential as those that lay down a suitable code of conduct.

Uwepekere, or stories of this kind — while not of the more serious nature of *Yukar* — play an important role in the socialisation and education of Ainu children. As Inez Hilger points out in *Together With the Ainu*, 'Children were taught social virtues: they learned life's values and mores as their elders understood them.'[110] Work, in particular, provided the opportunity to learn by doing things; hence being 'lazy' is equivalent to not learning. Says Hilger:

> A properly trained girl knew the time for cutting bulrushes and collecting bark for weaving. She knew how to weave attractive designs in mats, how to be skilful in embroidering, how to make carrying straps used in transporting babies and wood, and also straps used by men to suspend a sword from the shoulder. The girl, too, was trained to work in gardens and fields which was, and still is, woman's responsibility.[111]

The following Ainu story on the Seven Sisters of the Pleiades is quite humorous and charming. Female elder Numate Miyamoto and male elder Kunishiro Kushihara told this particular Uwepekere to early Japanese colonialists in Hokkaido. It was recorded in the historical document known as the *Sosekiso* and is retold in Tomio Sueko's *Ainu no Hoshi* (Ainu Stars). As the book is not available in English, Georgie Stevens, a young Australian lawyer living in Japan who works with Ainu and Japanese people, made the following translation:

> Long ago, there lived seven lazy, obstinate daughters.
>
> A hardworking young man who lived nearby often warned the seven daughters. 'Rather than playing idly every day, why don't you till your fields? No matter how much makeup you wear, no one will take a lazy girl for a wife.'
>
> The daughters would complain in chorus just like this:
>
> 'If we till the fields we'll get hot and sweaty, and our hands will get dirty.'
>
> When the youth replied again light-heartedly, 'If you get dirty you can wash in the river', the daughters, not to be beaten, replied, 'If we wash in the river we might be swept away and meet a terrible fate, so we don't want to.'
>
> 'If you are about to be swept away you can grab hold of the base of a plant.'

'We'll cut our hands and be hurt.'

'If your hand gets cut, you can treat it by binding it with cloth.'

'Even if we bind our wound, our hearts will be pounding and we will be upset. Oh we wish we could become stars. Because if you are a star you wouldn't have to do anything and that would be so nice.'

During all the banter back and forth the young man had become truly angry, and so the daughters got in a boat to run away from him. The man got in a boat and chased them, but could not seem to catch them.

The Kamui (God) who watched this turned the seven lazy daughters and the man who chased them, into stars and put them up into the sky.[112]

The seven daughters who became stars are known as *Arwannociw*. They appeared late in autumn when the work in the fields had finished, shining in the cold winter night sky, but the youngest daughter was ashamed of her actions and covered her face with both of her hands. As a result, people today can see no more than six stars. So, the name *Iwannociw* (six stars) was created much later, says Sueko. The young man who chased the daughters also became the star named *nospakur nociw*, and because he is basically hardworking, he circles the night sky almost all year round without rest. Interestingly, says Sueko, when Arwannociw appear it seems that nospakur nociw will catch up with them, but as winter progresses Arwannociw get faster, and by the time the snow begins to thaw nospakur has been completely outrun. In other words, he is completely separated from them. This is a funny tale, say the elders, because it happens the same way every year. Although Sueko does not identify which star is meant by nospakur nociw, it is likely that he may be one of the stars in Ursa Major because of the reference to him 'circling the night sky.' In astronomy these stars are referred to as 'circumpolar' stars because they never appear to set. Whether a star is circumpolar or not depends on two factors; their actual position in the sky, and the observer's latitude. For example, Ursa Major is circumpolar in London, but not in Cairo. At the North or South Poles all stars are circumpolar, whereas none are circumpolar at the equator.

The content of the story of Iwannociw (the six daughters of the Pleiades) changes little from the above narrative. In the north-eastern part of Ainu *moshiri* (Ainu country), nociw is also called *riop* (star); and so Iwannociw, as recorded in the Sosekiso, is called *Iwanrikop* (the string of six stars). Names for this constellation directly express the oral folklore on Arwannociw, such as *torannenociw*, meaning 'idle or lazy stars', and *toy ta sawot nociw*, 'fields tilling run away from stars' or 'the stars who hated farming.' It comes from *toy* meaning field, *ta* being tilling, *kur* being a person, usually a farmer or peasant, *sawot* meaning to run away from and *nociw* being star. This literally translates to: 'The star that ran away because he hated being a farmer.' The toytasawonociw mentioned here are thought to be star names, which originally referred to this same Arwannociw and to the stars of Ursa Major.

Here we have a situation where the stars of the Pleiades and Ursa Major are equated with one another interchangeably as seven stars. The reference to runaway stars is absorbing in light of the Hindu and Babylonian traditions of one of the Seven Sisters going to live with her husband in Ursa Major.[113] Sueko says that as interest in the stars was gradually lost with the passing of each era, it became more difficult to identify which star corresponded to the star names passed down from long ago. In particular, this tendency is often seen in people from the southern regions who had contact with Wajin (Japanese) from early times. It could also be said that the case of toytasawonociw is probably an example of the oral tradition of 'seven stars', 'seen in winter', being hastily referred to as the Ursa Major (Big Dipper). One of the possible reasons for this confusion about stars and their configurations is the deterioration of the living environment surrounding those who handed down (oral) folklore after contact with Wajin. For example, because of harsh labour and poor nutrition during their youth, many older women had poor eyesight, which made it difficult to locate stars.

Searching for the lost Sister

As previously noted, the Ainu did not attempt to explain why there were first seven stars, and then six. Being practically minded, they simply changed the star cluster's name to six idle girls rather than seven lazy daughters. Whether there is any significance of the interplay between six and seven in Ainu cultures is not known for certain. Not much is written about the symbolism of numbers in either Japanese or Ainu traditions, at least not in English. What we can know may be possibly gleaned from their mythologies. For instance, Reverend John Batchelor in *The Ainu and Their Folk-Lore* suggests that Ainu people, in particular, regarded the number six, as a sacred number. Thus 'we hear of a boat having six oars, rowed by six men, and having six gods as passengers,' and so on.[114] But by the same token he acknowledges that for some Ainu, 'seven appears to be thought the perfect number.'[115] Although the legends do not refer to the Pleiades in their zoological form as birds as in other cultures such as ancient Greece and Aboriginal Australia, they may have done so in the past as evidenced by their sacred shaved sticks. Of particular interest are the ones they call *Chikappo-Chikomesup*, which means 'Little Carved Birds'.[116] These are 'sticks of elder, or *cladrastis*, about four feet in length,' which they make into a kind of charm to protect themselves from disease and other poisons of the spirit.[117] Ainu shave the sides of the stick downwards so that they stand out from the stem and then make a split at the top. 'The shavings left on the sides of the sticks are intended to represent feathers or wings' and 'the split top the bird's mouth.'[118] The number of shavings left on the sticks depends on the theological beliefs of the creator, says Batchelor. 'Some have six left on them, and are therefore called *iwin rapushbe inao*, (or) "six-winged inao"; others have seven, and are called *arawan rapushbe inao*, (or) "seven-winged inao".'[119] Unfortunately Batchelor does not explore the symbolism behind this practice, but it clearly indicates interplay between both numbers which may or may not be related to the Pleiades.

The theme of the missing sister, as we have seen throughout this book, is common to many cultures and several writers have attempted to explain the persistence of this theme. Madame Blavatsky, for instance, has offered an esoteric explanation in *The Secret Doctrine* that relates to the various cycles of time and biblical prophecies.[120] Robert Graves' explanation is simply that the ancient Greeks invented it. Thus, in *The Greek Myths*, he says that sometime 'towards the end of the second millennium BC,'[121] one of the seven stars virtually 'disappeared' and 'her absence had to be explained.'[122] The Greek myths give us a choice of three of the Seven Sisters as a likely candidate for the Lost Pleiad. They include Electra, 'who veiled her face at the burning of Troy, appearing to mortals afterwards only as a comet; or Merope, who was shamed for marrying a mortal; or Celaeno, who was struck by thunderbolt.'[123] While Graves' explanation has an air of practicality, it still does not explain why some cultures that are much older than the Hellenic — such as Aboriginal Australia — have similar stories about a lost or missing sister. In Bandaiyan (Australia), these Dreamtime legends of the Seven Sisters go back in time to even before the commencement of the Greek civilisation.

There may, however, be a more scientific explanation to account for its disappearance, suggests American astronomer Edwin Krupp in *Beyond the Blue Horizon*, and this is known as the star variability theory. He believes it is possible that one of the stars in the Pleiades did disappear during the time that Graves has claimed, as stars are known to vary in their brightness and vanish in our night skies only to 're-appear' at some later date. In fact, this can happen on a regular or irregular basis. Such stars are referred to in astronomy as 'variable' stars.[124] The most likely contender in this situation, says Krupp, would be the star Pleione for these reasons:

> Pleione, now the seventh-brightest star of the Pleiades, is an eruptive variable star. Every now and then it flings away an outer layer of gas into an expanding bubble. It did so in 1970, 1938, and 1888, and perhaps earlier too. Observations of it in the 1930s by Harvard Observatory astronomer William A. Calder recorded a decline of about one-sixth magnitude over the three seasons of watching. Given its modest record of changing brightness, Pleione could have been at least a bit more conspicuous in the past.[125]

Along with the star variability theory, other scientific theories include the argument that 'one of the stars has gone dim,' or that the veils of dust surrounding the cluster reduced the starlight of one of its members, or else one of the adolescent stars 'might still be experiencing growing pains.'[126] By a process of valid scientific deduction, Krupp eliminates all but one of the four theories proposed — that of the stellar variability argument. But even though he concedes it may appear to 'have a felicitous modern scientific confirmation of ancient wisdom' which may confirm the Lost Pleiad legend on a scientific basis, he remains not entirely convinced.[127]

Despite this perfectly plausible explanation, remarkably Krupp's scientific judgment is swayed by an ancient Greek astronomy text, the *Katasterismoi*, which claims only six stars were visible in the cluster, despite mythic and cultural references to Seven Sisters.[128] He also enlists the aid of the ancient Greek poet Aratus, who boldly proclaimed a century later that the Pleiades 'are seven in the songs of men but only six to the eye' and maintained that none of the stars had ever perished throughout known Greek history.[129] This comment is a grandiose statement coming from a poet, and it is quite peculiar that an astronomer of Krupp's calibre would dismiss the logical explanation of stellar variability on the basis of what most scientists would traditionally perceive as purely 'subjective' and 'non-scientific' evidence. From a Marxist theoretical perspective, Krupp's argument could easily be characterised as engaging in a form of scientific 'false consciousness.' In other words, despite the fact that the Greek legends specifically state there were once seven maidens, one of whom either hid her face in shame or else left the family unit altogether because of some misdeed, ancient Greeks never really 'believed' this story, they simply made it up.

This begs the question: why would the legends speak of seven stars in the first instance when there were supposedly only ever six stars that were visible throughout history, as Aratus claims? It makes no sense mathematically or symbolically, because whereas six may be drawn from seven, seven cannot be drawn from six. Simply put, either there were seven stars or seven daughters to begin with, or there were six.

Stellar variability suggests there could have been more or less stars visible in the cluster at different periods throughout history but it need not diminish the veracity of the tale. Whatever the true explanation, there can be no doubt that the mystery of the Lost Sister of the Pleiades still remains an enigma. And while some have offered all sorts of explanations, spiritual and scientific, one thing is certain — that the stars of the Pleiades will continue to fascinate and keep us guessing. The next time you spot a Subaru, however, spare a thought for this Lost Maiden, whoever she may be, and pray that she will return some day, safely home to her sisters and loving family.

Notes

1. Kitao, *Starlore of Japan*, pp. 29 and 55. See also Juliet Piggott, *Japanese Mythology*, p. 8.
2. Sueko, *Ainu no Hoshi*, p. 199.
3. Kitao, *Starlore of Japan*, p. 28.
4. See, 'A brush daub on the heavens', which appeared in *Archaeastronomy and Ethnosastronomy News*, Quarterly Bulletin #19, March 1996. The article is available at <http://www2.gol.com/users/stever/subaru.htm>.
5. Blavatsky, *The Secret Doctrine*, vol. 2, p. 140.
6. Hackin et al., *The Mythologies of the East*, p. 385.
7. Aston, *Nihongi*, p. 69 his footnote #5.
8. Burnham et al., *Astronomy: The Definitive Guide*, p. 337.
9. Sueko, *Ainu no Hoshi*, p. 199.
10. Ibid.
11. Anesaki, *History of Japanese Religion*, p. 1.
12. See *Kayano et al. v. Hokkaido Expropriation Committee* (The Nibutani Dam Decision), 1598 Hanrei Jiho — 33; 938 Hanrei Times 75, Sapporo District Court, March 27, 1997. For an English commentary on the Japanese case, see the internet article by Mark Levin, Assistant Professor of Law at the University of Hawaii. See also 'An Act for the promotion of Ainu Culture, the spread of knowledge relevant to Ainu Traditions and an education campaign' which the Japanese government enacted on 1 July 1997 following the outcome of the Nibutani Dam case. This Act does not acknowledge Ainu indigeneity *per se* but at least there is the recognition

that Ainu are culturally different from the Japanese and have special needs based on their perceived differences.

13. Ibid, p. 9.
14. Piggott, *Japanese Mythology*, pp. 45–46.
15. See the concepts and arguments espoused by De Santillana and Von Dechend in *Hamlet's Mill.*
16. Piggott, *Japanese Mythology*, p. 46.
17. Graves, *The Greek Myths*, pp. 165 and 775.
18. Piggott, *Japanese Mythology*, p. 46.
19. Ibid.
20. Jobes, *Dictionary of Mythology, Folklore and Symbols*, vol. 1, p. 203.
21. Ibid.
22. Walker, *The Women's Encyclopedia of Myths and Secrets*, p. 1067.
23. Ibid.
24. Ibid.
25. Jobes, *Dictionary of Mythology, Folklore and Symbols*, vol. 1, p. 203. Juliet Piggott also makes reference to Benten's two sisters in *Japanese Mythology* at p. 28.
26. Krupp, *Beyond the Blue Horizon*, p. 248.
27. Piggott, *Japanese Mythology*, pp. 134–35.
28. Ibid, p. 135.
29. Ibid.
30. See the story of 'Where the frost comes from' in *Wise Women of the Dreamtime* by Johanna Lambert at pp. 44–50.
31. Piggott, *Japanese Mythology*, p. 134.
32. Ibid, p. 135.
33. Ibid.
34. Coulter and Turner, *Encyclopedia of Ancient Deities*, p. 182.
35. Philippi, *Songs of Gods, Songs of Humans*, p. 69.
36. Ibid.
37. Coulter and Turner, *Encyclopedia of Ancient Deities*, p. 182.
38. See the Sioux legend of the Seven Star Girls in the booklet by Rathbun, *First Encounters: Indian Legends of Devils Tower*, p. 5.
39. Hancock, *Heaven's Mirror*, p. 66.
40. Orbell, *The Illustrated Encyclopaedia of Maori Myth and Legend*, p. 53.
41. Jobes, *Dictionary of Mythology, Folklore and Symbols*, vol. 1, p. 203.
42. Piggott, *Japanese Mythology*, pp. 136–38.
43. Ibid, p. 136.
44. Ibid, p. 137.
45. Ibid, p. 138.
46. Ibid.
47. Ibid.
48. Ibid.

49. Ibid.
50. Ibid, p. 60.
51. Campbell and Noble et al., *Japan: An Illustrated Encyclopedia*, vol. 1, p. 106.
52. Ibid.
53. Walker, *The Woman's Dictionary of Symbols and Sacred Objects*, p. 163.
54. Ibid, p. 135.
55. Ibid, p. 419.
56. Ibid, p. 151.
57. Hackin et al., *The Mythologies of the East*, p. 446.
58. Ibid.
59. Graves, *The White Goddess*, p. 182.
60. Walker, *The Woman's Dictionary of Symbols and Sacred Objects*, pp. 205, 349, 372 and 465.
61. Allen, *Star Names*, p. 399. See also *Isis Magic* by Isidora Forrest, p. 132.
62. Philippi, *Songs of Gods, Songs of Humans*, p. 99.
63. Ibid.
64. Ibid.
65. Ibid. See the entire *Song of the Goddess of the Waters* at pp. 99–107.
66. Hilger, *Together With the Ainu*, pp. 10–12.
67. Batchelor, *The Ainu and Their Folk-Lore*, p. 60.
68. Ibid.
69. Ibid, p. 63.
70. Philippi, *Songs of Gods, Songs of Humans*, p. 175. See the entire *Song of the Old Boat Goddess* at pp. 175–84.
71. Levy, *Skywatching*, p. 150.
72. Best, *The Astronomical Knowledge of the Maori*, p. 60.
73. See *Hawaiki, Ancestral Polynesia* by Kirch and Green.
74. Makemson, *The Morning Star Rises*, pp. 232 and 266.
75. Batchelor, *The Ainu and Their Folk-Lore*, p. 20. For a photograph of an Ainu woman with her equivalent *moko* see p. 21.
76. 'Maori Origins', Australian Broadcasting Corporation TV Science Program *Catalyst*, broadcast on Thursday 27 March 2003. See the following website for further information: <www.abc.net.au/catalyst/stories/s823810.htm>. See also the research findings of geneticist Professor Bryan Sykes in *The Seven Daughters of Eve* in his individual chapters on 'The Puzzle of the Pacific' and 'The Greatest Voyagers'.
77. As Uchida's *Dialects of the Stars and Cultures* is not available in English, I have had to rely on the interpretation of Steve Renshaw and Saori Ihara in their archaeoastronomy article. See, 'A brush daub on the heavens', which appeared in *Archaeastronomy and Ethnosastronomy News*, Quarterly Bulletin #19, March 1996. The article is available at: <http://www2.gol.com/users/stever/subaru.htm>.
78. Kitao, *Starlore of Japan*, p. 45.

79. Renshaw and Ihara, 'A brush daub on the heavens', *Archaeastronomy and Ethnosastronomy News*, Quarterly Bulletin #19, March 1996.
80. Ibid.
81. Kitao, *Starlore of Japan*, p. 28.
82. Renshaw and Ihara, 'A brush daub on the heavens', *Archaeastronomy and Ethnosastronomy News*, Quarterly Bulletin #19, March 1996.
83. Ibid.
84. Kitao, *Starlore of Japan*, p. 29.
85. Renshaw and Ihara, 'A brush daub on the heavens', *Archaeastronomy and Ethnosastronomy News*, Quarterly Bulletin #19, March 1996.
86. Gimbutas, *The Language of the Goddess*, p. 247.
87. Kitao, *Starlore of Japan*, p. 25.
88. Ibid.
89. Ibid, p. 45.
90. Renshaw and Ihara, 'A brush daub on the heavens', *Archaeastronomy and Ethnosastronomy News*, Quarterly Bulletin #19, March 1996.
91. Kitao, *Starlore of Japan*, p. 45.
92. Ibid, p. 46.
93. Ibid, pp. 45–46.
94. Ibid, p. 46.
95. Ibid.
96. Gimbutas, *The Language of the Goddess*, p. 81.
97. Renshaw and Ihara, 'A brush daub on the heavens', *Archaeastronomy and Ethnosastronomy News*, Quarterly Bulletin #19, March 1996.
98. Ibid.
99. Ibid.
100. Kitao, *Starlore of Japan*, p. 46.
101. Ibid.
102. Allen, *Star Names*, p. 393.
103. Tyler Olcott, *Starlore of All Ages*, p. 412.
104. For more information on the status of minorities in Japan see the report *Japan's Minorities* by Professor George A. De Vos and William O. Wetherall commissioned by the Minority Rights Group of Britain.
105. Krupp, *Skywatchers, Shamans and Kings*, p. 169.
106. Anesaki, *History of Japanese Religion*, p. 33.
107. Ibid.
108. Ibid.
109. This information is taken from a Japanese website. It was translated by Georgie Stevens. See <http://aragusuku.tripod.co.jp/home.htm>.
110. Hilger, *Together With the Ainu*, p. 178.
111. Ibid, p. 190.
112. From *Ainu no Hoshi* (Ainu Stars) by Tomio Sueko at pp. 198–202 as translated by Georgie Stevens.

113. Krupp, *Beyond the Blue Horizon*, p. 248.

114. Batchelor, *The Ainu and Their Folk-Lore*, p. 306.

115. Ibid, p. 106.

116. Ibid, p. 103.

117. Ibid.

118. Ibid, p. 104.

119. Ibid, pp. 105–106.

120. Blavatsky, *The Secret Doctrine*, vol. 2, pp. 618–19.

121. Graves, *The Greek Myths*, p. 154.

122. Ibid, p. 220.

123. Ibid.

124. Krupp, *Beyond the Blue Horizon*, p. 245.

125. Ibid, p. 245.

126. Ibid, p. 243.

127. Ibid, p. 245.

128. Ibid, pp. 242–43.

129. Ibid, p. 245.

CHAPTER NINE

The Pleiades Calendar
Keeping Time with the Seven Sisters

The calendar is a very important piece of paper for every woman . . .
so the Seven Sisters Dreaming to us is important because the stars,
the moon, the sun controls the seasons. So right up until I was a
young girl they were still working by the seasons and the stars. I
would say that a diary and a calendar would be the two most
important things that the Seven Sisters Dreaming would be able to
tell Ngarrindjeri women over thousands of thousands of years . . .
they moved in certain circles. And in Spring when everything is
being born that's when the Sisters return, when swimming is
restricted, when the waters are full of life, too dangerous for women
to enter.

— Doreen Kartinyeri, Ngarrindjeri Elder of South Australia.[1]

All over the world people looked to the stars of the Pleiades to signal
the beginning and end of the seasons, to tell them when to collect
and gather food, when to sow and to harvest, when to expect rains
and when to engage in festivities and ceremonies. This effectively
meant the star cluster acted as a calendar to indicate all of these
things, including the end of the old and the beginning of the New
Year. Although only a handful of writers refer to a Pleiadian Calendar,

nonetheless several have alluded to its existence in one form or another. Anthony Aveni and Susan Milbrath both refer to a Pleiadian cycle in the Aztec, Inca and Mayan agricultural calendars. Maud Makemson writes specifically of the 'Pleiades Year' in several Polynesian cultures in *The Morning Star Rises*. Others like Helena Blavatsky, Richard Allen, Gertrude and James Jobes, and Giorgio De Santillana and Hertha Von Dechend have all noted ancient references to the cycle of precession of the equinoxes as the 'Great Year of the Pleiades.' This cycle marks the circle carved out in our night skies by our planetary wobble which covers some 26,000 years, or 25,900 years to be more precise. Although Madame Blavatsky does not specifically mention a Pleiadian Calendar as such, there is no doubt she considers the star cluster plays an extremely important role as Keepers of Time, Fate and Destiny in our part of the universe.

Likewise, Barbara Hand Clow not only draws parallels between the precession of the equinoxes, the Pleiades and the Mayan Calendar, but also asserts that these stars are primarily responsible for setting in motion time, as we know it, including the commencement of the Age of Aquarius. An ancient artefact found among ruins on the Greek isle of Crete in the Mediterranean may possibly be evidence that such a calendar existed in the Western world. The clay calendar, known as the Phaistos Disk, dates back to the early Minoan civilisation some 4,000 years ago during the Bronze Age. This artefact, says Robert Burnham Jr in *Burnham's Celestial Handbook*, has a number of symbols imprinted on it, including a repetitive pattern of 'seven dots enclosed in a circle.'[2] He is firmly of the view that the seven-dot motif 'can hardly represent anything other than the cluster of the Pleiades.'[3] Ultimately, 'all calendars were influenced by the Pleiades,' say Gertrude and James Jobes, who suggest that a Pleiades Calendar 'may have preceded the lunar and solar calendars.'[4] If proven true it would make the Pleiades Calendar one of the oldest calendars in the world.

Mana from the heavens

Almost everywhere the Pleiades were associated with abundance or scarcity of food supplies. Such was their magnitude that people would chant, sing, dance and pray to these stars to invoke their blessings, like the Maori people of Aotearoa in the southern Pacific, whose language is laden with proverbs and phrases that celebrate this celestial connection. In many places, however, the availability of food was dependent on whether the Pleiades brought with them the valuable rains to enable their crops to grow or bush foods to flourish in the wild. As a consequence, many different cultures marked their seasons by the rise and setting of this star cluster, including the division of the old and the New Year. For instance, in the Murray Islands of the Torres Straits between Papua New Guinea and Australia, the rising of the Pleiades or *Usiam* in the east at sunset signals the start of the turtle season in late October.[5] Their returned presence in the evening skies of the Southern Hemisphere during the late spring and early summer tells the islanders that it is time to begin clearing their gardens.

Other environmental signs include the flowering of certain plants and trees, but the cluster remain one of their principal markers, along with the stars of Orion's belt, which they call *Seg*.[6] In the Southern Hemisphere, in both the Torres Strait and in Tonga, these stars rise in the east at sunset. The Tongan people divided their year into two seasons and named both of them after the Pleiades — *Matarii i nia* (Pleiades above) and *Matarii i raro* (Pleiades below).[7] The first season began 'when the stars appeared on the horizon in the evening and continued while they remained above after dusk' and the second season 'began when they ceased to be visible after the sun set and continued until they again appeared above the horizon in the evening.'[8] Their name for the cluster is similar in sound to other Polynesian names, such as the Maori *Matariki* and the Native Hawaiian *Makali'i*.

The importance of these particular stars to the survival of people in ancient civilisations in the Americas such as the Mayans, Aztecs and Inca was second to none, where they featured prominently in all matters of daily life from astronomy to agriculture, engineering, architecture, mathematics and matters of the divine. The Aztec and Inca terms for the Pleiades have an important meaning in their connection with food, for the Aztec term *Tianquiztli* refers to the 'marketplace'[9] and the Inca term *Collca* translates as 'storehouse',[10] which is precisely where food is stored, sold, exchanged and distributed. The Quiché Maya refer to both the Hyades and Pleiades as *motz*, meaning 'handful', with the Hyades signifying a handful of beans and the Pleiades a handful of maize, whose setting 'corresponds to the time when these plants are sown.'[11] Beyond their victual significance lies a deeper, esoteric meaning whereby these handfuls of seeds, 'which, when planted, grow into the fertile World Tree seen passing prominently through the north-south overhead zone a few hours after the Pleiades have crossed the zenith.'[12] Likewise their relationship with honey in many of the world's legends is especially revealing in light of the biblical reference to their 'sweet influence.' Apart from bestowing immortality in the form of soma or ambrosia, the food of the gods, honey provided mortals with other desirable virtues and qualities such as the lauded gift of eloquence. In addition to these attributes were the many benefits of its healing properties, whose medicinal application was considerable.

This Pleiadian connection with food is further emphasised by the reference to Maia the eldest of the Greek sisters as the 'Great Mother' and the correlation of these stars with the cow in ancient Egypt.[13] Like dutiful mothers everywhere she fulfils her maternal obligations by providing food for all her children in all her various incarnations. In ancient Egypt she took the form of Hathor the Mother Cow goddess, who not only nurtured the living but provided sustenance and comfort to the souls of the dead in the Egyptian underworld.[14] The placement of these stars nearby or within the constellation of Taurus the Bull in several ancient cultures such as India, Mesopotamia, Greece and Egypt, would have enhanced these images of abundance and fertility that led to the idea of the Cornucopia, or Horn of Plenty. Now seen as decorative items in houses and temples, it originally symbolised 'the horn of the Great Mother in her cow or goat incarnation,' says Barbara Walker in *The Woman's Dictionary of Symbols and Sacred Objects*.[15] And from it, 'all good things poured forth.'[16] Given this connection of the star cluster with food, it is not surprising that many festivities and cultural gatherings soon gave rise to celebrate the gifts of mana from the heavens generally, and from the stars of the Pleiades specifically. As Sir James Frazier noted in *The Golden Bough*, among the Khoisan peoples of southern Africa, mothers would lift their babies up to the sky. This was done so that children might stretch their hands towards the stars of plenty and when they appeared above the horizon, 'they would begin to dance, and pray that they might give them abundance of food.'[17]

Elsewhere in the Pacific, Maori women would greet them in the mornings with song and dance. This Polynesian tradition survives in the Native Hawaiian New Year festival *Makahiki* that is still celebrated each year in late October.[18] Quite clearly, people from so many different cultures around the world love and venerate these stars for all sorts of reasons, including the provision of food so vital to their survival. The myriad number of festivals that derive from the ancient worship and observation of the Pleiades reflects the importance of this small cluster of stars and how it has influenced the human psyche. They include our New Year celebrations, May Day, the Feast of Lanterns, the Beltane Fires and the various Feasts of the Dead, including Halloween. In addition to these well known festivals, the cluster may have given rise to *Yom Kippur*, the Jewish Day of Atonement.[19] At this time of the year Jewish people 'still retain the ancient custom of blowing the *shofar* (ram's horn) at the conclusion of Yom Kippur to clear away evil influences.'[20]

Halloween and other Pleiadian festivals

Most people are familiar with the Halloween festival celebrated every year on the last day of October, if not by personal experience then most certainly by its immortalisation in many Hollywood films. Lanterns made from hollowed-out pumpkins and children dressed in supernatural garb in the guise of witches, hobbits and ETs who go knocking from door to door shouting 'trick or treat', are popular images of this particular Feast of the Dead. Although largely a North American cultural tradition, the origins of Halloween lie in the old Celtic feast of Samhain, or All Hallows' Eve, that was later Christianised as All

Saints' Day.[21] While it is commonly understood these festivals cele-brate the dead, it is a lesser known fact that they ultimately derive from ancient beliefs and observations of the Pleiades. At midnight, on or about the first of November, these stars reach their highest point in the night skies of the Northern Hemisphere.[22] Prayers for the dead were recited at midnight, for people believed the spirits of the dead would rise from their resting places on this precise date and time to visit family and friends to make merry and feast. Hence the tradition of 'trick-or-treat customs descended from a belief that the family dead would bring gifts or goodies to the children during their temporary return from the other world.'[23]

The original Celtic Feast of the Dead was actually a three-day festival regulated by the cluster and not three separate, unrelated feast days, says William Tyler Olcott in *Starlore of All Ages*.[24] This seems to fit in with the ancient Egyptian Feast of Hathor and Isis that took place over a three-day period, the only difference being that it commenced around the seventeenth of November, supposedly their special day.[25] As goddesses of the underworld, Isis and Hathor would have been appropriate role models for an Egyptian Feast of the Dead. It may seem that the Egyptian festival is late in comparison to the Celtic and other northern European festivals, but in our current epoch the cluster actually reach their highest point at midnight closer to the seventeenth of November. The difference in feast dates in the past is a direct result of precession. The same date was significant in ancient Persia (now modern day Iraq) where supposedly petitions presented to Persian kings on that day were not denied. Interestingly, their former name for November was *Mordad*, whose meaning 'the Angel of Death' was aptly suitable.[26]

A common pagan custom at these feasts of the dead was to extinguish 'every fire in the land,' say the Jobes.[27] This was done to ensure that the ghosts of the dead could travel to their final resting place in the Far West. However, once the stars of the Pleiades 'passed the meridian, the Druids lit new fires which were carried by fast runners the length and breadth of the land, and in this way each village started its fire with a sacred flame.'[28] A similar eastern tradition is found in the various Feasts of the Dead in India, China and Japan, known as the Feast of Lanterns.[29] Early in November, bonfires are kindled along riverbanks and seashores as paper sailboats carrying lighted candles are launched on the waters to ferry the dead souls to their eternal abode. This brings us to the fires of Beltane and to other celebrations involving the cluster, such as the annual May Day festival and its famous dances around the maypole, a thinly disguised Tree of Life and an ancient symbol for the cosmic axis. The spring festival celebrating life in the Northern Hemisphere 'when the Pleiades rose on the first day of May' after the long, cold winter was celebrated by the fires of Beltane that gave rise to such traditions as lighting bonfires and dancing around the maypole.[30] Occurring 'exactly midway through the year,' the lighting of the fires symbolised the death of winter and the 'rebirth' or the approach of warmer weather, when the stars of the Pleiades 'rose at dawn.'[31] More popularly known as Mayday, 'the festival was named after the Celtic Fire God, Bel.'[32]

Although officially named for a male god, it celebrated the powers of the feminine through the goddess Flora in her 'flower aspect' as 'harbinger of fruit to come,' says Walker in *The Woman's Dictionary of Symbols and Sacred Objects*.[33] Another name for the Green Goddess was Walpurga, or Maya, whose many variations include 'Maj, May, Maia, and the Maiden.'[34] Maia, as you will recall, is the name of the eldest of the seven sisters in Greek mythology, for which the month of May is named. This celebration of the male and female principles would explain the alternate names for the festival. Many people wore green on this day 'in honour of Earth's new green garment' representing 'nature's fertilisation.'[35] A pole representing the god's phallus was 'planted' in the womb of the Earth to ensure its fecundity and people would dance and sing around it. The maypole, as it came to be called, was painted and decorated with coloured ribbons that participants would hold as they encircled the pole.[36]

The parallel to celebrating the creative powers of nature translated in cultural terms as a time of sexual freedom for those constrained by the bonds of marriage. This 'sexual licence' continued in the rural areas of Europe and elsewhere until as late as the sixteenth century, says Walker, who argues that the early Christian fathers labelled May Eve as the work of witches.[37] Morality aside, whatever the Christian church may have thought, the sad fact is that there was a sinister side to the celebrations in the involvement of human sacrifice to the god Bel. Candidates were selected by drawing lots from oatcakes that were toasted in the flames of the fires. These were broken into small pieces, of which one was completely blackened as charcoal, and the person 'who drew the burnt piece' became the chosen victim.[38]

This gory Celtic custom of sacrificing humans at festival fires was unfortunately practised by other cultures around the world, including the Aztecs and Mayans. One major difference is that the Celtic and Inca feasts were celebrated on an annual basis, whereas the other two occurred just once every fifty-two years.[39] The Aztecs, Mayans and Incas, in particular, believed the world would end as prophesied by the priests if the gods were not placated by 'an endless supply of hearts and blood.'[40] The feast of the New Fire took place in November when the stars of the Pleiades crossed over in the skies at midnight. As the crossing time of the stars drew near, the priests would ascend the Hill of the Star to anxiously watch the movement of the cluster across the night skies. When they saw that the stars had safely passed by they knew the world would not end, and now successfully bound by the sacrificial offerings, was guaranteed its survival for another fifty-two years.[41] Mayan scholar John Major Jenkins has argued the fifty-two-year feast was an important precessional tool that enabled the Aztecs and Mayans to calibrate their calendars.[42]

Whatever their reasons for this horrendous cultural practice, human sacrifices are entirely repugnant and its association with this particular star cluster is not only an affront to these beautiful, loving, celestial beings but an outright slander. Perhaps some comfort may be drawn from the insightful analysis of Laurette Sejourn in *Burning Water*, who describes the Aztec repressive regime as having been 'founded on a spiritual inheritance which it betrayed and transformed into a weapon of worldly power.'[43] Or, as Graham Hancock states, 'theirs was a religious travesty of basic spiritual truths that enabled them to exploit the old prophecies of the death and rebirth of world ages to strike terror into the hearts of their citizens to bolster and maintain the fascist state.'[44] This form of social control has sadly been used all too frequently throughout history and has produced mindless, amoral, robotic human beings, while *true* spiritual warriors are burned at the stake as heretics, if not in literal terms then most certainly symbolic.

On a much brighter note, many Polynesian and other cultures celebrated the New Year when the stars of the Pleiades rose or set in the morning or evening skies. The actual dates differed, however, because of the change in latitude that determines the time when these stars are seen. This explains why the New Year was celebrated in late May in Aotearoa (New Zealand) because that is when the cluster rises in the north-east with the Sun just before dawn.[45] The Native Hawaiians, on the other hand, celebrate their New Year in late October when the stars of Makali'i rise in the east in the evening sky just after sunset.[46] The stars heralded the harvest festival that celebrated the annual return of Lono, the patron god 'of agriculture, fertility, peace and healing.'[47] The festival is a time of much celebration and feasting. In ancient times it involved the collection of taxes and an Olympian-styled event known as the Makahiki Games that involved a number of sports competitions, mock battles and ritual bathing to symbolise purification.[48] Native Hawaiian artist and author Herb Kane says that each island celebrated Makahiki in their own way.[49] On the Big Island of Hawaii, for example, the festival was announced by flying flags from the temples and during the festival 'a procession went around the island carrying the Lono standard.'[50] Known as the *akua loa* it was made of 'a tall pole with a small image from which were hung long banners of white *kapa* (bark-cloth), feather pennants, ferns and imitation birds.'[51] This avian standard is hardly surprising, given the common affiliation of birds with the Pleiades.

A closer linguistic examination shows that the connection between these stars and the Makahiki festival goes beyond their mere role as celestial semaphores to reveal a much deeper, important relationship. An obvious point of reference is the relationship between Makali'i and Lono. We know that the Native Hawaiian universe is multi-layered and comprises nine heavens, the highest of which is *Kahiki*, home to Lono, one of four primary Hawaiian gods.[52] Like the Pawnee Native people of the American plains, the Makahiki festival began 'when Lono's voice was first heard in the thunder.'[53] Regarded as the heavenly 'eyes of the chief', the Pleiades are considered 'high-born stars' — thereby indicating their high-ranking status — an appropriate recognition given their role in marking the Makahiki festival.[54] It is fascinating that both the Maori people and Native Hawaiians refer to the cluster as an 'assembly' of the Pleiades.[55] The Native Hawaiian expression *Na Hui Hui a Makali'i* and the Maori phrase *Te Hui Hui o Matariki* therefore imply they enjoy equal high status with other celestial chiefs — thereby evoking similar images of a heavenly body of lawmakers like that of the Greek pantheon on Mount Olympus. The festival's official name, *Makali'i*, provides an additional linguistic clue to their superior status. Closely connected linguistically, conceptually and culturally to *Makahiki*, it is just one step removed from *Kahiki*[56] or *Hawaiki*,[57] the fabled Native Hawaiian and Maori homelands, the place where time began.

This connection of the Pleiades with the creation of the beginning of time is reflected by the festival's theme of the 'first fruits', as it is commonly referred to throughout Polynesia.[58] Hence the *Kumulipo* (the Hawaiian hymn of creation) extols the virtues of the stars of Makali'i in this momentous event. As Edwin Krupp explains in *Shamans, Skywatchers and Kings*, the world first began with the same event that commences the New Year — the rise of the stars of the Pleiades.[59] This was the exact moment of the First Creation when the 'earth grew warm to the touch' and 'when the sky first began to turn.'[60] At this time, the Moon was 'first seen, thanks to the shadowed light of the sun.'[61] And out of the cosmic darkness the first couple emerged, born into the first light. This time of the First Creation is remarkably similar to the ancient Egyptian beliefs of Zep Tepi, or the First Time of the creation of the world. Here, as in the Polynesian myths, the first couple was created as the result of their separation from one another. The primordial couple in ancient Egyptian mythology was Sheb and Telfnut, whose mythic counterparts are

Apsu (Marduk) and Tiamat in the Babylonian creation epic, Rangi and Papa in Maori cosmology and Papa and Wakea in Hawaiian starlore, and so on. The universal mythical separation of cosmic parents, say the authors of *Hamlet's Mill*, symbolises 'the obliquity of the ecliptic' and 'the beginning of measurable time.'[62] In other words, it represents the separation of the axes of the equator from the ecliptic. 'Before this separation,' they say, 'time did not exist.'[63]

Counting time by the Pleiades

Many cultures commenced their New Year and marked the seasons by observing the stars of the Pleiades. In India these divisions of time extended to their son Karttikeya, who was often depicted with six heads to signify the six seasons of the Hindu calendar.[64] In *Hindu Myths*, Wendy Doniger O'Flaherty says the mythology surrounding Karttikeya and his mothers is rich in astronomical symbolism. Karttikeya's conception, for instance, which is described as having transpired 'when Agni is in the Krittika,' suggests the fateful event occurred during a conjunction of Mars and the Pleiades.[65] Similarly, his birth by the Krittika represents 'the birth of the New Year . . . during the new moon at the spring equinox, when the Sun is in the Pleiades.'[66] Doniger O'Flaherty's comment needs to be somewhat qualified as the new moon does not always fall precisely at the time of the spring equinox, so we must read this as meaning when the new moon is *nearest* to the spring equinox, not at it. Other cultures, Aboriginal Australians and Native Americans, for instance, divided the year into six or more seasons with these stars playing an important role as heavenly indicators.[67]

Although our modern calendar is based on the solar cycle whose four stations of the Sun (the two solstices and two equinoxes) give us our traditional four seasons, it also ultimately derives from ancient Greek observations of the 'special relationship' between the Sun and the Pleiades, say the Jobes.[68] Thus winter began when the star cluster set in autumn and ended with the spring equinox when they accompanied the Sun; summer commenced when they appeared in the night skies with the star Arcturus and autumn when they became visible in the night skies.[69] This notion of the stars of the Pleiades being used to determine the equinoctial and solstice Sun is not uncommon, even where it appears that the observers may be following a solar or lunar calendar. Even in those civilisations traditionally represented as worshipping the Sun (such as the Mayan, Aztec and Inca) we still see the Pleiades, either directly or indirectly, playing a major role in their calculations and measurements of time. The agricultural calendar of the Inca civilisation, for instance, remains wholly dependent on this star cluster in determining the equinoxes and solstices of the Sun.

The lunar calendar has proved useful and practical to many civilisations; however its major shortcoming is that it falls some eleven days short of the official year of 365.242199 days.[70] Obviously, if we followed this method of counting without making appropriate adjustments, the lunar calendar would throw our seasons into disarray. Furthermore, in the short period of just sixteen years it would totally reverse the summer and winter solstices, thus creating mayhem and an intolerable situation.[71] Despite this failing, many Indigenous peoples continue to use the lunar cycle, but they do so with great imagination and ingenuity through the use of their various techniques of *intercalation*, the technical term that describes the process whereby the lunar and other calendars are made to work as closely as possible to the solar one.[72] A well known example of intercalation is the insertion of the leap year into our Gregorian calendar every four years.[73]

However, despite our various techniques of intercalation, even our latest technological timepiece the atomic clock is no match against nature. For although it is 'accurate within a billionth of a second a year,' this clock — that works by measuring the atomic oscillations of caesium — is not entirely infallible, according to David Duncan.[74] Noting our fanatical 'compulsion for order and perfection,' he wryly states that 'Nothing is perfect, particularly nature — something we relearn every hurricane season, and whenever the latest Theory of Everything falls short.'[75] The lesson here perhaps might be that it does not matter whether people follow a lunar or solar calendar, or which one is cleverer or more accurate; rather that human beings have a sense of *sacred* time. That they use their intelligence and ingenuity to find ways of honouring all of the cosmos by acknowledging our kindredness with the stars, moons and planets for — as we have previously seen — there are cycles within cycles of time, and all that matters ultimately is not the rhythm or the beat but the *dance* itself.

Notwithstanding this, Indigenous peoples employed numerous intercalation techniques in their various calendars. A frequent practice among some Native American tribes, for instance, was to leave out a moon or two — that is, to simply not count it or give it a name in a calendar year. This process, says Evan Hadingham in *Early Man and the Cosmos*, is known as the 'disorderly month' technique.[76] Unsure whether it resulted from 'confusion or carelessness' by Native astronomers or whether it represented the remnant of some more ancient, defined system, Hadingham confesses to being none the wiser as to this particular method.[77] This appraisal fails to acknowledge Indigenous people's sense of 'the flow of seasonal rhythm,' and to whom time is an *experiential* activity as opposed to 'a mechanical device designated by the ticks of a clock.'[78]

In any event, the development of this particular form of intercalation was not the result of confusion, carelessness or decay but something quite deliberate and ingenious. For, as Anthony Aveni points out in *Stairways to the Stars*, the non-counting of a whole Moon was a simple method of attempting to consolidate the solar and lunar calendars. This was done by creating a cyclic pattern of 12-12-13-12-12-13, which could be improved by 'inserting an extra thirteenth into the sequence once the shortfall between first crescent and solstice builds up to a full month.'[79] Another mode of intercalation was to observe the rising and setting of the Pleiades in order to adjust the lunar calendar with the solar one, as among the Pawnee Indians of the American plains.[80] Like many other cultures of the Northern Hemisphere, their New Year was determined by the reappearance of the stars of the Pleiades in the early morning skies. The star cluster was used 'to set the moon count straight, although a twelfth month of *kaata* or darkness (occasionally paired with an extra thirteenth month), ensued before the year officially began.'[81] Their medicine men and women waited to hear the sound of thunder low on the distant horizon, which they perceived as the Voice of Heaven, 'for only with this sign could the elaborate spring creation rituals begin.'[82]

Unhinging the cosmic wheel, the end of the world's ages

The stars of the Pleiades not only played a significant role in determining the seasons and commencement of the New Year but together with the stars of Ursa Major — the Great Bear constellation — they are universally recognised as being jointly responsible for turning the huge celestial hub in the sky that determines the world's ages. The fact that the Ancients referred to the great cycle of precession of the equinoxes as the Great Year of the Pleiades rather than that of the Great Bear suggests theirs is by far the more important of the two.[83] Walker's description of the Halloween season, when these stars culminate at midnight in the northern night skies during November as marking the period when the cosmos opens up, is an intriguing one.[84] Even more so when you consider that many cultures saw this time as representing the annual destruction and reconstruction of the world. The star cluster figured prominently in this process — 'a myth so universal,' says Tyler Olcott in *Starlore of All Ages*, 'as to suggest a foundation of fact.'[85]

Hence the Masonic reference to the 'Great Architect', who not only built the fabled Solomon's Temple but who in turn fixes and repairs the dilapidated world as needs be. The connection of this cluster with a cataclysm of some kind, usually a deluge, thereby unhinging the cosmos (or opening it up) and setting in motion the processes of devastation and destruction of the world followed by its reconstruction, is the central theme and premise of *Hamlet's Mill* that goes to the very heart of the Pleiades Calendar. Like Tyler Olcott, the authors argue that there are far too many traditions connecting the stars of Ursa Major and the Pleiades with this or that kind of catastrophe to be overlooked.[86] Their thesis resonates with Madame Blavatsky's theories that they are the principal stars (in collaboration with those of Ursa Major) that govern time, including the destruction and recreation of the world's ages.[87]

This idea is not as strange as it may seem for this is a prevalent belief in several world mythologies, including the ancient Greeks, Hindus and Germans. For instance, the Greek philosopher Proclus wrote that 'the fox star nibbles continuously at the thong of the yoke which holds together heaven and earth.'[88] German folklore adds, the world will end when the fox star succeeds. This star is none other than the faint star Alcor in the constellation of Ursa Major the Great Bear, the other half of the splendid double star Mizar, or Zeta Ursae Majoris. In Hindu mythology this is none other than Arundhati, one of the Seven Sisters who left her heavenly abode in the star cluster to live with her husband, one of India's holy men known as the Seven Rishis. In ancient Babylonia there was a similar legend where Arundhati was known as *Elamitic Narundi*, the 'sister' of the *Sibitti*, their mythological counterpart to the Seven Sages.[89]

In astronomical terms, the stars of the Pleiades and Ursa Major are at vast distances from one another and at first there appears to be no apparent link between the two. However, if we were to step outside our biased, restrictive geocentric viewpoint to look at things from a wider perspective as they may appear from some other location in space, the patterns and connections may become more apparent. The obliquity of the ecliptic is one example of the joint relationship between these stars, for the Pleiades mark the ecliptic just as the stars of Ursa Major mark the north celestial pole during certain periods of the cycle of precession. What I find enormously curious about human mythology, apart from the pure entertainment value and spiritual wisdom it contains, is to what extent it reveals hidden knowledge about the universe in all of its realms and dimensions. Whatever that may be, these legends lend substance to Blavatsky's theories that connect with the great Hindu Cycles of Time, including the end of the Kali Yuga and the Fifth Sun as prophesied in the Mayan Calendar — both events identified as coinciding with the dawn of the Age of Aquarius.

Quite apart from the claim that the stars of the Pleiades and Ursa Major are the catalysts for the destruction and reconstruction of the world's ages, yet another absorbing aspect to this process is, essentially, the use of a net to assist in tearing down the house. You may recall from the introductory chapter that the association of cords, knots, webs and nets is one of several familiar themes identified with stories of the Pleiades. A Maori legend recounts how the powerful Samson-like hero Whakatau on a vengeance mission tore down a *marae* that led to a tribe's demise by first passing a rope round its posts before tying it around the house and pulling it with all his strength. The house toppled, crushing and killing all within. He then set the house alight, 'so that the whole tribe perished.'[90] Giorgio De Santillana and Hertha Von Dechend cite other examples of similar stories from around the world, including a similar Tahitian hero who uses a net to avenge the torture of his father by catching the perpetrators as they ran out of the house headlong into the trap. Tahaki then lights a fire and throws his ensnared prisoners upon it, where they burn to their death.[91]

The authors identify this net with the Pleiades star cluster by comparing these Polynesian stories to the Native Hawaiian tale of yet another hero, Kaulu, who flies up 'to Makali'i the great god' to borrow his nets — the Pleiades and the Hyades — to entangle a 'she-cannibal' before burning down her house.[92] It is clear who owns the nets, say De Santillana and Von Dechend. This hero is none other than the legendary Greek hunter Orion who is portrayed in the Farnese Globe with the stars of the Pleiades in his right hand. This, say the authors — together with the lesser-known fact that the cluster 'used to be called the *lagobolion*,' meaning 'hare net' — is proof enough of who is running the show.[93] An interesting surmisal but most unlikely, given that the Native Hawaiian term Makali'i does not refer to a male god as such but rather to the goddesses of the Pleiades. Ownership aside, the suggestion made in the introductory chapter is that these references to the knotted nature of the Pleiades and their depiction as cords, nets and webs may be read on many levels, including the spiritual and the scientific.

On a spiritual level knots, cords, webs and nets relate to the notion of destiny, fate and prophecy. From a scientific perspective, they are connected with navigation and creation in all of its facets from chemistry, biology and genetics, to physics, mathematics, astronomy and astrophysics. Ultimately our creation, and that of our universe, reflects the very web or matrix of creation, which many diverse cultural traditions ascribe to a female creator goddess such as the Greek Ariadne or Athene in her spider incarnation as Arachne, whose Native American counterpart, Spider Woman, spun the world and all of creation into existence. It is said that she destroyed the world several times only to remake it all over again. 'The world will end when her web is finished,' says Walker, who sees some similarities between Spider Woman and the goddess Kali.[94] In particular, she says, 'the six arms and two legs' illustration of Indian goddesses may be associated 'with the eight-legged spider.'[95] Furthermore, although the legends do not state which particular species symbolises Spider Woman, a black spider would almost certainly connect it with Kali, the Hindu goddess of death and destruction.

It is significant that Reginald Lewis refers to the Egyptian goddess Isis as the Spider Goddess in *The Thirteenth Stone*.[96] Weaving, knitting and knotting have always been traditionally associated with 'women's work.' A deeper examination reveals these activities 'were once considered magically able to control winds and weather, birth, death and fate,' says Walker.[97] Little wonder that knots were regarded with awe and that many men, in particular, came to fear women's 'knot magic.'[98] As a consequence, many rules and cultural prohibitions grew out of this fear that forbade women to thread or twirl their spindles in certain areas or locations such as near planted fields lest they should 'bind' crops[99] or even worse, the male libido. Given that DNA forms knots inside our genes that are unravelled by certain chemicals within our body in order to replicate to create new cells — including the unravelling and replication process that occurs in sexual reproduction — this can only serve nowadays to heighten this ancient feeling of awe.[100] The simple fact is, as mathematical knot theory tells us, we just could not exist in our three-dimensional world if we were not so knotted.[101]

In many cultures the telling of fortunes was once the exclusive domain of women because it was the primary function of the Goddess to decide the fates of men.[102] Robert Graves explains her role as Seamstress in *The Greek Myths* as follows. A prevalent belief is that the Goddess would tie 'the human being to the end of a carefully measured thread, which she paid out yearly, until the time came for her to cut it and thereby relinquish his (or her) soul to death.'[103] As the Triple Goddess that foretold fate she was known variously as the Fates, Norns, Moerae, Valkyries and as the Weird Sisters in Saxon mythology.[104] Graves tells us their names in Greek mythology were Clotho the 'spinner', Lachesis 'the measurer' and Atropos the most dreaded of the Three Fates whose name means 'she who cannot be turned, or avoided.'[105] Although the etymology of the words 'cloth' or 'clothes' can only be traced as far back as the Middle English period, its Germanic roots could possibly relate back to Clotho the Spinner, one of the Greek fates, especially given that cloth is spun from various materials, measured and cut.[106]

Within this context clothes become a fitting metaphor of our fate and destiny, like the Aboriginal 'skin' and Scottish tartan that determines kin relationships and social obligation. Certainly the pathology term 'atrophy' derives from one of these sisters of fate. Its Latin meaning of 'the wasting away of the body or an organ' actually derives from the Greek *atropos* for 'lack of nourishment,' which leads to one's demise.[107] Is this the Dreaded One who withholds sustenance, physical, emotional and spiritual? Who can say for certain? Given the connection between the Pleiades and the Fates through their widespread reputation as seers — and the particular affiliation of these stars with food supplies — this linguistic affiliation provides a fascinating insight into these relationships. The ability to *see* into the future, as with any soothsayer, makes some people uncomfortable, especially those who are fearful what the future may hold in store; others relish their powers and actively seek clairvoyant guidance through various oracles. In either case, they continue to be regarded equally with awe.

Another association of cords with the Pleiades and sacred sites that combines the spiritual and material worlds is that which binds mother and child, the umbilical or navel cord. In *Ngarrindjeri Wurruwarrin*, Diane Bell writes of its special significance among the Ngarrindjeri people of South Australia. The *miwi* is a revered object whose esoteric meaning is multivalent and, like all things consecrated, whenever the Ngarrindjeri women 'speak about miwi and the navel cord, they do so in hushed tones' as a mark of respect.[108] Imbued with a sense of kindredness and social obligation, the umbilical cord was tied with a bunch of feathers as some sort of talisman that could be exchanged to solidify alliances and enhance trade relationships.[109] In recognition of its binding powers a mutual greeting was to welcome one another 'with a hand gesture from the stomach' to indicate that one is of the 'same stomach, or same soul.'[110] Beyond the social aspects, the miwi denoted the highest order of initiation.[111] It told men and women when it was wise to speak and when it was not.[112] Essentially, says Bell, the miwi defined 'sacred moments and sacred relationships.'[113]

The Seven Sisters Dreaming features prominently in the teachings of Ngarrindjeri and other Aboriginal women, of which weaving is an essential part of their ritual because it tells a story of people, country and relationships. As Ngarrindjeri Elder, Daisy Rankine explains: 'The weaving is about our history . . . as we weave from the centre out, we weave the Ngarrindjeri world, like our miwi.'[114] Says Bell, 'Women's weaving draws together the threads of Ngarrindjeri lives into circular designs which, like cycles in the natural world, should not be broken.'[115] Ngarrindjeri women refer to the umbilical cord as a lifeline whose power is felt in all facets of family and tribal life, especially in ritual and in weaving.[116] Like weaving, the miwi, or navel cord, draws people, land and cosmos together that reflect the artistic and spiritual patterns of our relationships with one another. It is in this sense that weaving and our navel cords connect all of us to the Dreamtime where past, present and future are one.

Floods, pole-shifts and other catastrophes

Bringing down the house through the use of a net, or the removal of a plug or some other action undertaken by a mythical hero that symbolises the unhinging of the cosmic axis, invariably involves some major cataclysm such as floods, fires, earthquakes, pole-shifts or some other catastrophe. In the majority of these stories, the stars of the Pleiades are overwhelmingly implicated as being the primary cause.

Their relationship with rain is renowned and their connection with the various deluges that have plagued humanity throughout history which led to the destruction of numerous civilisations, including the lost lands of Atlantis, Lemuria, Ys, Mu and Lyonesse, to name a few, is legendary. De Santillana and Von Dechend tell of a strange tale from Borneo of a 'whirlpool island' that has a special tree that people can 'climb up into heaven' and bring back plant seeds from 'the land of the Pleiades.'[117] This aquatic theme, in particular the whirlpool aspect, provides an excellent visual image of the churning of the sea caused by the displaced axis pole of the celestial mill that is linked to the precession of the ages. The authors argue this 'whirlpool at the bottom of a tree' archetype is familiar to many cataclysmic myths that share the same basic themes.

Edwin Krupp succinctly summarises these themes in *Beyond the Blue Horizon*. Essentially, he says, these stories contain the Tree of Life or 'some other symbol of the world axis,' whose disturbance causes the water to swirl down the drain and lastly, that the whirlpool represents the underworld — 'the realm of the dead.'[118] The precise significance of these themes is borne out of Greek mythology, in particular the legend of Heracles' Eleventh Labour that involved collecting golden apples from the Hesperides whose gardens were tended and protected by their father Atlas and the serpent Ladon. The essential clue to this explication lies in the notion of the shifting pole as it becomes unhinged — a process commonly referred to as 'pole-shifts'. There are many stories of this phenomenon from different parts of the world, primarily among Native Americans, whose prophecies — like those of the Inca, Aztecs and Mayans — speak of these events that accompany the various cataclysms that occur during the change of world ages known as suns.

As we have already seen, many civilisations referred to the precession of the equinoxes as the Great Year of the Pleiades. The ancient Egyptians are reputed to have recorded two such cycles producing a grand total of 51,736 years, for the priests had informed Herodotus that the Sun 'had twice risen where it then set, and twice set where it then rose.'[119] In his article on pole-shifts, David Pratt says this does not 'necessarily mean that the sun used to rise in the west and set in the east, because as long as the earth rotates on its axis from west to east, as it does at present, the sun will always rise in the east and set in the west, even when the poles are inverted.'[120] The one exception would be when the inclination of the earth's axis exceeded 90 degrees such that the poles would be reversed and renamed 'so that the earth could then be said to rotate from east to west.'[121] Could this have possibly happened in our remote past? If so, how might this affect us during the next changing of the ages that is rapidly approaching? We will consider these matters in the section on the Mayan Calendar that prophesies all sorts of world changes and transformations. There still remains the question, which pole are these prophecies referring to? Is it the axis pole of the planet (the imaginary line running through the earth's centre to the north and south poles) or is it the axis pole of the ecliptic (the imaginary line encircling the middle of our globe like a child's hoop that marks the path of the Sun, Moon and planets)? The conclusion to this question reveals the hidden astronomical symbolism of Atlas and Hercules in the Greek legends.

Krupp tells us that De Santillana and Von Dechend traced the stream that gurgled 'down the cosmic drain hole to an intriguing water reclamation plant — Tartarus and the Depths of the Sea.'[122] Tartarus, as you may recall, is a section of the Greek underworld of Hades where the wicked were punished. Its name is derived from the Greek *tartaruga* for tortoise, and is therefore connected to the world-supporting turtle or tortoise of Eastern and Native American philosophies.[123] The father of the Greek Pleiades, Atlas, descended into Tartarus as a consequence of the heavy weight of the globe upon his shoulders as his twin-self Heracles temporarily relieved Atlas of his burden during the Eleventh Labour. Thus Atlas, who becomes Heracles, becomes Atlas, is the human personification of the world-supporting tortoise in anthropomorphic form. Tartarus has been identified as Canopus in the constellation of Carina the Keel of the Argo — the legendary ship that Jason and the Argonauts sailed in their search for the Golden Fleece.[124] Many have speculated on the possibility of the *Argo* as the ark that the biblical Noah or his Greek counterpart, Deucalion, sailed in to escape the deluge. The term may derive from Sanskrit, for Richard Allen tells us in *Star Names* that Hindus called the star *Agastya*, after one of the Seven Rishis, helmsman of their *Argha* and 'a son of Varuna, the goddess of the waters.'[125] Arks were 'magical ships,' says Robert Temple, in which sit 'those who come out of the womb,' in the sense that they repopulate the world after the Deluge.[126]

According to Krupp, the significance of Canopus lies in the fact that 'it is located fairly close to the south pole of the solar system, or the south ecliptic pole.'[127] In actual fact, it was much closer some 13,000 years ago than it is today. Perhaps what is more significant about Canopus is that it is the second brightest star in the night skies after Sirius and the brightest star near the South Pole.[128] Krupp suggests that Canopus must have appeared to the Ancients as a set, unmovable star because it is circumpolar (meaning it never sets).[129] In actual fact, as noted previously, whether a star is circumpolar or not depends on two factors — the latitude of an observer and the star's position in the sky. This means that Canopus would only appear to be circumpolar to observers at 38° S latitude (those locations parallel to Melbourne, Australia). Therefore Canopus could only have appeared as very low on the southern horizon to observers in the Northern Hemisphere. Also, although it would have appeared to be heavily weighed down, it would eventually set in the night skies.

Notwithstanding this minor technicality, its Arabic name Al Wazn means 'weight', which could possibly refer to an anchor of some sort, and therefore reflect the notion of Atlas being burdened with the heavy weight of the globe sinking into the depths of Tartarus.[130] The sacred writings, the *Avesta* mentions the star as 'pushing the waters forward,' which conjures up images of a celestial plug that keeps the deluge at bay.[131] Another reason for its significance, says Krupp, relates to its steadfast position in the sky in terms of the precessional cycle. As the precessional path carves out its circle around the North and South Poles (which affect the stars we identify as the Pole Star) it does so by tracing circles around the stars Canopus in the southern skies and Thuban or Alpha Draconis, the leading star of Draco the Serpent-Dragon in the Northern Hemisphere. Thus Draco appears to guard the northern ecliptic pole.[132] This dragon is none other than the serpent Ladon who vigilantly guarded the gardens of the Hesperides before Heracles killed it with an arrow shot from his bow. The *Argonautica* tells us Draco guarded 'the emblem of eternal vigilance in that it never set.'[133]

Another translation for Canopus from the Coptic language of Egypt is 'Golden Earth', which leads us to the hero Heracles and his quest to obtain the three golden apples of the Hesperides.[134] The ecliptic axis, says Krupp, holds the key to understanding the 'spiritual geography' underlying the constellations of Hercules (Heracles) and Draco, and reveals 'just why the golden apples of Hesperides were so valuable.'[135] The Ancients identified the two ecliptic poles 'as divine and transcendent destinations of souls.'[136] Essentially, the serpent Draco marks the site of Paradise, the realm of eternity in the north, whereas the hero Hercules stands over the underworld, the realm of the dead in the south ecliptic pole. Both the serpent and the hero guard their respective poles but Draco, in particular — despite being killed by Hercules on the earthly plane — continues to guard the golden apple tree and the fruits of eternal life in the heavenly sphere. This tree is not the polar axis but the sturdy ecliptic axis, says Krupp, and the fruit that hangs from its boughs are none other than 'the immortal stars and divine planets.'[137] This explains the hero's quest for immortality.

However, the most important aspect to this imagery, says Krupp, is the depiction of Hercules and Draco in traditional drawings and in their position in the heavens. In the Northern Hemisphere the superhero is shown kneeling on one foot with a portion of the serpent 'locked to the heavenly grindstone of the world mill' under his foot.[138] If Hercules (Heracles) guards the southern ecliptic pole, then that other strongman — Atlas — must by definition be guarding the southern pole of the cosmic axis. This explains why the two ironmen took turns in holding aloft the celestial globe because they represent both axes. In other words, both superheroes are holding up the two poles. Given their amazing scientific correlation these mythic images would therefore have to be among the most profound ever.

The starring role of the Pleiades in the Inca calendar

The ancient civilisations of the Inca, Maya and Aztecs of Central and South America are renowned for many achievements, including their splendid architectural masterpieces of pyramids, temples and observatories. The full extent of their astronomical, mathematical and scientific knowledge is yet to be uncovered and realised. What we do know is that although their cultures achieved extraordinary technological and other feats, it was at the same time like any other human society, fraught with the realities of the human condition of power struggles, emotional and spiritual vulnerabilities, and the need for social control. And while we may marvel at many of their achievements, we cannot forget the desperate situation of many of its citizens caught up in their empire's struggle to achieve fame and glory, or the vast number of lives that were lost or sacrificed in the name of 'spirituality' or 'progress.' Having said that, the stars of the Pleiades provided the key to all of the calendars that marked their important feast days and the various world ages, including the shorter fifty-two-year cycle of the Aztecs and Mayans, which presaged the end of the world.[139]

Like many other civilisations around the world the Pleiades commenced the beginning of the Inca agricultural and New Year as well as marking the seasons.[140] Even today their descendants still rely on these stars in contemporary farming practices where they are used to make crop predictions. The farmers say that when the cluster reappear large and bright the crops will flourish, while a small and dim rising portends a bad yield.[141] A recent scientific study into Indigenous weather forecasts based on these stars in Bolivia and Peru has confirmed that 'not only does the technique work,' but 'it turns out that the farmers have in effect been forecasting El Niño for at least 400 years, a capability modern science achieved less than 20 years ago.'[142] This close connection between the Pleiades and the provision of food supplies, as previously noted, is generally referred to as a storehouse, but perhaps more revealing is the reference to the cluster as 'mother' by some Inca.[143]

High in the Andes Mountains of Peru in the ancient capital of Cuzco, the Inca observed the night skies from the Coricancha or Temple of the Sun housed in the city centre that marks the present site of the church of Santo Domingo.[144] Its Spanish translation, 'golden enclosure', hides its true significance for — as the Temple of the Ancestors — it was dedicated to the more important celestial bodies worshipped by them, 'the Sun, Moon, Venus, and the Pleiades.'[145] The Sun was especially revered because the Inca believed they were the children of Inti — the Sun god, whose festival Intiraymi was celebrated in June at the winter solstice.[146] Given this belief and the temple's name, we could reasonably assume it was directly aligned to the solstices but, as Anthony Aveni demonstrates in *Stairways to the Stars*, the Coricancha is more 'closely aligned to the Pleiades.' In his revised edition of *Skywatchers of Ancient Mexico*, now simply titled *Skywatchers*, he says this slight deviation in the building's alignment was quite deliberate as the heliacal rising and setting of the Pleiades on the Cuzco horizon were used to determine the summer and winter solstice and the all-important Feast of the Sun.[147]

This combined use of artificial structures and natural features of the landscape to observe the movements of heavenly bodies is a universal astronomical practice in many cultures. Evidence of these ground-based observatories can be seen at various archaeological sites — mounds, stone arrangements, standing stones and the like — the most celebrated of these being Bighorn Medicine Wheel in the mountains of Wyoming in the USA and Stonehenge on Salisbury Plains in the southwest of England. Working from a similar premise and employing the principles of sacred geometry, the Inca devised an intricate astronomical scheme known as the *ceque* system that forms the basis of the Inca Calendar that is as clever and complex as the more famous of the two — the Mayan Calendar. The term ceque comes from the Quechua language and means 'ray' or 'sight line.'[148] Scientists have since discovered from the city's layout that it was purposely built to operate as a virtual star map — an earthly planisphere that aligned natural and artificial structures with certain constellations, individual stars and planets — especially the stars of the Pleiades. Its ray-like organisation is a reflection of the rays of the Sun as it passes over the zenith.

Like a huge mandala, these ceque lines radiated out from the walls of the Coricancha in the city centre to the countryside that effectively divided the city into 'pie-like wedges' in much the same way that the city of Paris is oriented with a series of roads stretching outwards from the central Arc de Triomphe.[149] Each wedge was filled with *huacas*, a series of sacred sites on which were housed temples, shrines, geographic features or some other object that marked the invisible ceque lines.[150] Other huacas were made visible through their incorporation into the city's architecture and road works that created straight visual lines that may still be observed from the Coricancha. There were 41 ceque lines in all, which divided the city, and 328 huacas.[151] The number of huacas may have represented 'twelve sidereal lunar months,' says Aveni.[152] Therefore, by adding the 37 days in the year when the Pleiades disappear from the Cuzco skies — the so-called 'dead time' in 'the seasonal calendar when the fields lay fallow'— this produces the sum of 365 days, which represents the solar year.[153]

This ground-based compass incorporated a basic moiety division of the city along an east–west axis (aligned to the Pleiades) that created an upper (northern) and lower (southern) half before further dividing the city along a north–south axis, thereby creating a quartered effect.[154] The Inca term for quarter in this region is *suyus* and each possessed different numbers of ceques in a counting sequence that reflected 'both lunar synodic and sidereal periods.'[155] This means the Inca were counting the year by the Moon's movements together with the stars of the Pleiades, which effectively means their calendar was a combination of lunar and Pleiadic cycles. The quartering system may possibly have derived from observations of the four quarters of the Moon — full, half, quarter and new. This might explain why there were different numbers of ceques in their quarter. In any event, not surprisingly they named their city *Tahuantinsuyu* to acknowledge this quadripartite scheme, for it means 'Four-Quarters.'[156]

This tradition of quartering our environment is a universal phenomenon that relates back to archetypal notions of the fourfold division of the universe such as the four corners of the world, four horsemen of the Apocalypse, four winds and so on. On their foundation, most cities were first divided into halves before being further divided into quarters, such as the city of New Orleans in the American state of Louisiana which is divided into four quarters, including the well known French Quarter which hosts the annual Mardi Gras festival. Walker suggests the quartered division — especially its replacement of 'the round village' (mark of the feminine principle) — appears to be associated with the rise of patriarchy.[157] The cross, in particular, that is formed by quadrants represents the male god's sacrifice that was assumed to have taken place 'at the centre of things.'[158] Consequently, the turned cross on its side came to mark 'the usual location of the Tree of Life, *axis mundi*, omphalos . . . centre of the world body, and other interpretations of the X that marks the spot.'[159]

Certainly the sacred geometrical design of Cuzco marked it as another navel of the world, but the fact that it combined both the quadrant with the circle suggests it incorporated the male and female creative principles, like that of the Celtic cross known to Hindus as a sign of sexual union between the phallus (cross) and the yoni (circle). Thus, Robert Lawlor argues in *Sacred Geometry*, that squaring of the circle 'contains many symbolic keys for the contemplation of universal creation.'[160] In fact, in older times Cuzco was once known as the navel of the world for it marked the spot where the Children of the Sun thrust a golden rod into the ground to found the Inca civilisation and empire.[161] The twins, a brother and sister, carried the golden sceptre of Viracocha their father from Lake Titicaca to Cuzco. He had instructed them to build their court at the spot where the rod disappeared entirely 'with one single thrust.'[162] This action may be emblematically read as a means of establishing an 'umbilical link,' says Graham Hancock in *Heaven's Mirror*.[163] When the Temple of the Sun was built to commemorate this momentous event, it housed an open courtyard that surrounded the sacred field. At its centre a stone coffer was laid to mark the precise spot where the rod had been planted.

At one time it was entirely 'covered with 55 kilograms of gold.'[164] In fact, the entire temple was once covered in gold, including its roof and eave troughs. And in the empty field was planted symbolic corn 'fashioned out of pure gold.'[165] Needless to say the temple roof, walls, stone coffer and gold corn were harvested and pillaged by the greedy Spanish conquistadors, who took the golden booty back to Spain.

The Mayan calendar and Pleiadian prophecies

The mere mention of the Mayan Calendar excites many a reader because of its acclaimed ingenuity. Enormously admired and respected for the vastness of its cycle, the complex mathematical calculations involved and the marvels of its exactitude, it remains a shining example of one of the pinnacles of human achievement. To speak of *the* Mayan Calendar is a bit of a misnomer for it is made up of several calendars, not just one, in fact seventeen in all, each comprising a series of long and short cycles involving a number of planets, individual stars, clusters and constellations.[166] Not surprisingly, one of these cycles is based on their observations of the Pleiades.[167] Ever since the Mayan Calendar was discovered, many writers have emphasised its Venusian rather than its Pleiadian cycle because this planet represented their supreme creator god, Kukulcan, the feathered rainbow serpent whom the Aztecs called Quetzalcoatl.[168] However more recent research — particularly that of John Major Jenkins — suggests these stars played a far more significant role in their calendar than previously realised. Not only did they signal the date of their most important calendar event — that of the New Fire ceremony that was held once every fifty-two years in November[169] — but, as he argues in *Maya Cosmogenesis 2012*, their conjunction with the Sun's passage in the zenith heralds the fabled return of their plumed Serpent God at the end date of the Mayan Calendar.[170] By their calculations this is soon to occur on the summer solstice of 21 December 2012 when the Mayan celestial clock strikes 13.0.0.00, only to return the odometer count to zero to begin the Long Count of the precessional cycle once again.[171]

Driven to understand why the Mayans chose this exact date to mark the end of their calendar led Jenkins on an amazing, revolutionary journey that provided him with profound insight into Mayan cosmology, which has enormous implications for our spiritual growth and transformation. On this designated date the summer solstice Sun will mark a special alignment between the ecliptic and the centre of the Milky Way galaxy, an event so rare it only occurs once every 25,800 years![172] Mayan scholar Linda Schele first identified this crossing of the ecliptic between the constellations Sagittarius and Scorpio as marking the exact spot of the Mayan Cosmic Tree or Tree of Life.[173] There are similar tales in other mythologies of the world axis being guarded by a serpent or dragon, such as the hundred-headed serpent Ladon in the Garden of the Hesperides. Serpents, dragons, crocodiles and scorpions effectively mean the same thing; therefore it is not surprising that the Mayan equivalent of the Tree of Life was called the Crocodile Tree, an ancient Mayan creation symbol from their holy book, the *Popul Vuh*.[174] The crossroad formed by the celestial alignment points to the source of creation and rebirth to that region within the central bulge of the Milky Way galaxy. More specifically, the alignment represents the heavenly union of the Cosmic Mother and Father who created the world.[175] This, says Jenkins, is the true meaning of the cryptic message inscribed on Pacal's sarcophagus at Palenque. Given the precision of Mayan astronomical calculations, those astronomers who see no value or relevance in astrology may now choose to rethink their intellectual bias. It seems that the Ancients knew far more than modern day astronomers, for the predicted solstice Sun that will take place near the centre of our galaxy can be precisely marked by drawing a straight axis line from the Archer's arrow to the Scorpion's upturned tail.[176]

The significance of this rare astronomical event led Jenkins to ponder on its symbolic meaning to the ancient Mayans. In their mythology, the Milky Way represents the celestial pathway to the Mayan Underworld whose entrance was marked by the Crocodile Tree, which points in the general direction of the centre of our galaxy.[177] The four lines of the cross, which radiate out from the axis point or cosmic navel correspond to the four cosmic roads that are referred to in the *Popul Vuh* as the place of creation where the 'fourfold siding' or 'stretching' of the umbilical cord occurred.[178] This quartering of the cosmos is reflected in Mayan landscaping like that of the Inca city of Cuzco. Like other cultural traditions the four 'Year Bearers' hold up the four corners of the world as depicted in 'The World' tarot card of the Rider Waite deck. Jenkins tells us the four bearers at Izapa are '*Ahau* (Solar Lord), *Chiccan* (Serpent), *Oc* (Dog) and *Men* (Eagle).'[179] Quite clearly they are the Mayan version of the Four Horsemen of the Apocalypse of St John's Revelation who, in astrological terms, are the four signs of the Fixed Cross — Aquarius, Scorpio, Leo and Taurus. In the Egyptian chapter we saw that these four fixed signs act as a virtual cosmic fulcrum that turns the precessional wheel that creates the various ages. These four Year Bearers play a vital role in the Calendar Round that combines the *tzolkin* and *haab* cycles of the Mayan Calendar because, as Jenkins explains, 'they are the only day-signs on which the New Year can fall.'[180] The *Popul Vuh* tells us that one of the four bearers 'was killed' and 'his head was placed in the crook of a calabash tree' along one of the crossroads.[181]

As previously noted, crosses, trees, stones and other such items that mark the spot of the male god's sacrifice — which took centre stage 'where all forces came together' — represent the Cosmic Axis.[182] The notion of the god who dies to save humanity lies at the heart of several great religions. This sacrificial god, the Wandjina or *One Hunahpu*, is none other than *One Ahau*, the Aquarian Solar Man-God or Jesus Christ. Apart from their names is there anything that can tell us where these four sky-bearer beings come from? Jenkins is silent on the subject but Mayan elder Don Alejandro Peres has plenty to say in *The Mystery of the Crystal Skulls* by Chris Morton and Ceri Louise Thomas. He claims that long ago four celestial beings or prophets from the Pleiades collectively known as *Mia* brought with them a number of crystal skulls to assist humanity in our spiritual growth.[183] Furthermore, he maintains, there were not thirteen crystal skulls as commonly believed but fifty-two in all.[184] Curiously, one of many Aboriginal Australian names for the Pleiades is Meamai or Maimai, which sounds remarkably like the eldest of the Greek daughters, Maia.

Remarkably this number reflects the exact number of years of the New Fire ritual that was marked by these stars. Also, a numerological interpretation reveals an interesting picture when the numerals are added together. Thus 52 equals 5 plus 2, which equals 7, which stands for the Seven Sisters; and 13 equals 1 plus 3, which equals 4, that represents the four beings that were sent down from the cluster. Although Peres does not specify their gender it may be asked, what are four male gods doing there in the first place, given this region's overwhelming identification with women? The answer to this question lies in the twofold cosmic relationship between the Pleiades and Kukulcan (the feathered rainbow serpent, whose astronomical manifestation is the planet Venus) and between the Pleiades and the Sun.

The rattling of the Pleiades

The modern descendants of the Mayan people live mostly in the Yucatan Peninsula of Mexico and in neighbouring regions and countries such as Belize, Honduras and Guatemala. Spanish-speaking Mayans refer to the cluster as *Las Siete Cabrillas*, or 'The Seven Kids,'[185] which refers to young goats as opposed to children. The ancient Mayan name for the Pleiades was *Tzab*, which referred to the rattlesnake's tail, an easily recognised hieroglyph in all of the Mayan codices.[186] This was no ordinary snake but the supreme god Kukulcan. A popular icon in the Americas, the anthropomorphic form of the feathered serpent symbolised the bearded, white saviour god of the Golden Age who died, but as legend says, will return one day.[187] Not only is the mythology surrounding this deity similar to Celtic traditions of a saviour king-god who — like Arthur — lies sleeping until his people need him once again, but there is an uncanny phonetic resemblance to the Irish god *Cu Chulainn*.[188] This once and future king, known to the Aztecs as the serpent-king-god Quetzalcoatl, was celebrated in the annual rebirth rituals of the Inca. Among the Mayans and Aztecs it took place just once every fifty-two years in the infamous 'Binding of the Years' ceremony involving huge numbers of human sacrifices, including those of children.[189]

This visual image of the rattlesnake and its tail confirms the close relationship between Venus and the Pleiades; their various representations — together or on their own — reveal a hidden language encoded with information on astronomy that has an impact on meteorology and agriculture. When featured together as feathered serpent with its rattle tail intact, it not only heralded the rainy season — which therefore made it a likely sign of fertility — but as a form of astronomical shorthand it may have indicated a conjunction between Venus and the star cluster.[190] Conversely, the depiction of these stars on their own as 'feathered rattlers' indicated their disappearance from the skies, hence their dislocation from the rattlesnake. Both images can be seen at numerous Mayan sites oriented toward the setting of the Pleiades such as Teotihuacán and Xochicalco, and the Caracol (observatory) and Upper Temple of the Jaguars at Chichén Itzá.[191]

Their close ties with a supreme male deity are a familiar theme in world mythology. Of special note is the similarity between Kukulcan, or Quetzalcoatl, with that of Zeus, king of the Greek gods whose special insignia was the eagle.[192] The notion of a feathered serpent combines two very powerful symbols of spiritual transcendence, the bird and the serpent.[193] By shedding its skin the serpent symbolises life, death and rebirth that encapsulates our concept of time. Like Aboriginal Australians who worshipped the feathered Rainbow Serpent, the Mayans venerated the rattlesnake, the sign of their supreme creator. But this was not just any rattlesnake but a particular species known as *Crotalus durissus durissus* and its subspecies, which they called the *Ahau Can* — Great Lordly Serpent.'[194] Not only did the patterns on the snake's back influence Mayan art and architecture, but some believe the serpent may have taught the ancient Mayans about time as well.[195]

Several writers, for instance, have noted Mayan observations of the rattlesnake's physical traits such as the shedding of its skin or the replacement of its rattle, as having influenced their perceptions of time. The authors of *The Mayan Prophecies*, Adrian Gilbert and Maurice Cotterell, note a direct correlation between the base equivalent of the Mayan *uinal* in the vigesimal counting system (the twenty base system) and the fact that this particular species of rattlesnake 'loses and replaces its fangs every twenty days.'[196] Furthermore, as they point out, 'the standard glyph for this looks like the open jaws of a snake with two very obvious and prominent fangs.'[197] Unlike other rattlesnakes, this particular species sheds its skin only once a year 'in mid-July, around the time in the Yucatan when the Sun reaches the highest point in the sky for the second time in the year.'[198] Gilbert and Cotterell suggest this observation led to a 'natural correspondence between Sun and serpent, that annually renew themselves together.'[199] The association of the Pleiades with the creative powers of the Sun and the serpent is borne out by their heliacal rising in the morning sky on the twenty-fifth of April in the Yucatan that announces 'the first annual passage of the sun across the zenith, a phenomenon said to be responsible for the fertilization of the seeds.'[200] Here, as in other parts of the world, the early morning rise of the cluster portents 'the coming of the rains and fixes the first day of the planting.'[201]

Elsewhere in other regions of Mesoamerica, the Pleiades rise in conjunction with the Sun closer to the twentieth of May. The connection between these stars, the feathered serpent and our Sun is commemorated in the Pyramid of Kukulcan at Chichén Itzá. Thousands flock to this site on the twenty-first of March during the spring equinox to watch the famous shadow play of the descending serpent on the northern stairway that occurs just once every year on this day only.[202] A closer examination of the esoteric relationship between the star cluster and the feathered rainbow serpent at Chichén Itzá by Mayan scholar John Major Jenkins shows yet another important astronomical alignment, which until now has previously gone unnoticed by astronomers.[203] By focusing on the zenith passages of the Sun and the Pleiades and their observation from the Castillo pyramid, Jenkins has discovered a coming astronomical event that is of enormous symbolic significance. He notes that on 20 May 2012 there will be a conjunction of the Sun, Moon and the Pleiades accompanied by a solar eclipse 'that will sweep across central and western North America.'[204] This, says Jenkins, will be the sign in the heavens that portends the return of Quetzalcoatl.[205] Here we have confirmation yet again that these stars play a major role in the changing of the world ages.

As previously noted, the transition from one age to another brings enormous changes and cataclysms of which the stars of the Pleiades play a significant role in its occurrence and announcement. Given their connection with a number of musical instruments, including the 'Voice of Heaven' thought to emanate from that celestial region, the serpent's rattle takes on particular significance. That sound plays an important role in universal creation, destruction and reconstruction is not unknown to science. Like Joshua's trumpet, whose reverberations brought down the city walls of Jericho; the rattling of the Pleiades is sure to have a profound effect, whether real or symbolic. The esoteric symbolism between this star cluster and the feathered serpent is further heightened by the widespread belief that the rattlesnake grew 'an extra rattle each time it moulted.'[206] Curiously, the Pythagorean theory of the Harmony of the Spheres maintained that with each new age, 'new instruments, new strings, or, as in the case of Odysseus, a new peg' are called for, 'thereby creating new harmonies.'[207]

Within this context, the feathered rattle of the Pleiades might hold the key to understanding why the Aztecs celebrated the fifty-two-year cycle along with the Mayans. The Florentine Codex of Sahagún contains an Aztec drawing of the star cluster that resembles a triangular shape with a rounded edge at its base. Although Anthony Aveni refers to their portrayal as 'an arrow-shaped configuration,' there is nothing to suggest it was necessarily perceived as such, and so its specific meaning is left open to conjecture.[208] In my view, the shape looks more like a rattle than an arrow, which may account for its loose beads. Imbibed with specific meaning, rattles are used by many Native peoples as sacred musical instruments in ritual and ceremonies and are an important tool for shamans. Furthermore, the affiliation of the Pleiades with spiritual and divinatory matters ensured that their trademark glyph became 'the special insignia' of the Inca priests, 'who carried a short stick with rattlesnakes' tails attached to it.'[209]

Many mistakenly fear the Mayan prophecies because of their apparent message of doom and gloom, but I believe it heralds exciting times ahead for all of humanity. As we rapidly approach the end of the Age of the Fifth Sun — that happens to coincide with the Age of Aquarius and the end of the Kali Yuga — and if prophecy is correct, we can expect enormous changes on a personal and global level. Perhaps the cataclysms will be on the physical, environmental level such as floods and other catastrophes that have visited our planet in times past. Or else they may refer to the inner cataclysms within us — those of a psychological or spiritual encounter that we may never have experienced before. Whatever the stars have in store for us remains to be seen and whether the prophecies come true or not, one thing is certain — that the stars of the Pleiades will continue to shine, guide and instruct us on this fragile but many splendoured journey we call life. We need only trust these gentle, glowing celestial beacons for we *are* kin, and we share an ultimate, common destiny with these fellow stars of our galaxy who are so near and yet so far.

Notes

1. As quoted in *Ngarrindjeri Wurruwarrin* by Diane Bell at p. 580.
2. Burnham Jr, *Burnham's Celestial Handbook*, p. 1869.
3. Ibid.
4. Jobes and Jobes, *Outer Space*, p. 336.
5. Frazer, *Aftermath: A Supplement to the Golden Bough*, p. 394 and *Stars of Tagai* by Nonie Sharp, p. 4.
6. Ibid.
7. Allen, *Star Names*, p. 400.
8. Ibid.
9. Aveni, *Skywatchers*, p. 32 and *Stairways to the Stars*, p. 100.
10. Aveni, *Stairways to the Stars*, p. 46.
11. Milbrath, *Star Gods of the Maya*, p. 38.
12. Aveni, *Stairways to the Stars*, p. 50.
13. Allen, *Star Names*, p. 405.
14. Storm, *Egyptian Mythology*, p. 36.
15. Walker, *The Woman's Dictionary of Symbols and Sacred Objects*, p. 90.
16. Ibid.

17. Frazier, *Aftermath: A Supplement to the Golden Bough*, p. 340.
18. Krupp, *Skywatchers, Shamans and Kings*, p. 169.
19. Jobes and Jobes, *Outer Space*, p. 336.
20. Walker, *The Woman's Dictionary of Symbols and Sacred Objects*, p. 139.
21. Ibid, p. 180.
22. Jobes and Jobes, *Outer Space*, p. 338.
23. Walker, *The Woman's Dictionary of Symbols and Sacred Objects*, p. 180.
24. Tyler Olcott, *Starlore of All Ages*, p. 413.
25. Jobes and Jobes, *Outer Space*, p. 336 and Tyler Olcott, *Starlore of All Ages*, p. 412.
26. Ibid, *Outer Space*, p. 339 and ibid, *Starlore of All Ages*, p. 413.
27. Jobes and Jobes, *Outer Space*, p. 338.
28. Ibid.
29. Ibid, pp. 339–40.
30. Ibid, p. 338.
31. Ibid, p. 339.
32. Ibid, p. 338.
33. Walker, *The Woman's Dictionary of Symbols and Sacred Objects*, p. 186.
34. Ibid.
35. Ibid.
36. Ibid.
37. Ibid.
38. Jobes and Jobes, *Outer Space*, p. 339.
39. Jenkins, *Maya Cosmogenesis*, p. 21.
40. Hancock, *Heaven's Mirror*, p. 15.
41. Aveni, *Skywatchers*, p. 33.
42. Jenkins, *Maya Cosmogenesis*, p. 24.
43. Quoted in *Heaven's Mirror* by Hancock, p. 15.
44. Ibid.
45. Best, *The Astronomical Knowledge of the Maori*, pp. 33–34 and 47.
46. Kane, *Ancient Hawaii*, p. 44.
47. Ibid.
48. Ibid, pp. 44–45.
49. Ibid, p. 44.
50. Ibid.
51. Ibid.
52. Krupp, *Skywatchers, Shamans and Kings*, p. 169.
53. Ibid.
54. See the website of the Polynesian Voyaging Society at: <http://pvs.hawaii.org/navigate/stars.html>.
55. Best, *The Astronomical Knowledge of the Maori*, p. 53 and Kyselka, *An Ocean In Mind*, pp. 9 and 48.
56. Krupp, *Skywatchers, Shamans and Kings*, p. 169.

57. Orbell, *Maori Myth and Legend*, pp. 50–51.
58. Krupp, *Skywatchers, Shamans and Kings*, p. 169.
59. Ibid, p. 171.
60. Ibid.
61. Ibid.
62. De Santillana and Von Dechend, *Hamlet's Mill*, p. 135.
63. Ibid, p. 153.
64. Doniger O'Flaherty, *Hindu Myths*, p. 104.
65. Ibid.
66. Ibid.
67. See *Early Man and the Cosmos* by Evan Hadingham, pp. 101–105 and the Native American tale in the children's book, *Thirteen Moons on Turtle's Back* by Bruchac and London. Anthony Aveni also mentions this calendric system in *Stairways to the Stars* at p. 32. For information on the seasons in the calendar of Aboriginal Australians see *Where the Ancestors Walked* by Phillip Clarke at pp. 131–35 where he talks about the six seasons of the Nyungar people of Western Australia. He also discusses some other calendars of different Aboriginal nations throughout the book.
68. Jobes and Jobes, *Outer Space*, p. 336.
69. Ibid.
70. Duncan, *The Calendar*, p. 17.
71. Ibid.
72. Aveni, *Stairways to the Stars*, p. 54.
73. Ibid, p. 55.
74. Duncan, *The Calendar*, p. 321.
75. Ibid, p. 323.
76. Hadingham, *Early Man and the Cosmos*, p. 101.
77. Ibid.
78. Aveni, *Stairways to the Stars*, p. 32.
79. Ibid, p. 55.
80. Hadingham, *Early Man and the Cosmos*, p. 108.
81. Ibid.
82. Ibid.
83. Allen, *Star Names*, p. 393.
84. Walker, *The Woman's Dictionary of Symbols and Sacred Objects*, p. 180.
85. Tyler Olcott, *Starlore of All Ages*, p. 414.
86. De Santillana and Von Dechend, *Hamlet's Mill*, p. 386.
87. Blavatsky, *The Secret Doctrine*, vol. 2, p. 549.
88. De Santillana and Von Dechend, *Hamlet's Mill*, p. 385.
89. Ibid, p. 385.
90. Ibid, p. 174.
91. Ibid, p. 175.
92. Ibid.

93. Ibid.
94. Walker, *The Woman's Dictionary of Symbols and Sacred Objects*, p. 163.
95. Ibid, p. 419.
96. Lewis, *The Thirteenth Stone*, p. 22.
97. Walker, *The Woman's Dictionary of Symbols and Sacred Objects*, p. 142.
98. Ibid, p. 143.
99. Ibid, p. 142.
100. Ridley, *Genome*, pp. 3–9.
101. Casti, *Five More Golden Rules*, pp. 14–19.
102. Walker, *The Woman's Dictionary of Symbols and Sacred Objects*, p. 36.
103. Graves, *The Greek Myths*, p. 204.
104. Walker, *The Woman's Dictionary of Symbols and Sacred Objects*, p. 36.
105. Graves, *The Greek Myths*, p. 204.
106. Ayto, *Word Origins*, p. 118.
107. *The Macquarie Concise Dictionary*, p. 61 (3rd edn, 1998).
108. Bell, *Ngarrindjeri Wurruwarrin*, p. 499.
109. Ibid, p. 491.
110. Ibid.
111. Ibid.
112. Ibid, p. 490.
113. Ibid.
114. Ibid, p. 542.
115. Ibid, p. 544.
116. Ibid, p. 498.
117. De Santillana and Von Dechend, *Hamlet's Mill*, p. 213.
118. Krupp, *Beyond the Blue Horizon*, p. 298.
119. As quoted in *The Secret Doctrine* by Blavatsky, vol. 1, p. 435.
120. David Pratt, 'Poleshifts: Theosophy and Science Contrasted', January 2000, p. 4 of internet article at: <http://ourworld.compuserve.com/homepages/dp5/pole5.htm>.
121. Ibid.
122. Krupp, *Beyond the Blue Horizon*, p. 298.
123. Walker, *The Woman's Dictionary of Symbols and Sacred Objects*, p. 392.
124. Krupp, *Beyond the Blue Horizon*, p. 298.
125. Allen, *Star Names*, p. 71.
126. Temple, *The Sirius Mystery*, p. 180.
127. Krupp, *Beyond the Blue Horizon*, p. 298.
128. Burnham et al., *Astronomy: The Definitive Guide*, p. 351.
129. Krupp, *Beyond the Blue Horizon*, p. 298.
130. Temple, *The Sirius Mystery*, p. 163.
131. Allen, *Star Names*, p. 71.
132. Krupp, *Beyond the Blue Horizon*, p. 298.
133. Allen, *Star Names*, p. 204.

134. Ibid, p. 68.
135. Krupp, *Beyond the Blue Horizon*, p. 299.
136. Ibid.
137. Ibid.
138. Ibid.
139. Aveni, *Skywatchers*, pp. 3–31.
140. Ibid and p. 166.
141. Ibid.
142. Kurt Sternlof, 'Science and Folklore Converge in Andean Weather Forecasts Based on The Stars,' *Columbia University News*, 10 January 2000, p. 1. See the website at: <www.columbia.edu/cu/pr/00/01/pleiades.html>.
143. Bauer and Dearborn, *Astronomy and Empire in the Ancient Andes*, p. 104.
144. Aveni, *Skywatchers*, p. 310 and *Stairways to the Stars*, p. 150.
145. Ibid, p. 310.
146. Ibid, p. 319 and *Stairways to the Stars*, p. 168.
147. Aveni, *Skywatchers*, p. 319.
148. Ibid, p. 316 and *Stairways to the Stars*, p. 157.
149. Aveni, *Stairways to the Stars*, p. 156.
150. Ibid, p. 158.
151. Ibid.
152. Ibid, p. 170.
153. Ibid.
154. Aveni, *Skywatchers*, p. 317.
155. Ibid.
156. Aveni, *Stairways to the Stars*, p. 155.
157. Walker, *The Woman's Dictionary of Symbols and Sacred Objects*, p. 46.
158. Ibid.
159. Ibid.
160. Lawlor, *Sacred Geometry*, p. 74.
161. Hancock, *Heaven's Mirror*, p. 274.
162. Ibid, p. 273.
163. Ibid, p. 274.
164. Ibid, p. 277.
165. Ibid.
166. Hand Clow, *The Pleiadian Agenda*, p. 47. She claims there are seventeen cycles in all that make up the Mayan Calendar.
167. Ibid. Indigenous Mayan author Hunbatz Men refers to this particular cycle as *Calendario del Tzek'eb o Pleyades*.
168. Aveni, *Skywatchers*, pp. 186 and 274–75.
169. Also known as the 'Binding of the Years' ceremony, see Hancock, *Heaven's Mirror*, p. 19.
170. Jenkins, *Maya Cosmogenesis*, p. 78.

171. Ibid, p. 23.
172. Ibid, p. xxxviii.
173. See references to Linda Schele's work in the bibliography of *Maya Cosmogenesis*.
174. Ibid, p. 122.
175. Ibid, pp. 120–21.
176. Moore, *Stars of the Southern Skies*, p. 12.
177. Jenkins, *Maya Cosmogenesis*, pp. 10, 61, 112 and 118.
178. Ibid, pp. 57 and 117–18.
179. Ibid, p. 58.
180. Ibid, p. 20.
181. Ibid, p. 57.
182. Walker, *The Woman's Dictionary of Symbols and Sacred Objects*, p. 46.
183. As quoted in *The Mystery of the Crystal Skulls* by Chris Morton and Ceri Louise Thomas, p. 388.
184. Ibid, p. 392.
185. Aveni, *Skywatchers*, p. 34.
186. Ibid.
187. Hancock, *Heaven's Mirror*, p. 19.
188. See Barbara Walker's *The Women's Encyclopedia of Myths and Secrets* at pp. 195–96 for further information on the Irish god *Cu Chulainn*.
189. Hancock, *Heaven's Mirror*, p. 19.
190. Milbrath, *Star Gods of the Maya*, p. 259.
191. Ibid, p. 263.
192. Walker, *The Woman's Dictionary of Symbols and Sacred Objects*, p. 400.
193. Campbell, *The Power of Myth*, pp. 18 and 45–47.
194. Gilbert and Cotterell, *The Mayan Prophecies*, p. 133. See also *Beyond the Blue Horizon* by Edwin Krupp, p. 97.
195. Gilbert and Cotterell, *The Mayan Prophecies*, pp. 134–38.
196. Ibid, p. 138.
197. Ibid.
198. Ibid.
199. Ibid.
200. Aveni, *Skywatchers*, p. 34.
201. Ibid.
202. Krupp, *Beyond the Blue Horizon*, p. 98.
203. Jenkins, *Maya Cosmogenesis*, p. 14.
204. Ibid, p. 79.
205. Ibid.
206. Gilbert and Cotterell, *The Mayan Prophecies*, p. 138.
207. De Santillana and Von Dechend, *Hamlet's Mill*, p. 369.
208. Aveni, *Skywatchers*, p. 33.
209. Milbrath, *Star Gods of the Maya*, p. 258.

REFERENCES

Adamson, Stephen (ed.) 1997, *Mother Earth, Father Sky: Native American Myth,* Time–Life Books, Amsterdam.

Allen, Richard Hinckley 1963, *Star Names: Their Lore and Meaning,* Dover Publications, New York.

Alpers, Anthony 1977 (1964), *Maori Myths and Legends: Retold by Anthony Alpers,* 4th edn, Longman Paul, Auckland.

Anesaki, Masaharu 1930, *History of Japanese Religion,* Kegan Paul, London.

Ashe, Geoffrey 1992, *Atlantis: Lost Lands, Ancient Wisdom,* Thames and Hudson, London.

Aston, W. G. 1956 (1896), *Nihongi: Chronicles of Japan from the Earliest Times to AD 697,* George Allen and Unwin, London.

Audouze, Jean and Israel, Guy (eds) 1994, *The Cambridge Atlas of Astronomy,* 3rd edn, Cambridge University Press, Cambridge.

Ayto, John 1991, *Dictionary of Word Origins,* Arcade Publishing, New York.

Aveni, Anthony 1993, *Ancient Astronomers,* Smithsonian Books, Washington DC.

Aveni, Anthony 1997, *Stairways to the Stars: Skywatching in Three Great Cultures,* John Wiley and Sons, New York.

Aveni, Anthony 2001 (1980), *Skywatchers: A Revised and Updated Version of Skywatchers of Ancient Mexico,* University of Texas Press, Austin.

Batchelor, John 1901, *The Ainu and Their Folk-Lore,* Religious Tract Society, London.

Bauer, Brian and Dearborn, David 1995, *Astronomy and Empire in the Ancient Andes: The Cultural Origins of Inca Sky Watching,* University of Texas Press, Austin.

Bauval, Robert and Gilbert, Adrian 1994, *The Orion Mystery: A Revolutionary New Interpretation of the Ancient Enigma,* Crown Trade Paperbacks, New York.

Bauval, Robert and Hancock, Graham 1996, *Keeper of Genesis,* Heinemann, London.

Beaver, Pierce (ed.) 1994, *The World's Religions,* Lion Books, Oxford.

Bell, Diane 1998, *Ngarrindjeri Wurruwarrin: A World That Is, Was, and Will Be*, Spinifex Press, North Melbourne.

Berlitz, Charles 1972, *Mysteries from Forgotten Worlds: Rediscovering Lost Civilizations*, Souvenir Press, London.

Berlitz, Charles 1994, *Atlantis: The Lost Continent Revealed*, Fontana Paperbacks, New York.

Berndt, Ronald and Berndt, Catherine 1989, *The Speaking Land: Myth and Story in Aboriginal Australia*, Penguin Books, Australia.

Berndt, Ronald and Berndt, Catherine 1992 (1977), *The World of the First Australians*, Aboriginal Studies Press, Canberra.

Best, Elsdon 1922, *The Astronomical Knowledge of the Maori*, Monograph No. 3, Dominion Museum, Wellington.

Bird Rose, Deborah 2000, *Dingo Makes Us Human: Life and Land in an Australian Aboriginal Culture*, 2nd edn, Cambridge University Press, Sydney.

Blavatsky, Helena P. 1969, *Collected Writings 1882–1883*, Volume 4, Theosophical Publishing House, London, compiled by Boris Zirkoff.

Blavatsky, Helena P. 1977, *The Secret Doctrine: The Synthesis of Science, Religion and Philosophy*, Volumes 1 and 2, Theosophical University Press, Pasadena (facsimile of the original edition of 1888).

Blavatsky, Helena P. 2002 (1877), *Isis Unveiled*, 2 volumes, Quest Books, Wheaton, Illinois.

Bowker, John 1997, *World Religions: The Great Faiths Explored and Explained*, Dorling Kindersley, Surry Hills, UK.

Bradley, Marion 1988, *The Mists of Avalon*, Sphere Books, New York.

Brotherston, Gordon 1992, *Book of the Fourth World: Reading the Native Americas Through Their Literature*, Cambridge University Press, New York.

Brown, Joseph Epes (ed.) 1989 (1953), *The Sacred Pipe: Black Elk's Account of the Seven Rites of the Oglala Sioux*, Civilization of the American Indian Series, vol. 36, University of Oklahoma Press, Norman.

Brown, Joseph Epes 1992, *Animals of the Soul: Sacred Animals of the Oglala Sioux*, Element, Rockport, Massachusetts.

Bruchac, Joseph and London, Jonathan 1997, *Thirteen Moons on Turtles Back: A Native American Year of Moons*, Puttnam and Grosset, New York.

Buchanan, Noel (ed.) 1993, *Discovering the Wonders of Our World*, Reader's Digest Association, London.

Bullfinch, Thomas 1993, *Bullfinch's Mythology: The Age of Fable*, Modern Library, New York.

Burnham Jr, Robert 1978, *Burnham's Celestial Handbook: An Observer's Guide to the Universe Beyond the Solar System*, 3 volumes, Dover Publications, New York.

Burnham Jr, Robert et al. 2001, *An Australian Geographic Guide to Space Watching: The Amateur Astronomer's Guide to Starhopping and Exploring the Universe*, Australian Geographic, Sydney.

Burnham Jr, Robert, Dyer, Alan and Kanipe, Jeffrey 2003, *Astronomy: The Definitive Guide*, Fog City Press, San Francisco.

Cahir, Sandy 2002, *Mythology Series: Livewire Classic Australian Titles*, Cambridge University Press, Australia.

Cameron, Dorothy 1981, *Symbols of Birth and Death in the Neolithic Era*, Kenyon-Deane, London.

Campbell, Alan and Noble, David S. et al. 1993, *Japan: An Illustrated Encyclopedia*, 2 volumes, Kodansha, Tokyo.

Campbell, Joseph 1988, *The Power of Myth*, Doubleday Publishing, New York.

Casti, John L. 2000, *Five More Golden Rules: Codes, Chaos, and Other Great Theories of 20th Century Mathematics*, John Wiley and Sons, New York.

Cayce, Hugh Lyn (ed.) 1969, *Edgar Cayce on Atlantis*, Howard Baker, London.

Cayce, Hugh Lyn (ed.) 1986, *The Edgar Cayce Collection*, Wings Books, New York (4 volumes in one).

Chatwin, Bruce 1987, *Songlines*, Jonathan Cape, London.

Clarke, Phillip 2003, *Where the Ancestors Walked*, Allen and Unwin, Sydney.

Collins, Andrew 2000, *Gateway to Atlantis: The Search for the Source of a Lost Civilisation*, Headline Books, London.

Cooper, Jeanne 1978, *An Illustrated Encyclopaedia of Traditional Symbols*, Thames and Hudson, London.

Cotterell, Maurice 1998, *The Supergods*, Thorsons (imprint of Harper-Collins), London.

Coulter, Charles and Turner, Patricia 2000, *Encyclopedia of Ancient Deities*, McFarland and Company, Jefferson, North Carolina and London.

Couper, Heather and Henbest, Nigel 1988, *The Stars: From Superstition to Supernova*, Pan Books, London.

Dalai Lama 1995, *The Power of Compassion: A Collection of Lectures by His Holiness the XIV Dalai Lama* (transl. Geshe Thupten Jinpa), Thorsons (imprint of Harper-Collins), London.

Deloria, Vine 1977, *Red Earth, White Lies: Native Americans and the Myth of Scientific Fact*, Fulcrum Publishing, Colorado.

De Santillana, Giorgio and Von Dechend, Hertha 1999, *Hamlet's Mill: An Essay on Myth and the Frame of Time*, 5th edn, Nonpareil, Boston.

De Vos, George 1974, Japan's *Minorities: Burakumin, Koreans and Ainu*, Report no. 3, Minority Rights, William Wetherall, Minority Rights Group, London.

Doniger O'Flaherty, Wendy 1975, *Hindu Myths*, Penguin Books, London.

Doniger O'Flaherty, Wendy 1981, *The Rig Veda*, Penguin Books, London.

Drayer, Ruth 2003, *Numerology: The Power in Numbers,* Square One Publishers, New York.

Duncan, David Ewing 1998, *The Calendar,* Fourth Estate, London, 1998.

Dunsford, Cathie 2001, *Song of the Selkies,* Spinifex Press, North Melbourne.

Eliade, Mircea 1978, *A History of Religious Ideas,* 3 volumes, University of Chicago Press, Chicago.

Eliade, Mircea 1991a, *Images and Symbols: Studies in Religious Symbolism,* Princeton University Press, Princeton.

Eliade, Mircea 1991b, *The Myth of the Eternal Return,* Princeton University Press, Princeton.

Elphick, Jonathon (ed.) 1995, *The Atlas of Bird Migration: Tracing the Great Journeys of the World's Birds,* Marshall Editions, London.

Fenton-Smith, Paul 1999, *The Tarot Revealed: A Simple Guide to Unlocking the Secrets of the Tarot,* 4th edn, Simon and Schuster, Roseville.

Fenton-Smith, Paul 2000, *Mastering the Tarot: A Guide to Advanced Tarot Reading and Practice,* Simon and Schuster, Roseville.

Flem-Ath, Rand and Wilson, Colin 2000, *The Atlantis Blueprint,* Little, Brown and Company, London.

Forrest, Isidora 2001, *Isis Magic: Cultivating a Relationship With the Goddess of 10,000 Names,* Llewellyn Publications, St Paul.

Frazier, James 1936, *The Golden Bough: A Study in Magic and Religion,* 3rd edn, 13 volumes, Macmillan and Company, New York.

Frissell, Bob 1994, *Nothing in This Book is True But It's Exactly How Things Are,* Frog Limited, Berkeley.

Gilbert, Adrian and Cotterell, Maurice 1996, *The Mayan Prophecies: Unlocking the Secrets of a Lost Civilization,* Element Books, Shaftesbury.

Gimbutas, Marija 1991, *The Language of the Goddess,* Harper Collins, New York, 1991.

Goble, Paul 1988, *The Lost Children,* Aladdin Paperbacks, New York.

Graves, Robert 1993, *The Greek Myths,* Penguin Books, USA (originally published as 2 volumes in 1955).

Graves, Robert 1997 (1948), *The White Goddess: A Historical Grammar of Poetical Myth,* 4th edn, Carcanet Press, London.

Gunn Allen, Paula 1992, *The Sacred Hoop: Recovering the Feminine in American Indian Traditions,* Beacon Press, Boston.

Hackin, J. et al. 1996, *The Mythologies of the East,* 2 volumes, Aryan Books, New Delhi (first published in 1932 as *Asiatic Mythology*).

Hadingham, Evan 1984, *Early Man and the Cosmos,* Walker and Company, New York.

Hancock, Graham 1998, *Heaven's Mirror,* Crown Publishers, New York.

Harden, M. J. 1999, *Voices of Wisdom: Hawaiian Elders Speak*, Aka Press, Kula.

Harney, W. E. 1959, *Tales from the Aborigines*, Robert Hale, London.

Harris, Geraldine and Pemberton, Delia 1999, *The British Museum Illustrated Encyclopaedia of Ancient Egypt*, British Museum Press, London.

Hart, George 1986, *A Dictionary of Egyptian Gods and Goddesses*, Routledge and Kegan Paul, London.

Hand Clow, Barbara 1995, *The Pleiadian Agenda: A New Cosmology for the Age of Light*, Bear and Company Publishing, Santa Fe, New Mexico.

Hand Clow, Barbara 2001, *Catastrophobia: The Truth Behind Earth Changes In the Coming Age of Light*, Bear and Company Publishing, Rochester, Vermont.

Hendrickson, Robert 1997, *Encyclopedia of Word and Phrase Origins*, Facts on File Inc., New York.

Heyerdahl, Thor 1966 (1950), *The Kon-Tiki Expedition: By Raft Across the South Seas*, Allen and Unwin, London.

Heyerdahl, Thor 1968, *Sea Routes to Polynesia*, Allen and Unwin, London.

Hilger, Inez 1971, *Together With the Ainu: A Vanishing People*, University of Oklahoma, Norman.

Hoad, T. F. (ed.) 1986, *The Concise Oxford Dictionary of English Etymology*, Clarendon Press, Oxford.

Husain, Shahrukh 1997, *The Goddess: Power, Sexuality and the Feminine Divine*, Duncan Baird Publishers, London.

Icke, David 1999, *The Biggest Secret: The Book That Will Change the World*, Bridge of Love Publications, Scottsdale, Arizona.

Irwin, Keith 1965, *The 365 Days: The Story of Our Calendar*, George G. Harrap and Company, London.

Isaacs, Jennifer (ed.) 1981, *Australian Dreaming: 40,000 Years of Aboriginal History*, Lansdowne Press, Sydney.

Jenkins, John Major 1998, *Maya Cosmogenesis 2012: The True Meaning of the Maya Calendar End-Date*, Bear and Company, Rochester, Vermont.

Jung, Carl 1978 (1964), *Man and His Symbols*, Pan Books, London.

Jobes, Gertrude 1962, *Dictionary of Mythology, Folklore and Symbols*, 3 volumes, Scarecrow Press, New York.

Jobes, Gertrude and Jobes, James 1964, *Outer Space: Myths, Name Meanings, Calendars — From the Emergence of History to the Present Day*, Scarecrow Press, New York and London.

Kane, Herb 1976, *Voyage: The Discovery of Hawaii*, Island Heritage Press, Honolulu.

Kane, Herb 1997, *Ancient Hawaii*, Kawainui Press, Honolulu.

Keats, John 1927, *Endymion: A Poetic Romance*, Oxford University Press, London.

Kitao, Kouichi 2002, *Starlore of Japan: The Starscape of a People*, Ama River Publishing Company, Amherst, Massachusetts.

Kirch, Patrick Vinton and Green, Roger 2001, *Hawaiki, Ancestral Polynesia: An Essay in Historical Anthropology*, Cambridge University Press, Cambridge.

Kraay, Colin 1966, *Greek Coins*, Thames and Hudson, London.

Krupp, Edwin 1991, *Beyond the Blue Horizon*, Oxford University Press, Oxford.

Krupp, Edwin 1997, *Skywatchers, Shamans and Kings: Astronomy and the Archaeology of Power*, John Wiley and Sons, New York.

Kyselka, Will 1987, *An Ocean in Mind*, University of Hawaii Press, Honolulu.

Lambert, Johanna 1993, *Wise Women of the Dreamtime: Aboriginal Tales of the Ancestral Powers*, Inner Traditions, Rochester, Vermont.

Lame Deer, Archie Fire and Erdoes, Richard 1992, *The Gift of Power: The Life and Teachings of a Lakota Medicine Man*, Bear and Company Publishing, Sante Fe, New Mexico.

Lamy, Lucie 1981, *Egyptian Mysteries*, Thames and Hudson, London.

Langloh Parker, Katherine 1978, *Australian Legendary Tales*, Bodley Head, Sydney.

Lattimore, Richmond 1959, *Hesiod*, University of Michigan Press, Ann Arbor.

Lawlor, Robert 1982, *Sacred Geometry: Philosophy and Practice*, Thames and Hudson, London.

Lawlor, Robert 1991, *Voices of the First Day: Awakening in the Aboriginal Dreamtime*, Inner Traditions, Rochester, Vermont.

Lesko, Barbara 1999, *The Great Goddesses of Egypt*, University of Oklahoma Press, Norman, Oklahoma.

Lewis, David 1994, *We, the Navigators: The Ancient Art of Landfinding in the Pacific*, University of Hawaii Press, Honolulu.

Lewis, James 1994, *The Astrology Encyclopedia*, Gale Research Inc., Detroit.

Lewis, Reginald 1997, *The Thirteenth Stone*, 2nd edn, Fountainhead Press, Plymouth and Fremantle.

Levy, David 1995, *Skywatching*, RD Press and the Nature Company Guides, Sydney.

Lippincott, Kristen et al. 1999, *The Story of Time*, Merrell Holberton, London.

Lurker, Manfred 1980 (1974 in German), *The Gods and Symbols of Ancient Egypt*, Thames and Hudson, London.

Mackenzie, Donald 1996, *Legends of China and Japan*, Senate Books, London.

Mackenzie, Donald 1996, *South Seas: Myths and Legends*, Senate Books, London.

Makemson, Maud 1941, *The Morning Star Rises: An Account of Polynesian Astronomy*, Yale University Press, New Haven.

Massola, Aldo 1968, *Bunjil's Cave: Myths, Legends and Superstitions of the Aborigines of South East Australia*, Lansdowne Press, Sydney.

Mathews, Janet 1994, *The Opal That Turned Into Fire*, Magabala Books, Broome, Western Australia.

McDonald, John 1996, *House of Eternity: The Tomb of Nefertari*, Thames and Hudson, London.

McKenna, Terence 1992, *Food of the Gods: The Search for the Original Tree of Knowledge*, Bantam Books, New York.

Milbrath, Susan 1999, *Star Gods of the Maya: Astronomy in Art, Folklore, and Calendars*, University of Texas Press, Austin.

Michell, John and Rhone, Christine 1991, *Twelve-Tribe Nations and the Science of Enchanting the Landscape*, Phanes Press, Grand Rapids.

Milton, John 1923, *On the Morning of Christ's Nativity*, Milton's hymn with illustrations by William Blake and a note by Geoffrey Keynes, Cambridge University Press, London,

Moore, Patrick 1994, *Stars of the Southern Skies*, Penguin Books, New York.

Moore, Patrick (ed.) 2002, *Astronomy Encyclopedia*, Oxford University Press, Sydney.

Morton, Chris and Thomas, Ceri Louise 1998, *The Mystery of the Crystal Skulls*, 2nd edn, Harper Collins, London.

Motz, Lloyd and Nathanson, Carol 1987, *The Constellations*, Doubleday, New York.

Mountford, Charles 1970a, *The Dawn of Time: Australian Aboriginal Myths*, Rigby Ltd, Sydney.

Mountford, Charles 1970b (1965), *The Dreamtime: Australian Aboriginal Myths*, Rigby Ltd, Sydney.

Mountford, Charles 1971, *The First Sunrise: Australian Aboriginal Myths*, Rigby Ltd, Sydney.

Mowaljarlai, David and Malnic, Jutta 1993, *Yorro Yorro: Aboriginal Creation and the Renewal of Nature*, Inner Traditions, Rochester, Vermont. (First published 1993 by Magabala Books.)

Nakayama, Shigeru 1969, *A History of Japanese Astronomy: Chinese Background and Western Impact*, Harvard University Press, Harvard.

Nasr, Seyyed Hosein 1993, *An Introduction to Islamic Cosmological Doctrines*, State University of New York Press, Albany.

Neumann, Erich 1955, *The Great Mother: An Analysis of the Archetype*, Routledge and Kegan Paul Ltd, London.

Nitsch, Twylah 1991, *Other Council Fires Were Here Before Ours*, Harper Books, San Francisco.

O'Brien, May 1990, *The Legend of the Seven Sisters: A Traditional Aboriginal Story from Western Australia*, Aboriginal Studies Press, Canberra.

Oliver, Paul 2001, *World Religions*, Hodder Headline, London.

Orbell, Margaret 1996, *The Illustrated Encyclopaedia of Maori Myth and Legend*, University of New South Wales Press, Sydney.

Peat, David 1995, *Blackfoot Physics: A Journey into the Native American Universe*, Fourth Estate, London.

Philippi, Donald 1982, *Songs of Gods, Songs of Humans: The Epic Tradition of the Ainu*, North Point Press, San Francisco.

Piazzi Smyth, Charles 1978, *The Great Pyramid: Its Secrets and Mysteries Revealed*, Gramercy Books, New York (originally published in 1880 as *Our Inheritance in the Great Pyramid*).

Piggott, Juliet 1983, *Japanese Mythology*, Peter Bedrick Books, New York.

Pinkola Estes, Clarissa 1992, *Women Who Run With the Wolves: Contacting the Power of the Wild Woman*, Rider (imprint of Random House), Sydney.

Ransome, Hilda 1986 (1937), *The Sacred Bee in Ancient Times and Folklore*, Bee Books New and Old, Burrowbridge.

Rathbun, Shirley 1982, *First Encounters: Indian legends of Devils Tower* (pamphlet), Sand Creek Printing, Wyoming, USA.

Reed, A. W. 1972, *Maori Legends*, A. H. and A. W. Reed, Wellington.

Rees, Martin 1999, *Just Six Numbers: The Deep Forces That Shape the Universe*, Weidenfeld and Nicolson, London.

Renshaw, Steve and Ihara, Saori 1996, 'A brush daub on the heavens', *Archaeoastronomy and Ethnosastronomy News*, Quarterly Bulletin no. 19, March. The article is available at: <http://www2.gol.com/users/stever/subaru.htm>.

Rice, Patty 1980, *Amber: The Golden Gem of the Ages*, Van Rostrand Reinhold Company, New York.

Ridley, Matt 1999, *Genome: The Autobiography of a Species in 23 Chapters*, Fourth Estate, London.

River, Lindsay and Gillespie, Sally 1992, *The Knot of Time: Astrology and the Female Experience*, 2nd edn, Women's Press, London.

Roberts, Jean and Roberts, Ainslie 1989, *Dreamtime Heritage*, Art Australia, Adelaide.

Roland, Paul 1995, *Revelations: The Wisdom of the Ages*, Carlton Books, London.

Romer, John and Romer, Elizabeth 2001, *The Seven Wonders of the World: A History of the Modern Imagination*, Seven Dials, London.

Rucker, Rudy 1984, *Infinity and the Mind: The Science and Philosophy of the Infinite*, Palladin Books, London.

Ryan, P. M. 1997, *The Reed Dictionary of Modern Maori*, 2nd edn, Reed Books, Auckland.

Sams, Jamie and Carsons, David 1988, *Medicine Cards: The Discovery of Power through the Ways of the Animals*, Bear and Company, Santa Fe, New Mexico.

Schimmel, Annemarie 1993, *The Mystery of Numbers*, Oxford University Press, Oxford.

Schwaller de Lubicz, R. A. 1998, *The Temple of Man*, 2 volumes, Inner Traditions, Rochester, Vermont (originally published in French under the title *Le Temple de l'homme* in 1957).

Segal, Justin 1998, *The Amazing Space Almanac*, Lowell House, Chicago.

Sejourne, Laurette 1956, *Burning Water: Thought and Religion in Ancient Mexico*, Thames and Hudson, London.

Sesti, Giuseppe 1991, *The Glorious Constellations: History and Mythology*, Harry N. Abrams, New York (translated from the Italian by Karin Ford).

Shah, Idries 1977, *The Sufis*, 4th edn, Octagon Press, London.

Sharma, Brijendra Nath 1978, *Festivals of India*, Abhinav Publications, New Delhi.

Sharp, Nonie 1998, *Stars of Tagai: The Torres Strait Islanders*, Aboriginal Studies Press, Canberra.

Simon, Margaret 2003, *The Meeting of the Waters: The Hindmarsh Island Affair*, Hodder Books, Sydney.

Sitchin, Zecharia 1998, *The Cosmic Code: Book VI of the Earth Chronicles*, Avon Books, New York.

Sjoo, Monica 1991, *The Great Cosmic Mother: Rediscovering the Religion of the Earth*, Harper Books, San Francisco.

Sora, Steven 2003, *Secret Societies of America's Elite: From the Knights Templar to Skull and Bones*, Inner Traditions, Rochester, Vermont.

Spence, Lewis 1915, *Myths and Legends of Ancient Egypt*, Harrap, London.

Spence, Lewis 1995, *History of Atlantis*, Studio Editions (imprint), London.

Storm, Rachel 1999, *The Encyclopedia of Eastern Mythology*, Lorenz Books, New York.

Storm, Rachel 2002, *Egyptian Mythology*, Hermes House (imprint), London (first published 1999 and 2002 as part of a larger compendium, *The Encyclopedia of Eastern Mythology*).

Sueoka, Tomio 1979, *Ainu no Hoshi* (Ainu Stars), Asahikawa Series, Volume 12, Public Corporation for Asahikawa Promotion, Asahikawa, Japan.

Sykes, Bryan 2001, *The Seven Daughters of Eve*, Bantam Press, Sydney.

Taub, Liba 2003, *Ancient Meteorology*, Routledge, London.

Temple, Robert 1989 (1975), *The Sirius Mystery: New Scientific Evidence of Alien Contact 5,000 Years Ago*, Arrow Publications, London.

Temple, Robert 2000, *The Crystal Sun: Rediscovering a Lost Technology of the Ancient World*, Random House, London.

Tomas, Andrew 1972, *Atlantis: From Legend to Discovery*, Robert Hale and Company, London.

Tompkins, Peter 1978, *Secrets of the Great Pyramid*, Harper and Row, New York.

Tripp, Edward 1972, *The Handbook of Classical Mythology*, Arthur Baker Ltd, London.

Tyler Olcott, William 1911, *Starlore of All Ages*, C. P. Puttnam's Sons, New York.

Uchida, Takeshi 1973, *Dialects of the Stars and Cultures*, Iswasaki Bijitsu Company, Tokyo.

Unaipon, David 2001, *Legendary Tales of the Australian Aborigines*, Melbourne University Press, Melbourne. (The edited version by Stephen Meuke and Adam Shoemaker of Unaipon's unpublished manuscript of 1924–25.)

Van Zandt, Eleanor and Stemman, Roy 1976, *Mysteries of the Lost Lands*, Aldus Books, London.

Versluis, Arthur 1994, *Native American Traditions*, Element Books, Rockport.

Von Däniken, Erich 1973, *In Search of Ancient Gods*, Souvenir Press, London.

Walker, Barbara 1983, *The Women's Encyclopedia of Myths and Secrets*, Harper Collins, New York.

Walker, Barbara 1988, *The Woman's Dictionary of Symbols and Sacred Objects*, Harper Collins, New York.

Wallis Budge, E. A. 1971, *Egyptian Mysteries*, Dover Publications, New York.

Wallis Budge, E. A. 1995a (1899), *Egyptian Ideas of the Afterlife*, 3rd edn, Dover Publications, New York.

Wallis Budge, E. A. 1995b (1899), *The Book of the Dead: The Hieroglyphic Transcript and English Translation of the Papyrus of Ani*, Gramercy Books, New Jersey.

Wasson, Robert 1971, *Soma: Divine Mushroom of Immortality*, Harcourt Brace Jovanovich, New York.

Watterson, Barbara 1996, *Gods of Ancient Egypt*, 2nd edn, Sutton Publishing, Phoenix.

West, M. L. (ed.) 1978, *Hesiod: Works and Days*, Oxford University Press, Oxford.

Wilkinson, Richard 2003, *The Complete Gods and Goddesses of Ancient Egypt*, Thames and Hudson, London.

Wilson, F. P. 1970 (1935), *The Oxford Dictionary of English Proverbs*, 3rd edn, Oxford University Press, Oxford.

Other books available from Spinifex Press

Trauma Trails, Recreating Song Lines: The Transgenerational Effects of Trauma in Indigenous Australia

Judy Atkinson

Shortlisted, *The Australian* Awards for Excellence in Educational Publishing, 2003

Providing a startling answer to questions of how to solve the problems of generational trauma, this book moves beyond the rhetoric of victimhood and provides inspiration for anyone dealing with Indigenous and non-Indigenous communities today. Beginning with issues of colonial dispossession, Judy Atkinson also deals with trauma caused by abuse, alcoholism and drug dependency.

"*Trauma Trails* is a remarkable book . . . a substantial reconciliative achievement that should encourage others to bridge the cultural divide in imaginative ways."

— Antonia Estin, *Journal of Australian Studies*

ISBN: 9781876756222

Kick the Tin

Doris Kartinyeri

Doris Kartinyeri was taken from a South Australian hospital, twenty-eight days after her birth while her Ngarrindjeri family mourned her mother's death. She was never returned to them.

A compelling and courageous journey into the soul of the individual to find meaning and substance after loss. This is a powerful memoir by a member of the stolen generation.

"[*Kick the Tin*] is a story of courage and survival, powerfully demonstrating how the human spirit can soar despite all the injuries and injustices which threaten to drag it down."

— Lowitja O'Donoghue

ISBN: 9781875559954

Ngarrindjeri Wurruwarrin

Diane Bell

Winner, NSW Premier's Gleebook Award for Cultural and
Literary Criticism
Finalist, *The Age* Book of the Year
Finalist, Queensland Premier's History Award
Finalist, Gold Medal of the Australian Literary Society
Finalist, Kiriyama Award

Diane Bell presents a finely drawn portrait of the struggle to
protect the sacred sites on Kumarangk/Hindmarsh Island, which
became the subject of legal investigations, a media frenzy and
mass anthropological argument. It contains the stories presented
by Ngarrindjeri women concerning their sacred sites and the
traditions upheld by the Ngarrindjeri community.

"A formidable [work] . . . It leaves the reader wondering whether
the outcome would have been different had the contents of the
book been known at the time of the events it describes."

— John Toohey, *Canberra Times*

"A magisterial work . . . every Ngarrindjeri person I have spoken
to applauds this book. Ethnographies of this sort are usually
avoided for any number of reasons, not the least of which is the
sense that attempting such a study is too hard. That Bell attempts
and succeeds in this without sacrificing scholarship or standards
is an magnificent achievement."

— Christine Nicholls, *Times Higher Educational Supplement*

ISBN: 9781742199184

Daughters of the Dreaming

Diane Bell

Finalist for the 1993 J. I. Staley Prize

"The energy, intelligence and sheer talent deployed in this work are formidable."

— Faith Bandler, *Sun*

"A landmark both of feminist scholarship and of anthropological fieldwork."

— *Women's Review of Books*

"*Daughters of the Dreaming* seems destined to take its place alongside the classics of Australian anthropological literature."

— Robin Lucas, *Sydney Morning Herald*

ISBN: 9781876756154

Daughters of the Pacific

Zohl dé Ishtar

Indigenous women have a voice in this book, speaking up for a nuclear-free Pacific from Hawai'i, the Marshall Islands, the Northern Marianas, Guam, Belau, Fiji, Australia, Aotearoa/New Zealand and Te Ao Maohi/Tahiti Polynesia.

"We are already dying from nuclear war while you are still thinking how to prevent it."

— Chailang Palacios, Northern Marianas

ISBN: 9781875559329

Moebius Trip

Giti Thadani

Giti Thadani has been driving her jeep around India for fifteen years. Blessed with a rare historical sensibility, including an eye for architectural detail, she ventures off-road in search of lost temples, sculptures and cosmological sites from Madhya Pradesh to Kanyakumari to Gujarat.

Her inner and outer journeys unfold each other. Her passion for architecture, sculpture, mythology, iconography and artistic heritage is infectious. She delves into the unknown histories of women. A travelogue unlike any other, *Moebius Trip* is at once a road journal, a collection of musings and a cry for respect and preservation of one of the world's oldest civilisations.

ISBN: 9781876756543

*If you would like to know more about Spinifex Press,
write for a free catalogue or visit our website*

Spinifex Press

Orders:
PO Box 5270
North Geelong, Victoria 3215

and

PO Box 105
Mission Beach, Queensland 4852
Australia

women@spinifexpress.com.au
www.spinifexpress.com.au